BIOASSAY TECHNIQUES AND

ENVIRONMENTAL CHEMISTRY

BIOASSAY TECHNIQUES AND ENVIRONMENTAL CHEMISTRY

Gary E. Glass, Editor

Environmental Protection Agency
National Water Quality Laboratory
Duluth, Minnesota

FOREWORD

In an effort to shed light on the complexities of environmental studies, a symposium of invited biologists and chemists was held at the National American Chemical Society meeting, Washington, D.C., 1971. The participating scientists agreed to publish their work in the form of a monograph since the principles and techniques used were similar even though their areas of specialty were widely divergent.

The papers are original contributions carefully refereed, include reports of research as well as reviews of principles and past work, and have not been published elsewhere.

All royalties resulting from the sale or use of this monograph are being contributed to the scholarship fund of the American Chemical Society, Lake Superior Section.

This monograph is addressed to biologists and
chemists concerned with the aquatic environment.
The complexities of both chemical and biological
measurements are illustrated by papers which represent the "state of art" as it is being practiced.
The need for close cooperation between chemists and
biologists in solving environmental problems can be
seen daily at the National Water Quality Laboratory,
where the protection of this nation's fresh water
uses through the development of water quality criteria is the goal. These criteria are being developed
for aquatic organisms, since they provide the most
accurate assessment of environmental quality. The
need for understanding and relating the biological
response in a changing environment to chemical
measurements made on that environment is without
bound, yet research in this area has just begun to
address the need.

Gary E. Glass
Duluth, Minnesota
January 1973

TABLE OF CONTENTS

SECTION V
APPLICATIONS

SECTION I

AN ILLUSTRATION OF
ENVIRONMENTAL COMPLEXITIES

CONTRIBUTOR TO SECTION I

Frank M. D'Itri. Institute of Water Research and Department
     of Fisheries and Wildlife, Michigan State University,
     East Lansing, Michigan.

1. MERCURY IN THE AQUATIC ECOSYSTEM

Frank M. D'Itri. Institute of Water Research and Department of Fisheries and Wildlife, Michigan State University, East Lansing, Michigan

INTRODUCTION

Mercury is a naturally occurring element that is normally found in minerals, rocks, soil, water, air, plants, and animals. The ubiquitous nature of mercury is due, in large part, to the high vapor pressure of the element and its compounds. Mercury can be divided into two major categories of organic and inorganic compounds. The inorganic mercury category contains the widely recognized elemental silvery liquid metal as well as a host of compounds wherein the mercury is present in one of its two most common forms, either as a mercurous ion (Hg^+) or mercuric ion (Hg^{++}). The second category containing organic mercury includes chemical compounds which contain carbon atoms that are covalently bound to a mercury atom. These relatively new mercury compounds were introduced in increasing numbers after the turn of the century. Organic mercury compounds, in turn, may be subdivided into two main categories: the aryl and the alkyl mercury compounds. Of the two, because of their resistance to degradation, the alkyl mercury compounds are generally considered to be a much greater threat to man and his environment than either the aryl or any of the inorganic mercury compounds.

Until recently, the main source of mercury in the environment has been the erosion and leaching of mercury containing geological formations by rainfall that also transports the mercury to streams and lakes by ground water runoff. Since the industrial revolution, however, increasing amounts of mercury have been lost to the environment from waste products

as a result of manufacturing processes that utilize
mercury or from the improper disposal of industrial
and consumer products which contain mercury compounds.
Mercury is consumed in the manufacture of electrical
apparatus, chlorine, caustic soda, industrial control
instruments, paint, pharmaceuticals, cosmetics,
paper, pulp, agricultural fungicides and dental
amalgams as well as in a great many minor industrial
and consumer products. Moreover, the amalgamative
and catalytic properties of mercury and its compounds
are responsible for many additional industrial uses.[1]
    In urban and industrial areas the sanitary sewer
systems serve as a convenient disposal system for
mercury-containing consumer products. These mercury
discharges are small enough to be considered insignificant individually. However, altogether, potentially large quantities of mercury are discharged
into the watercourses that receive the effluent from
sewage treatment plants. The total mercury content
of the sewage system is the result of all the inci-
dental uses of mercury-containing compounds by
individuals and businesses. These include the
mercury found in water based paints, paper products,
cosmetics, broken thermometers, mercury amalgam
tooth fillings, discarded pharmaceuticals, household
and laundry disinfectants, and the runoff of mercury
lawn and garden fungicides. On the average, the
mercury concentration of sewage effluent is one
order of magnitude greater than the watercourse that
receives it.[2] Klein[3] has studied amounts of mercury
in the receiving waters of the Grand Rapids and
Holland, Michigan, sewage treatment plants. His
findings indicate that the average range of mercury
concentrations in the raw influent and final
effluent of the Grand Rapids plant are 1.0-4.6 and
0.6-2.0 ppb respectively. For the Holland plant
the average mercury levels for similar water was
1.0-2.2 and 0.8-1.6 ppb. Therefore, the mercury
concentrations in the final effluent of these
sewage treatment plants averaged about 1 ppb. However, as these data indicate, substantially higher
values of mercury in the final effluent are possible
depending on the absorptive capacity of the activated
sludge in removing the mercury. Therefore, in an
urban area, from 400 to 1000 pounds of mercury per
million population are discharged into receiving
water every year.
    In addition, large quantities of mercury are
lost to the environment through burning or other
utilization of fossil fuels as well as through the

smelting of ores to recover such metals as copper, lead, and zinc. Coal typically contains small amounts of many of the known elements which are not commonly reported as part of the standard ash analysis. Furthermore, many coal and brown coal ashes contain exceptionally high amounts of certain trace elements relative to their average content in the earth's crust.[4-6] In a survey of the literature involving the occurrence of rare and uncommon elements in coal, Gibson and Selvig[7] reported that more than half of the known elements, including mercury, have been found in coal. V. M. Goldschmidt[8] explained the enrichment of mercury and other trace elements in coal and the uptake of nutrient and ballast elements by the living plants from the soil, subsoil, and water. These trace elements are ultimately translocated into the roots, twigs, and leaves. Goldschmidt offers another possible explanation that the normally soluble trace elements are precipitated as the respective sulfides by the special local chemical environment which develops the hydrogen sulfide associated with the anaerobic rotting of plant materials.

The available data concerning the concentration of mercury in coal are very limited. Goldschmidt and Peters[4] have reported the presence of mercury in different kinds of coal, peats, and shales from Germany and England while Inagaki[9] qualitatively found mercury present in some Japanese coals. In a report on the spectrographic analyses of 596 spot samples for 38 different elements from 16 coal seams in West Virginia, Headless and Hunter[10] found that the mercury content in coal ash ranged from less than 100 ppm to 280 ppm with an average value of 130 ppm. While these values appear inordinately high for coal ash, the presence of mercury is apparent. Recently, Ruch *et al.*[11] established that 55 raw coal samples from 10 coal seams in Illinois have a mean mercury concentration of 0.18 ppm. In addition, 11 coal samples from states other than Illinois have mercury concentrations within the same range as Illinois coals or slightly lower. Also, Joensuu[12] found the levels of coal highly variable with concentrations in the range of 0.1 to 1 ppm common.

The very limited available information indicates that most of the current data is in the Russian literature.[13-23] The distribution of mercury in coal in various parts of the USSR has been reported to range between 0.02 and 300 ppm.[13-19] In addition,

up to 3,000 ppm of mercury have been reported in the
solid bitumens of the Sevan-Akerin zone of the Lesser
Caucasus.[20]   While these high values of mercury in
coals have been reported, it is important to recognize that coal deposits containing more than one ppm
mercury are usually located in or adjacent to known
mercury mineralizations. When mercury containing
coals are used in the coking process, free mercury
and mercuric sulfide have been recovered as byproducts in the subsequent distillation of the coal
tar.[21-24]

Pakter[25-26] reported that coal charges from 80
percent of the byproducts of coking plants in the
USSR were analyzed for their mercury content, and
mercury was detected in practically all of them.
The overall average mercury concentration for all
the plants was 0.28 ppm.

The data are even more sparse for the mercury
content in crude oil than for coal. In the Wilbur
Springs District, California, which includes the
important Abbott mercury mine as well as four or
five smaller mercury deposits, White[27] reported
that hydrocarbons and methane are frequently associated with mercury deposits. Also, the Mt. Diablo
District of California is known for the association
of mercury with petroleum, hydrocarbon gases, and
saline oilfield waters. Bailey *et al.*[28] reported
that native mercury and possibly other forms of
mercury occur in petroleum, natural gas, and the
brine of the Cymric oil field, Kern County,
California. A quantitative spectrographic analysis
of the metal content was made on the ash of three
Cymric field crude oils. The analyses showed that
these crude oils contained remarkably large amounts
of mercury. The reported values ranged from 1.9 to
21 ppm. The latter value is one of the highest
reported for any natural fluid and compares with the
21 ppm that Trost[29] reported for Santa Barbara crude
oil. The Cymric oil field natural gas that separates
from the petroleum and brine upon release of pressure
is saturated with mercury vapor at the oil field,
but the mercury evidently combines with hydrogen
sulfide from the sour natural gases that are released
from the other oil fields. This mercury is precipitated in the pipe lines and some native mercury
separates from the crude oil at the pumping station.

The mercury content in brines associated with
oil producing wells has been reported to be between
0.02 to 0.2 ppm. It has been estimated that the
total quantity of mercury discharged from all fluids

during the existence of the Cymric oil field would probably be equivalent to many thousands of flasks of mercury at 76 pounds each.[30]

Although the mercury content of fossil fuels may appear to be insignificantly small (see Table 1), if the total amounts of coal and oil that are consumed in the world since the industrial revolution are considered, proportionately the amount of mercury released into the environment would conceivably exceed the amounts of mercury lost through the exploitation of mercury related technology. In order to appreciate the significance of this source relative to other sources of mercury contamination in the environment, it is interesting to note that in 1968 in the United States alone, the production of coal and crude oil was in excess of 500 and 488 million tons respectively.[31-33] Assuming that these fossil fuels contained between ten parts mercury per billion and one part per million, this represents from 20,000 to 2,000,000 pounds of additional mercury which could ultimately be lost into the environment of the United States each year. In addition, Klein[34] estimates a worldwide mercury input in excess of 6.0 million pounds per year into the environment from this source alone (see Table 1).

Finally, another potentially very large environmental mercury contamination source occurs in the recovery or use of raw materials which contain small amounts of mercury. Traces of mercury are present in many sulfide ores in concentrations which are not economically feasible to recover. When these ores are roasted (especially gold, copper, lead, and zinc), the mercury is vaporized and driven off with the evolved sulfur dioxide. Even though the gas is cooled, scrubbed and filtered, some of the mercury escapes into the atmosphere.[36] In special cases however, significant quantities of mercury have been recovered as a by-product of copper, lead, gold, and zinc production.[37] Klein[34] has identified smelting of sulfide ores of various metals as a major source of mercury to the atmosphere. If the ore concentrates mercury, copper, lead, and zinc to the same degree, Klein reasons that the relative quantities of each of these metals in the ore will be approximately the same as their relative crustal abundances. This assumption provides the upper limit of 60 million pounds of mercury per year as the result of smelting operations (Table 2). Klein arrives at a minimum value by

Table 1

Mercury Concentrations in Fossil Fuels in Parts Per Billion and Estimates of Maximum Possible Releases of Mercury to the Environment from this Source

Type of Fossil Fuel	*Range*	*Estimated Average Mercury Concentration*	*World Consumption tons/year*	*Mercury Released to Environment tons/year*	*Reference*
1. Peat	60-300	-	-	-	-
2. Coal	10-8500*	-	-	-	-
3. Coal in mecuriferous basins	20-300,000	-	-	-	-
Coal (all types)	10-300,000	~1000	3×10^9	3000	34
4. Bitumens asphalts	2000-900,000	-	-	-	-
Fuel Oil	2-4	-	-	-	35
Gasoline	2-17	-	-	-	35
5. Crude oil and Products	2-21,000**	40	2×10^9	80	34
6. Natural gas	†	40	0.6×10^9	20	34
Maximum possible release of mercury into environment				3100	34

* Highly variable depending on geographical location.

**The value of 21,000 ppb is from the Cymric oil field, Kern County, California.[30]

† The natural gas from the Cymric oil field is saturated with mercury vapor.[30]

Table 2

A. Mercury Content in Copper, Lead, and Zinc Ores[34]

Location, Type	Samples	Range (ppm)	Mean
Turkestan, general	1662	4.4-200	40
Achisai, Lead-Zinc, Sulfide	39	0.3-40	6.3
Oxidized lead	50	0.3-1000	130
Boliden, Chalcopyrite	5	2-10	5
Sphalerite	1	500	
"Sulfominerals"	5	4-100	33
North Sweden, general	27	1-80	13.7

Estimated mean for the 4 locations, 20 ppm.

B. Estimates of Maximum Possible Releases of Mercury into the Environment from Smelting of Copper, Lead, and Zinc Ores[34]

Metal	Crustal Abundance (%)	Relative Abundances	Tons of Ore Processed	Production (Tons/Year)	Estimated Mercury Produced with the Metal (Tons/Year)
Copper	4.7×10^{-3}	1.8×10^{-3}	9×10^8	6×10^6	11,000
Lead	1.6×10^{-3}	5×10^{-3}	6×10^7	3×10^6	15,000
Zinc	8.3×10^{-3}	1×10^{-3}	6×10^7	4×10^6	4,000
Mercury	8.3×10^{-6}	-			
				Total	30,000 tons/yr

assuming that the metal ores have an average mercury content of 3 ppm. This 3 ppm estimate is based on the "mercury halo" method of prospecting for sulfide ores. Mercury in an ore can diffuse into the surface soils where concentrations of 1 ppm indicate the presence of a sulfide ore. Therefore, Klein suggests that in order for an underground ore to maintain a surface concentration of 1 ppm mercury, a concentration of at least 3 ppm mercury is required in that ore body. A third estimate presented in Table 2 is based on the few analyses that are available for mercury in ores. Based on the estimates of average mercury concentrations of 3 and 20 ppm in ores and the data presented in Table 2, the yearly worldwide release of mercury from smelting ores ranges between 6 and 40 million pounds. Therefore, at a minimum, smelting of copper, lead, and zinc ores potentially could release as much mercury into the atmosphere as was industrially consumed in the United States in 1968, and smelting could be a tenfold greater source.[34]

World attention was initially focused on the environmental mercury problem when human beings were poisoned by eating mercury contaminated fish and shellfish in the middle and late 1950's. In a chemical plant located near Minamata, Japan, mercuric sulfate and chloride were used as the catalytic agents to convert acetylene into acetaldehyde and vinyl chloride.[38-39] As a result of these operations, the mercury catalysts were methylated in a side reaction to form the highly poisonous methylmercury chloride which was discharged as part of the wastewater effluent into Minamata Bay. Subsequently, the methylmercury chloride that accumulated in fish and shellfish was the toxic source of severe neurological disorders among inhabitants in the vicinity of Minamata Bay. Today the term Minamata Disease describes the chemical and pathological characteristics of this neurological disorder and has become synonymous with methylmercury poisoning specifically and alkylmercury poisoning in general. From 1953, when Minamata Disease first occurred, to the present, 121 cases of the disease resulting in 46 deaths have been confirmed in the Minamata area with an additional 47 confirmed cases and 6 deaths in Niigata, Japan.[40]

OCCURRENCE AND DISTRIBUTION OF MERCURY

Background concentrations of mercury in nature that have not been contaminated by man are difficult to determine. Wide and anomalous variations in the

mercury background concentrations exist for a variety
of reasons. For example, since mercury is mobile in
high temperature environments, igneous rocks should
contain less mercury than sedimentary formations such
as shale, but these values are highly variable. The
mercury concentrations of the various rock classifi-
cations are presented in Table 3. And because of
the volatile nature of mercury, dissemination
aureoles form around cinnabar deposits. These
aureoles can be quite extensive and serve as path-
finders in searching for mercury or any sulfide
metals associated with mercury deposits.[41-43]
Fredrick and Hawkes[44] reported that the soils underlying extensive systems of mineral rich ore veins
in the Pachuca-Real Del Monte District of Mexico
contained from 250 to 1900 ppb of mercury over a
background value of 50 ppb. Another factor that
makes it difficult to ascertain the background
mercury concentrations is that some areas containing
mercury mineralizations have been polluted with
mercury. And the amount of mercury from pollution
cannot be separated from the mercury background concentration levels which have been reported in the
literature as ranging from one to over 500 ppb.[44-55]
For example, Martin[54] reported the natural background
levels of some English soils to fall between the
values of 10 and 60 ppb whereas Andersson[55] reported
that the mercury content of 200 soil samples ranged
between 20 and 920 ppb with an average of 70 ppb.
Kimura and Miller[56] found the natural mercury content
of sandy loam soil from Puyallup, Washington, USA,
to be 0.166 ppm. The mercury content of Nevada and
Texas soils and non-mineralized soils elsewhere in
the United States usually falls between 0.20 and
0.040 ppm.[57]

Reports of what appear to be inordinately high
mercury soil concentrations, such as Andersson's
920 ppb, are especially significant in areas which
are devoid of mercury mineralizations. While the
mercury content of a given sample will vary in
different geographical locations, varying amounts
of humic materials and clays in the soil fractions
modify the soil's ability to accumulate mercury[48]
(see Table 1). Andersson[50-55] reported that the
humus or organic fraction of a soil has the greatest
affinity for mercury, and a direct relationship
exists between these two parameters. He also found
that the mercury content of the organic component
of a moraine soil was 1100 ppb while its mineral
component contained only 80 ppb of mercury.[55]
Furthermore, from these studies it appears that the

Table 3

Mercury Concentrations of Rocks in Parts Per Billion[45]

Rock Classifications	Range	Mean
A. *Igneous Rocks*		
1. Ultrabasic rocks (dunites, kimberlites, etc.)	7-250	168
2. Intermediate rocks		
Intrusives (diorites, etc.)	13-64	38
Extrusives	20-200	66
3. Basic rocks		
Intrusives (gabbros, diabase, etc.)	5-84	28
Extrusives (basalt, etc.)	5-40	20
4. Acid rocks		
Intrusives (granites, granodiorites, syenite, etc.)	7-200	62
Extrusives (rhyolite, trachyte, etc.)	2-200	62
5. Alkali-rich rocks (nepheline syneties, phonolites, etc.) (data mainly from Khibina and Lovozero massifs, USSR)	40-1400	450
B. *Sedimentary Rocks*		
1. Recent sediments		
Stream, river and lake sediments	10-700	73
Ocean sediments	<10-2000	100
2. Sandstone, arkose, conglomerate, etc.	<10-300	55
3. Shales and argillites		
Shales, mudstones, and argillites	5-300	67
Carbonaceous shales, bituminous shales, etc.	100-3250	437
4. Limestone, dolomite, etc.	<10-220	40
5. Evaporites		
Gypsum, anhydrite	<10-60	25
Halite, sylvite, etc.	20-200	30
C. *Metamorphic Rocks*		
1. Quartzites, etc.	10-100	53
2. Amphibolites	30-90	50
3. Hornfels	35-400	225
4. Schists	10-1000	100
5. Gneisses	25-100	50
6. Marble, crystalline dolomite, etc.	10-100	50

adsorption-desorption of mercury on the humus content
of a soil is pH dependent. Mercury is adsorbed on
the humus at low pH values, and as the pH increases,
larger amounts of mercury are adsorbed by the soil
minerals. Therefore, this mechanism plays a significant role in the translocation of mercury in
the environment. These data are summarized in Table
4.

The relatively little information available
with regard either to the qualitative or the quantitative aspects of atmospheric mercury is also
summarized in Table 4. In the early 1930's Stock
and Cucuel[48] reported the average of total mercury
in air to be 0.02 ng/m^3. And by analyzing the air
in 20 homes and their adjacent outdoor areas,
Goldwater[57] found that the mercury concentrations
ranged from one to 41 ng/m^3 for the indoor samples
and from 0.0 to 14 ng/m^3 for the outdoor samples.
In a Japanese study Fujimura[58] established that the
concentration of mercury in the air ranged between
zero and 14 ng/m^3 for various non-industrial locations. The amount increased to 18,000 ng/m^3 in air
in the vicinity of a busy superhighway. In addition,
Fujimura found that the mercury concentration in
street dusts ranged between 0.018 and 0.022 ppm.
In Japan another area[58] of high mercury concentration results because large amounts of organomercurial
fungicides are used to treat rice crops. The
atmospheric mercury concentration in these areas
has been reported as high as 10,000 ng/m^3. In
Russia Sergeyev[59] indicated that the concentration
of mercury in the air was less than 10 ng/m^3.

In the United States, Cholak[60] found the average
mercury content of suspended particulate matter to
be 0.10 and 0.17 ng/m^3 in Cincinnati, Ohio,and
Charleston, South Carolina,respectively, with an
overall range of 0.03 to 0.21 ng/m^3. And in an
analysis of the air over Denver, McCarthy *et al.*[61]
found the concentration of airborne mercury to range
from 2 to 5 ng/m^3. By using an airplane they also
investigated air samples collected in areas of known
mercury mineralization and found mercury levels as
high as 60 ng/m^3. In a recently completed two year
study Williston[46] disclosed that the atmosphere at
10,000 feet 20 miles west of San Francisco contained
less than 1 ng/m^3 while the concentration of mercury
vapor in the San Francisco Bay area varied according
to the season of the year. The summer values ranged
from 2 to 50 ng/m^3, and the winter values ranged
from 1 to 25 ng/m^3. Williston also noted a correlation

13

Table 4

Mercury Concentrations of Water, Soils, Glacial Materials and
Air in Parts per Billion Except Where Noted

Description	Range	Mean	Reference
A. *Natural Waters*			
1. Rainwater	0.05-0.48	0.2	45
2. Normal stream, river and lake waters	0.01-0.1*	0.03	45
3. Oceans and seas	0.03-5.0	0.03-0.05	--
4. Stream and river waters near mercury deposits	0.5 -100	variable	45
5. Coal mine water (Donets Basin, USSR)	1.0 -10	variable	45
6. Hot springs and certain mineral waters	0.01-2.5**	0.10	45
7. Normal ground waters	0.01-0.10	0.05	45
8. Ground waters and mine waters near polymetallic sulfide deposits	1.0 -1000	variable	45
9. Oil field and other saline waters	0.1 -230	variable	45
B. *Soils and Glacial Materials*			
1. Normal soils	20-150	70	45
2. Normal tills, glacial clay, sand, etc.	20-100	50	45
3. Soils, tills, etc. near mercury deposits, sulfide deposits, etc.	up to 250 ppm variable		45
4. Soil horizons (normal)†			
A. (humic)	60-200	161	45
B.	30-140	89	45
C.	25-150	96	45

5. Soil horizons (near mercury deposits)†

A. (humic)	200-1860	480	45
B.	140-605	275	45
C.	150-554	263	45

C. *Air, Volcanic Exhalations and Soil Air*
   in Nanograms (10^9 grams) per cubic meter

1. Atmosphere	2-10	‡	45
2. Air over mercury deposits	30-1600	highly variable	45
3. Soil air over mercury deposits	0-200	highly variable	45
4. Volcanic exhalations	100-9600	#	45

*An upper limit of 17 ppb has been reported while values above 0.1 ppb probably due to natural or industrial contamination.

**Fifty parts per billion is the probable value in hot springs presently precipitating mercury minerals.

†Examples from Clyde Forks Area, Ontario.

‡Higher values in industrial areas; lower values (0.5 ppm) over oceans.

#Data only from USSR.

between increased atmospheric mercury content and smogs caused by temperature inversion. Furthermore, he postulated that the higher summer values may be due to rising soil temperatures which allow some of the mercury in the soil to escape into the air, be dissipated by the wind and precipitated by the rain. It would be reasonable to assume that the concentration of mercury in air, and thus in rainwater, would be appreciably higher in industrialized areas, primarily because of the increased consumption of fossil fuels, evaporation, and mechanical losses from mercury containing industries. Analyses of soil samples collected in the St. Clair River - Lake St. Clair - Detroit River industrial complex suggest that this is true.[62] These data indicate that the mercury concentrations in soils that are distant or windward from industrialized areas averaged around 0.03 ppm while in or adjacent to the highly industrialized regions the mercury concentrations were as much as ten times higher. For comparative purposes, in 1967 the American Conference of Governmental Industrial Hygienists established that the toxic threshold limit values (TLV) for an 8-hour exposure in the air are 100 $\mu g/m^3$ for mercury and inorganic mercury compounds and 10 $\mu g/m^3$ for organic mercury materials.[63]

Since mercury vapor is present in the atmosphere, it also must be present in rainwater although very little information about this is available in the literature. In one study[48] of only 17 samples mercury was detected in 12 of them at concentrations ranging from 0.05 to 0.48 $\mu g/l$. Eriksson[64] has reported the concentrations of mercury in rainwater to range between zero and 0.20 ppb. By using low temperature neutron activation analysis techniques, Brune[65] found that the mercury content of rainwater (presumably in Sweden) was 0.33 ppb. While these values confirm the presence of mercury vapor in rainwater and air, the wide variability of the values must reflect the increasing pollution burden resulting from the relatively large quantities of mercury that escape from highly industrialized areas into the atmosphere.

The existence of mercury in the marine ecosystem was first indirectly documented by Proust.[66] Proust used sulfuric acid to convert salt from the sea into hydrochloric acid. At that time sulfuric acid was commonly produced from sulfur trioxide which was obtained by roasting sulfide ores. These ores frequently contained mercury impurities which, although

Proust concluded that the mercury originated in the
sea water, could also have been introduced from the
sulfuric acid. Later Garrigou,[67] Willm,[68] and
Bardet[69] confirmed the existence of mercury in some
mineral waters in France. Since the lakes, rivers,
and streams of the world are the primary method of
transporting mercury in the environment, it would
be useful to establish normal levels of mercury in
the surface, ground and ocean waters. In some
German water samples that were analyzed in the
1930's, Stock and Cucuel[48] and Stock[70] found a range
of from 0.02 to 0.07 ppb. Wilkander[71] established
a mean value of 0.05 ppb of mercury in his analyses
of river water and drainage water from cultivated
soils in Sweden. In a study of 300 natural water
samples from central Italy, Dall'Aglio[72] found a
range of from 0.01 to 0.05 ppb of mercury. In his
investigation of the groundwater of Tuscany, Dall'Aglio
reported the mercury levels to range from 0.01 to 0.05
ppb whereas waters from areas that contain mercury
mineralizations contained about 0.2 ppb. Dall'Aglio
was able to locate cinnabar deposits by tracing the
increasing mercury content of groundwaters. By
using low temperature neutron activation analyses,
Brune and Landstrom[73] established the mercury content
of a single Swedish groundwater sample to be 0.46 ±
0.06 ppb. In a series of replicate analyses on nine
natural water samples, Hinkle and Learned[74] found
that seven of the samples ranged between 0.2 and 0.7
ppb, and the remaining two samples contained 1.0 and
4.2 ppb. Wershaw[75] investigated 73 surface waters
from various parts of the United States and found
that they ranged from less than the detectable limit
of 0.1 ppb to more than 5.0 ppb mercury. Recently,
Chau and Saitoh[76] reported that a lake water sample
taken from Lake Ontario near Hamilton Bay contained
an average mercury concentration of 0.048 ppb.
Based on these few papers, "normal background levels"
of mercury in surface and groundwaters might reasonably be in the range of 0.02 to 0.7 ppb, depending
on the availability of mercury mineralization or
pollution in the respective area.

The normal background mercury levels vary widely
in unpolluted fresh water lakes, streams, and rivers
because of the complex ions that occur naturally in
these waters. Many parameters affect the movement
of mercury and other metal ions through the environment. These parameters include the pH,
temperature, redox potential, alkalinity, and
various naturally occurring chelating agents. The

variations among these and other unknown or misunderstood parameters can increase the mercury that is available to a system many times beyond what would be considered the "normal background levels."[77-81]

The values reported for the mercury content of ocean waters also vary greatly. In ocean water the reported values fall between 0.03 and about 3 ppb.[51,65,82-87] Hosohara, *et al.*[85] reported that the total mercury content of the waters of Minamata Bay, Japan, ranged between 1.6 and 3.6 ppb, probably reflecting the serious pollution of that area. The plankton was found to contain from 3.5 to 19 ppm of mercury which suggests that considerable amounts of mercury in sea water are consumed or adsorbed by plankton. The high values of mercury in the water are probably partly due to this increased mercury content of the organic constituents. Also, these data indicate that large quantities of mercury have been deposited in the bay as the result of manufacturing processes that utilize mercury as a catalyst.[38] It has been estimated that in excess of 600 tons of mercury have been deposited in the bay from these manufacturing processes.[88]

In another study Hosohara[86] found that the vertical distribution of mercury in the Ramapo Deep area of the Pacific Ocean increased with the depth, the values of 0.15 to 0.27 ppb of mercury were obtained at the surface and at 3000 meters respectively. Leatherland, *et al.*[89] reported the mercury concentration of four surface ocean water samples from the Northeast Atlantic to be between 0.013 and 0.018 µg/l. The subsurface water samples ranged between 0.003 and 0.020 µg/l. These values agree well with the early work of Stock and Cucuel,[48] but they are substantially lower than the 0.06 to 0.27 µg/l that Hosohara[86] reported in the deep ocean water of the Northwestern Pacific. These large variations may simply be a function of geographical location and reflect the hydrothermal and volcanic nature of the Northwestern Pacific basin. Harriss[90] determined the mercury content of deepsea manganese nodules taken from various locations in the Pacific, Indian and Atlantic Oceans. The results were highly variable, ranging from less than 1 ppb to 810 ppb. However, distinct regional differences in the mercury content of the nodules was apparent. For example, all the manganese nodules with mercury contents greater than 58 ppb were taken from regions of the ocean basins characterized by volcanic and hydrothermal features. Therefore, Harriss suggested that the major sources of mercury to deepsea manganese nodules are submarine volcanic and hydrothermal emanations. Based on the available data, the concentrations of mercury which can be assumed to be due to natural environmental mercury range from 0.03 to 5.0 ppb with a normal mean value varying near 0.03 ppb. However, these data demonstrate that variations can be much greater depending on both the depth and geographical location where the water sample is taken. The mercury concentrations of the oceans and seas are summarized in Table 4.

MERCURY IN SEDIMENT

Mercury accumulates in sediments both from natural sources such as geological weathering and from direct and indirect sources of man's activities. In waters polluted by high mercury concentrations such as Minamata Bay, levels as great as 2010 ppm on a dry weight basis have been reported.[40] Normally, river, lake and ocean sediments contain below 70 to 100 ppb because they do not receive substantial amounts of mercury from geological weathering or direct manmade pollution.[45]

Turekian and Wedepohl[91] estimated that sedimentary rocks have mercury levels in the range of parts per billion. In a study of mercury in pelagic sediments along a traverse of the East Pacific rise, Bostron and Fisher[92] found mercury concentrations ranging between 1.0 and 400 ppb. Burton and Leatherland[93] found that the surface of bottom mud from different parts of the Southhampton estuary in the English Channel had mercury levels ranging from 190 to 640 ppm. The mud from underlying anoxic layers in the same location contained 2200 to 5700 ppb. These amounts at 10 cm below the surface show how reducing conditions can increase the accumulation. D'Itri, *et al.*[94] studied the mercury levels in an oligotrophic lake and an eutrophic lake in a remote underdeveloped area of Northern Michigan. While the mercury content of the soils surrounding the two lakes was approximately the same, the eutrophic lake sediments (*vis.* 0.30 to 1.25 ppm dry weight) contained substantially more mercury than the oligotrophic lake sediments (*vis.* 0.03 to 0.12 ppm dry weight).

In areas that receive some form of manmade mercury pollution, these levels can be substantially higher. For example, an investigation of the mercury levels in the St. Clair River and Lake Erie indicate dry weight mercury levels ranging from less than 500 to 9200 ppb. The higher levels result from known mercury discharges.[95] The average total mercury content of Lake Erie in the vicinity of the Raisin River ranged from 190 to 530 ppb on a dry weight basis.[96] Kennedy, *et al.*[97] established that the mercury content in the uppermost layers of 132 unconsolidated sediment samples taken from Southern Lake Michigan averaged between 100 and 400 ppb. However, the base levels of mercury on a dry weight basis found at a depth 10 to 15 cm was only 30 to 60 ppb.

With known rates of sediment accumulation, Thomas[98] was able to study the distribution of mercury in the sediment of Lake Ontario. He found that the levels of mercury in the sediments began to rise above background levels of 358 ppb by the turn of the century. Between 1900 and 1940 the mercury levels increased gradually. Then between 1940 and 1952 the levels quadrupled over the mean background level. Since 1950, mercury levels in the sediments have fluctuated although they have shown a slow steady increase to the present time.

By studying the mercury content (dry weight) of various bottom sediments in Wisconsin, Konrad[99] found that background mercury levels ranged from less than 10 to 350 ppb. High mercury deposits were found below the discharges from factories that use mercury in their manufacturing operations (7500–50,000 ppb). Moreover, significant amounts of mercury were found below several sewage treatment plants, *i.e.*, 50 to 6800 ppb. The average mercury concentrations in the sediments taken from 30 yards below the East Lansing, Michigan, sewage treatment plant are approximately 140 ± 20 ppb.[2] Klein and Goldberg[100] substantiated these data when they showed that the mercury concentration in surface sediment samples taken from near municipal sewer ocean outfalls were eight to ten times higher than similar deposits farther from the outfall.

Dissolved mercury is commonly removed from the aquatic environment through adsorption on suspended organic and inorganic particulate matter which precipitates as sediment. Mercuric and monomethylmercuric cations can be adsorbed onto selected materials with organic, quartz, and/or silicate matrices. For example, Jernelov and Freyschuss[101] reported that fish taken from a lake treated with mine tailing sludge had mercury levels of less than 10 ppb whereas similar species of fish from control lakes contained mercury levels of between 90 and 1200 ppb.

In addition, water soluble organic complexing agents such as the humates and fulvates can complex soluble mercurial species to form both water soluble and insoluble mercury complexes.[102]   These insoluble mercury complexes precipitate directly from solution into the sediments of the lake or stream whereas the soluble mercury complexes are removed through absorption to whatever organic or inorganic particulate matter is present in the lake.   Then they are removed through sedimentation.   However as the sediments become increasingly anaerobic, the precipitated mercury compounds most likely are converted to mercuric sulfide.   This reduces the possibility that it will be recycled into the aquatic environment.[103]   This removal mechanism is absent in aquatic ecosystems that are aerobic year around, and therefore the mechanism for natural decontamination of a lake or stream is related to the absorptive capacity of the dissolved and particulate matter.   Recently, Cranston and Buckley[104] established that the adsorptive capacity and surface area of the particles determine how much mercury is adsorbed.   Fine suspended matter has the greatest ability to absorb dissolved mercury (*vis.* up to 34 ppm on a dry weight basis).   Moreover, the mercury content increases exponentially with the mean specific surface area of suspended particulate matter that is less than 60 microns in size.

Thus, the contributions of mercury to the atmosphere, hydrosphere, pedosphere, and biosphere occur not only as the result of the natural geochemistry cycle of mercury but also from the industrial, domestic and mining activities of man. The generalized geochemical cycle of mercury is shown in Figure 1.

BIOLOGICAL METHYLATION

From an ecological standpoint a most serious aspect of the environmental mercury problem is that selected microbes in an aquatic environment are capable of biologically synthesizing methylmercury from mercuric (II) ion.[103,105-109] These studies indicate that the populations of microbial species, organic pollution loading, mercury concentration, temperature, pH of the system as well as other unknown parameters affect the biological methylation process.

Until the late 1960's it was generally accepted that mercury compounds and especially the elemental

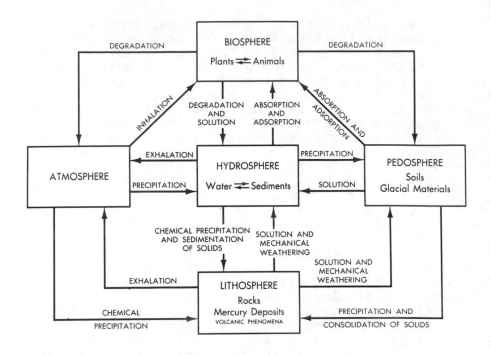

Figure 1. Generalized geochemical cycle of mercury.[45]

mercury released into the environment would be simply assimilated by the environment and, in effect, be diluted to the point where they were no longer a problem.[105] In 1967 this impression was proven incorrect by Jensen and Jernelov.[106-107] They established that inorganic mercury could be methylated in the bottom sediment contained in freshwater aquaria to form both mono- and dimethylmercury. As a possible explanation, they suggested that certain living organisms have the capacity to methylate any mercuric compounds that are present to yield mono- and dimethylmercury, and they postulated the following two reactions:

$$Hg^{++} + 2\ R\text{-}CH_3 \rightarrow CH_3HgCH_3 \rightarrow CH_3Hg^+ \tag{1}$$

$$Hg^{++} + R\text{-}CH_3 \rightarrow CH_3Hg^+ \xrightarrow{R\text{-}CH_3} CH_3HgCH_3 \tag{2}$$

These experiments provided the first evidence that the conversion of mercury compounds may involve anerobic microbes. It is interesting to note, however, that this mechanism was originally proposed as a hypothesis by Fujiki[110] in 1960 to explain the presence of thiomethylmercury in the shellfish of Minamata Bay, Japan. Fujiki suggested that the mercury could be alkylated by "plankton and other marine life." This hypothesis was rejected after it was discovered that a chemical plant using a mercury catalyst was responsible for the discharge of methylmercury into the bay. About a year later in the United States, Wood *et al.*[108] used cell-free extracts of a methanogenic bacteria to demonstrate that alkyl-B_{12} type compounds containing alkyl-cobalamin can serve as alkylating agents and convert inorganic mercury into monomethyl and dimethylmercury. In addition, they found that under mildly reducing conditions the alkyl-B_{12} compounds could serve as intermediates for a non-enzymatic methylation of the mercuric ion. The authors further postulated that if this non-enzymatic methyl transfer reaction is significant in biological systems, it would be enhanced by anaerobic conditions and by increasing the number of bacteria capable of synthesizing alkylcobalamin. Therefore, the studies of Wood *et al.*[108] substantiate a combination of the two mechanisms postulated by Jensen and Jernelov,[106-107] while at the same time describing a mechanism for a non-enzymatic conversion of the mercuric ion into dimethyl or monomethylmercury. Recently, a more complete series of possible reaction mechanisms for the enzymatic biological synthesis of monomethyl and dimethylmercury has been postulated by Wood.[109] These mechanisms are based on the reaction of mercuric ion with B_{12}-containing enzyme systems. The acetate synthetase mechanism with anaerobic organism *Clostridium thermoaceticium* and *Clostridium sticklandii* is given in Figure 23 where:

1. $NADP^+$ is oxidized nicotinamide-adenine dinucleotide phosphate
2. NADPH is reduced nicotinamide-adenine dinucleotide phosphate
3. THF is tetrahydroflolate
4. N^5CH_3 is N^5-methyltetrahydrofolate
5. B is usually the base 5,6-dimethylbenzimidazole, but a variety of basic ligands are able to coordinate to the cobalt atom.

Bertilsson and Neujahr[111] described the following mechanisms for the methylation of mercury:

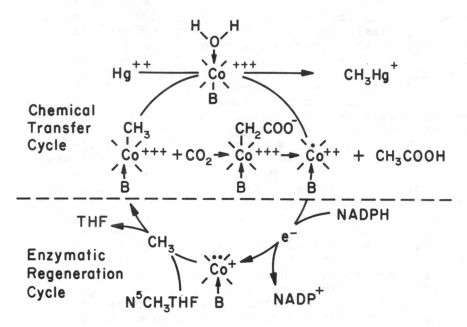

Figure 2. *An anaerobic biological synthesis mechanism for methylmercury.*[109]

$$CH_3^+ + Hg^\circ \to CH_3Hg^+ \qquad (3)$$

$$CH_3\cdot + Hg^+ \to CH_3Hg^+ \qquad (4)$$

$$CH_3:^- + Hg^{++} \to CH_3Hg^+ \qquad (5)$$

$$CH_3:^- + RHg^+ \to CH_3HgR \qquad (6)$$

where RHg^+ is an organomercury cation. Therefore, in order for either Hg^{++} or RHg^+ to become methylated a methyl carbanion donor is required. Presently the only naturally occurring compound capable of accomplishing this transformation is the vitamin B_{12} derivative, methylcobalamin.

These authors also describe the non-enzymatic methyl transfer from methylcobalamin to mercury compounds forming methylmercury as well as other methylated mercury compounds that have toxicological properties. Moreover, the reactivity of mercury chloride towards methylcobalamin is greater than the reactivity of several other organomercurials tested. Therefore, these data are inconsistent with the postulate of dimethylmercury formation as the

24

main methylation product because the reaction rate between methylcobalamin and monomethyl mercury is much slower than the reaction rate of methylcobalamin with mercuric chloride. Imura, *et al.*[112] have also reported on the chemical transformation from methyl-cobalamin to inorganic mercury under various conditions. They found that the transmethylation reaction proceeded at an unexpectedly high rate after methylcobalamin was mixed with inorganic mercury in neutral aqueous solutions in the absence of reducing conditions. Also, the initial product of the reaction is dimethylmercury, especially when equimolar or lesser amounts of mercuric chloride are used. Additional mercuric chloride converted the synthesized dimethylmercury to monomethyl mercuric chloride. The reaction mechanisms for these transmethylations appear to be:

$$2 \ CH_3-B_{12} \ + \ HgCl_2 \ \rightarrow \ (CH_3)_2Hg \tag{7}$$

$$(CH_3)_2Hg \ + \ HgCl_2 \ \rightarrow \ 2 \ CH_3-HgCl \tag{8}$$

Data which further confuse the mechanics of the methylation of mercury have been presented by Yamaguchi, *et al.*[113] who reported finding a compound resembling methylmercury in an area surrounding a caustic soda plant where only metallic mercury was used in the manufacturing operations. The methyl-mercury was isolated in the sludge taken from a pit where calcium carbide was added to neutralize mercury contaminated water from the electrolysis plant. Significantly, these authors also experimentally produced trace amounts of methylmercury in a reaction of inorganic mercury with amorphous carbon, *i.e.*, the material of which electrodes used in the continuous mercury cells are made. The non-enzymatic methylation of mercuric ion with methylcobalamin has many ramifications. However, more data will be required to evaluate the ecological impact of this mercury methylation mechanism.

It is important to note that the inorganic elemental or divalent mercury species is usually required before the biological or chemical methylation reaction to methylmercury will proceed. Frequently, however, the mercury compounds discharged into the environment are not in these forms. Therefore, the transformation reactions among the different compounds of mercury in nature are significant.[114] And while hundreds of different inorganic and organic compounds

have been synthesized in the past 100 years, the most important classes of mercury compounds lost to the environment can be categorized as:
1. metallic mercury, $Hg°$
2. inorganic divalent mercury, Hg^{++}
3. phenylmercury, $C_6H_5Hg^+$
4. alkoxyalkylmercury, $CH_3OCH_2CH_2Hg^+$ or alkylmercury in general.

The conversions of these types of mercury compounds into inorganic divalent mercury ions in nature are well known and have been schematically summarized and are presented in Figure 3.

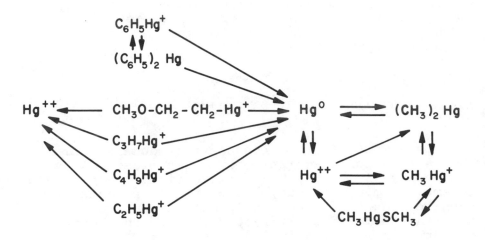

Figure 3. A model of the conversion of mercury in the aquatic environment.[105]

Moreover, it has been established that the oxidation potential in the water sediment interface of lakes and rivers is sufficient to oxidize metallic mercury ($Hg°$) to divalent mercury (Hg^{++}). This reaction occurs because mercuric ion has an absorptive affinity for organic mud. Werner[115] derived an equation to express the redox potential of the $Hg° \rightleftarrows Hg^{++} + 2e$ couple as a function of the absorptive affinity of the mercuric ion for organic mud (α) and the free mercuric ion concentration of the system.

$$E = 850 + 30 \log \frac{Hg^{++} \text{ total}}{\alpha}$$

Where

    E is the potential in millivolts required to oxidize
         metallic mercury to mercuric ion
    α is a measure of the affinity of mercuric ion for
         organic mud
    30 is the Nernst constant involving 2 electrons
    850 is the standard reduction potential (*vs.* the
         standard hydrogen electrode) for the
         $Hg° \rightleftharpoons Hg^{++} + 2e$ couple in millivolts.

According to Werner, α-coefficients for the binding strength of the mercuric ion-mud "complex" are of the order of 10^{21}. Therefore, if the free total mercuric ion concentration is 2 mg/l ($\sim 10^{-5}$M), less than 80 millivolts are required to effect the oxidation. Additionally, a decrease in the free total mercury ion concentration further decreases this potential. The redox potential generated by natural waters is a function of the dissolved oxygen concentration and the pH. In oxygenated waters at or near neutral pH values, a redox potential in the range of 520 millivolts is common.[116] This is far in excess of the 80 millivolts required for the oxidation of elemental mercury to mercuric ion according to Werner's equation.

Therefore, it is apparent that most mercury compounds released into the environment can be directly or indirectly transformed into the mercuric species required for either the chemical or biological methyl transfer reaction.[117] Furthermore, Bouveng[117] has suggested that there is a correlation between the elevated levels of mercury in fish living in polluted environments and the levels of phenylmercuric acetate released from paper and pulp factories or when the mercury is a result of a chloralkali operation. These data led to the postulation that methylmercury can be formed by the combination of organic carbon material with phenylmercuric acetate.

The studies of Jensen and Jernelov[106-107] and Wood, *et al.*[108] indicate that relative amounts of mono- and dimethylmercury compounds produced in a given system are a function of microbial species, organic pollution loading, mercury concentration, and the temperature and pH of that system. The data appear to indicate that at low mercury contamination levels dimethylmercury is the ultimate product of the methyl transfer reaction whereas if higher concentrations of mercury are introduced, monomethylmercury is produced. Moreover, it appears that neutral and alkaline environments favor the

27

formation of dimethylmercury[108,118] which readily
decomposes to monomethylmercury in mildly acidic
environments. Therefore, the pH and the mercury
levels can be correlated in a mercury contaminated
body of water. In summary, assuming that the flow
of mercury into a body of water has stopped, the
biological methylation process would favor the production of dimethylmercury in neutral or alkaline
waters which would tend to escape into the atmosphere because of its relatively high vapor pressure
or volatility. Conversely, if the body of water
has the slightest acidic character, the amount of
dimethylmercury that is produced and which subsequently evaporates into the atmosphere can be
drastically reduced because the transformation
equilibrium now favors the conversion of dimethylmercury to the less volatile monomethyl form. That
some lakes in remote areas with no sources of mercury
pollution still contain fish with extremely high
mercury levels may simply be a factor of their
acidity since the "natural" purging or release of
mercury into the atmosphere is reduced. The cycle
of mercury interconversions in nature which appear
to be evident is depicted in Figure 4 .

Additional Microbial Conversions
of Mercury Compounds in the Aquatic Ecosystem

It appears that some microbes are capable of
methylating mercury to form mono- or dimethylmercury
while other microbes are capable of reducing inorganic
divalent mercury to elemental mercury which can subsequently evaporate into the atmosphere. As early
as 1964, Magos, *et al.*[119] reported that *Pseudomonas
aeruginosa, Proteus* spp. and two more unidentified
microorganisms in a water supply were shown to
convert mercuric ion to elemental mercury. This
mercury was volatilized from such biological media
as tissue homogenates, plasma, and urine; and after
a latent period of about 10 hours the loss of mercury
was reported to be as high as 75 per cent. Furukawa,
et al.[120] established that *Pseudomonas* K-62, which
was obtained from the soil, decomposed various
types of organomercurials and produced metallic
mercury.
Investigating the changes that mercurial compounds undergo in the activated sludge system,
Yamada, *et al.*[121] reported that when various mercurial
compounds were added to a mercury-adsorbant sludge,
some fraction of the mercury was vaporized from the

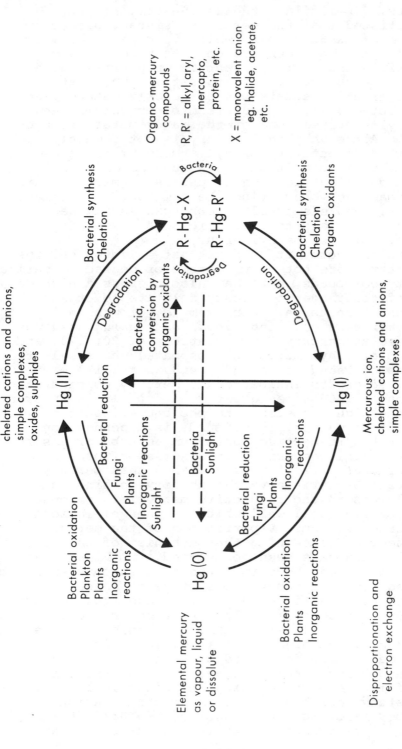

Figure 4. *The cycle of mercury interconversions in nature.*[45]

culture. In similar studies, Suzuki, *et al.*[122-123] demonstrated that the mercury-resistant bacterium *Pseudomonas* K-62 strain, which is capable of mercurial uptake and conversion, was used to remove mercurials that were present in industrial wastewater. Matsumura, *et al.*[124] established that phenylmercuric acetate can be quickly metabolized by selected soil and aquatic microorganisms, and diphenylmercury is one of the major metabolic products. Although phenylmercury is not commonly converted directly to methylmercury by microbes under aerobic conditions, Tonomura, *et al.*[125] reported that at least one species of microorganism was capable of converting phenylmercury to elemental mercury. And elemental mercury can easily be oxidized to yield the mercuric ions needed for the formation of methylmercury.

Furthermore, it appears that selected strains of bacteria are not only capable of degrading various mercury compounds, but they also possess a high degree of tolerance to mercury relative to other strains of bacteria. For example, growth inhibition of *Escherichia coli* and *Pseudomonas aeruginosa* was observed at mercury concentrations of 20, 120, and 450 ppm of ethyl mercuric phosphate, while selected strains of *Pseudomonas* and *Pseudomonas*-like bacteria exhibited growth inhibition at mercury concentrations as much as 1000 times greater.[126-128] It is apparent that the high resistance to the bacteriostatic activity of mercurial compounds is dependent on whether the respective bacterium adsorbs or absorbs the mercurial compound. In a series of papers, Tonomura, *et al.*[126-131] established that mercury-resistant bacteria have the ability to bind and loosely absorb the respective mercurial compound to the bacterial cell wall which is then biologically stimulated to cause vaporization of the mercury. Two theories concerning the mechanism of the mercury vaporization have been postulated.[128] First, the vaporization may be stimulated by a gaseous substance evolved from the bacterial surface on which the mercury compound is bound. Or second, the mercurial compound may be chemically converted into a substance such as elemental mercury which is more volatile. Conversely, the high bacteriostatic activity of mercurial compounds toward bacteria such as *Escherichia coli* and *Pseudomonas aerugenosa* may be the result of a different type of chemical or biological binding. In a study designed to determine the location of adsorbed mercuric ion in *Escherichia coli*, Harris, *et al.*[132] established that the mercuric ions are

not deposited on the outer surface of the cell.
Rather, they combine with the cytoplasmic portion
of the bacterium.

On the basis of these few studies, it is quite
probable that selected microbiological systems play
an important, but largely unknown, role in reducing
the toxicity of mercurials that are present in the
environment. This mechanism could be responsible
for the loss of mercury from stored water solutions
reported by Corner and Rigler[133] and emphasizes the
necessity of adequately preserving and analyzing
water samples as soon after collection as possible.
Furthermore, it is not beyond the realm of speculation that a balance exists among the strains of
bacteria as they relate to the mercurial compounds
present in an aquatic environment. Thus the bacteria
either convert the mercurial compounds into dimethylmercury or degrade them into mercurial species which
are volatilized into the atmosphere. This supposition
implies that bodies of water that contain fish with
high mercury levels have bacteria populations which
favor the synthesis of methylmercury.

Biological Magnification

Not only are the methylated forms of mercury
more highly toxic than other forms of mercury, but
they are also more biologically mobile. Aquatic
organisms are able to concentrate methylmercury
either directly from the water or through the food
chain.[134-137] Thus, two biological processes are
involved: the biological conversion of relatively
low toxicity inorganic mercury compounds into highly
toxic methylmercury and the biological magnification
of this material in the aquatic organism.

While mercury concentrations greater than 10-20
ppb can be toxic to aquatic organisms, sublethal
levels are also absorbed and biologically magnified.
The accumulation results from absorption of mercury
by ingestion or directly from water through the
organisms' outer surfaces (skin or epithelium) and/or
across gill membranes during respiration. Since
mercurials are at least one thousand times more
soluble in lipids than in water,[138] they are easily
extracted from water or food by contact with the
lipid portions of the tissues. That 85-95% of the
total mercury in contaminated fish[139-140] is in the
form of methylmercury is accounted for by its
affinity for sulfhydryl groups and lipids.

Moreover, the magnitude of biological accumulation of mercury by aquatic organisms is a function not only of the species and its exposure interval but also of the feeding habits (trophic level), metabolic rate, age or size of the organism, and the various water quality parameters as well as the degree of mercury pollution. While very few data are available on this biological magnification of mercury in the aquatic environment, this process appears to be a function of the particular organism at each trophic level. It has been demonstrated, for example, that pike (*Esox lucius* L.), which are at the top of the food chain, have levels of mercury in their muscle tissue that are 3,000 times greater than the level of the water in which they are taken.[134] Hannerz[135] reported that, all other factors being equal, the accumulation of mercury by aquatic organisms was more a function of their feeding habits and metabolic rates than of their trophic level in the food chain.

Furthermore, the amount of inorganic and organic dissolved and particulate matter in an aquatic environment may be significant in the accumulation of mercury by fish. D'Itri, *et al.*[94] established that rainbow trout (*Salmo gairdneri*) taken from an oligotrophic lake contained mercury levels three times greater than those of rainbow trout taken from a nearby eutrophic lake.

Jernelov[141] established that methylmercury is released from the sediments as a function not only of the depth at which it is buried but, more importantly, by the activity of the macrofauna that live in the sediments. Without macro-organisms the formation and release of methylmercury occurs almost entirely in the top centimeter of the sediment. However, if aggressive bottom feeding organisms such as *Anolonata* are present, as much as nine centimeters of the surface layer of sediment is disturbed; and methylmercury can be released into the water.

In a survey study of the literature, Chapman, *et al.*[142] reported that freshwater phytoplankton, macrophytes, and fish, potentially can biologically magnify mercury concentrations from water 1000 times while freshwater invertebrates have a 100,000 magnification factor. The freshwater invertebrate magnification factor was estimated from collateral data and is of questionable value because researchers have accumulated divergent data. Hannerz[135] reported an invertebrate biological magnification factor in the

range of 500-2000, 3300-8500, 900-4200, and 250-560
for methoxyethyl mercury hydroxide, methylmercury
hydroxide, phenylmercuric acetate, and mercuric
chloride respectively. Johnels, *et al.*[134] reported
a biological magnification factor of 3000 for northern pike while Underdal and Hastein[143] have
established biological magnification factors of
7000-10,000 for some Salmonidae and Percidae species
downstream from a wood pulp factory. The wide variation in the biological magnification is not unex-
pected considering the many diverse biological and
chemical parameters involved, and Hannerz[135] noted
as much as a tenfold variation in the mercury content
of fish of the same species that were exposed to
identical levels of mercury.

Rucker and Amend[144] established that rainbow
trout exposed to 60 ppb of ethylmercury for one
hour a day over ten days contained mercury levels
of 4,000 and 17,300 ppb in their muscle and kidney
tissues. Miettinen, *et al.*[137] have shown that
northern pike and rainbow trout are able to
assimilate and concentrate into their muscle tissues methylmercury bound to ingested food objects.
Also, several theories have been advanced to explain
the mechanism by which methylmercury is concentrated
at each trophic level of the food chain. One such
theory is that a series of transfers take place
during which the methylmercury from the water is
taken up by the phytoplankton, ingested by zooplankton, and then consumed by forage fish. Hamilton[145]
has recently shown that the levels of mercury in
fish food organisms increase at each trophic level
of the food chain. Another theory suggests that
methylmercury and the bacteria that produce it are
consumed by benthic organisms which are then preyed
upon by bottom feeding forage fish which, in turn,
fall prey to piscivorous fish. While the mechanism
of mercury accumulation is not clear, it is obviously
a function of one or more of the following: the
metabolic rate in individual fish, differences in
the selection of food objects as a fish matures, or
the fish's epithelial surface area.[135,146] In all
likelihood the actual mechanism is a complex combination of these parameters. The transfer of
mercurials in the aquatic food chain is presented
in Figure 5.

No clear trends are currently evident to explain
how the various species of fish accumulate mercury.
Generally, however, the predatory species, such as
muskellunge (*Esox masquinongy*), northern pike (*Esox*

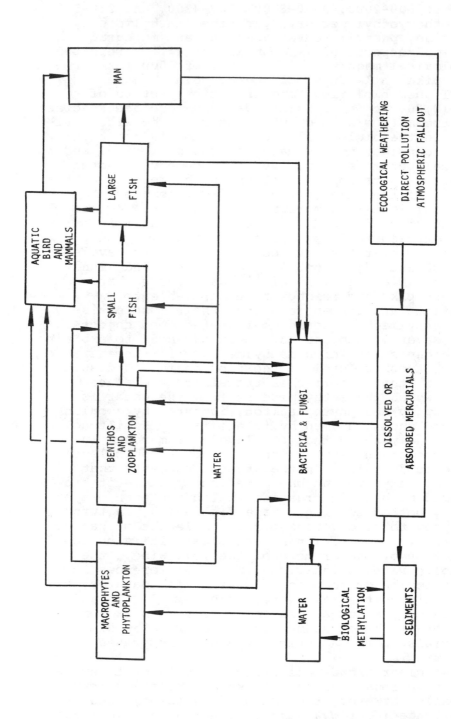

Figure 5. A generalized transfer scheme for mercury in the aquatic ecosystem to man.

lucius), walleye (*Stizostedion vitrium*), and lake
sturgeon (*Acipenser fulvescens*), accumulate the
highest quantities of mercury while herbivores,
such as the black bullhead (*Ictalurus melas*), carp
(*Cyprinus carpo*), gizzard shad (*Dorosoma cepedianum*),
and silver redhorse (*Moxostoma anisurum*), accumuate
the least. It is recognized that these are quite
broad categories because while the mechanism by
which mercury accumulates and concentrates in fish
has not been fully explained, at least seven factors
are involved: species of fish, exposure interval,
degree of mercury pollution, age and size of fish,
the metabolic rate of individual fish, differences
in the selection of food as the fish mature, and the
ephithelial surface area of the fish. Additionally,
the biological magnification process is also affected
by several aspects of water quality including
temperature, pH, organic pollution loadings, hardness, alkalinity, heavy metal loadings, and dissolved
oxygen. The accumulation mechanism is probably a
complex combination of all these biological and
chemical parameters.

Moreover, with respect to biological magnification, the respective mercury levels for different
species of fish can only be validly compared with
fish taken from the same body of water because of
the many different environmental and water quality
factors. Therefore, assuming that all other
parameters are roughly equal within a selected
aquatic environment, different species of fish can
be placed into broad categories according to their
ability to accumulate mercury. Recently, Bails[147]
arranged some species of fish taken in the St. Clair
River, Lake St. Clair, the Detroit River and the western basin of the Lake Erie area of the Great Lakes
into four broad categories based on their 1970
mercury levels. These categories and the species
of fish in each category are listed in Table 5.
The fish used in this study were taken from a large
geographic area which is not a closed system. Rigid
categories, therefore, cannot be established because
fish migration and mercury pollution levels vary
throughout the waterways. Also, some latitude must
be allowed for species to interchange adjacent
categories. In a closed ecological system such as
a lake where equilibrium conditions exist in the
mercury pollution loadings and exposure times, the
corresponding categories would be expected to be
more exact.

Table 5

Scientific Names and 1970 Mean Mercury Levels of Fish Taken from Lake St. Clair and the St. Clair River

Common Name	Scientific Name	Category	1970 Mean Mercury Levels (ppm)
Muskellunge	*Esox masquinongy*	I	>3.0
Lake Sturgeon	*Acipenser fulvescens*		
Northern Pike	*Esox lucius*		
Sauger	*Stizostedion canadense*	II	2.0-3.0
Smallmouth Bass	*Micropterus dolomieui*		
Walleye	*Stizostedion vitreum*		
White Crappie	*Pomoxis annularis*		
Channel Catfish	*Ictalurus punctatus*		
Drum (freshwater)	*Aplodinotus grunniens*		
Longnose Gar	*Lepisosteus asseus*		
Rock Bass	*Ambloplites supestris*	III	0.5-2.0
White Bass	*Morone chrysops*		
Yellow Perch	*Perca flavescens*		
Black Bullhead	*Ictalurus melas*		
Bluegill	*Lepomis macrochirus*		
Carp	*Cyprinus carpio*		
Quillback	*Carpiodes cyprinus*		
Gizzard Shad	*Dorosoma cepedianum*	IV	<0.5
Goldfish	*Curassius auratus*		
Largemouth Buffalo	*Ictiobus eyprinellus*		
Silver Redhorse	*Moxostoma anisurum*		
Rainbow Smelt	*Osmerus mordax*		
Yellow Bullhead	*Ictalurus natalis*		

Preserved Museum Fish as
Mercury Pollution Indicators

In Sweden Berg, *et al.*[148] used neutron activation procedures to analyze the mercury content of feathers on bird skins preserved in several Scandinavian University museums. They were able to correlate a sharp increase in the mercury content of the feathers of seed eating birds with the introduction of alkyl-mercury fungicides as grain seed treatments in the early 1940's. Moreover, the research group also demonstrated a slow but significant increase in the amount of mercury in the feathers of fish-eating birds after 1900. The authors speculated that this increase in the mercury levels of fish-eating birds paralleled Sweden's industrial growth and the attendant increased losses of mercury in the environment.

Recently, Evans, *et al.*[149] also used this method to test 57 preserved fish obtained from the University of Michigan, Michigan State and Wayne State Universities. Their data provide information on environmental mercury levels in various species of fish from the St. Clair River, Lake St. Clair, the Detroit River, and the western basin of Lake Erie, all part of the Great Lakes region. The museum samples were taken between 1920 and 1965. Figure 6 graphically compares the mean mercury levels of preserved fish specimens from the St. Clair River and Lake St. Clair with similar fish obtained in that area during 1970. The categories are given in Table 5. All the specimens collected before 1945 show the same low categorical mercury levels except muskellunge which registered high levels even prior to 1945. The samples taken in 1965 indicate that all specimens except those in category IV had elevated mercury levels in comparison with the 1945 levels, and the various concentration rates had begun to separate species by mercury levels.

Fish samples from the western basin of Lake Erie and the Detroit River indicated that mercury levels were approximately half those found for the same species of fish in Lake St. Clair and the St. Clair River during 1970 and 1971. This categorization of species according to their 1970 mercury content, given in Table 5 applied equally well to fish taken from Lake Erie and the Detroit River. Unfortunately, muskellunge were not taken from the western basin of Lake Erie or the Detroit River during 1970 or 1971, and Category IV is not represented. However,

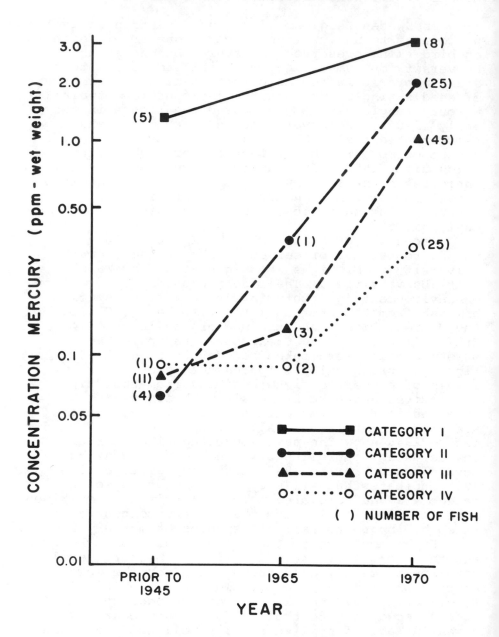

Figure 6.   *Comparison of mean mercury levels of preserved museum fish specimens from Lake St. Clair and the St. Clair River with mean mercury levels of fresh fish taken in 1970 from the same area.*

the other three categories were represented by a number of species during 1970 and 1971. In Category I, the western basin Lake Erie and lower Detroit River samples averaged less than 0.3 ppm mercury; the Category II samples averaged between 0.3 and 1.0 ppm; and Category III fish averaged between 1.0 and 2.0 ppm mercury.

Figure 7 compares the 1970-71 mercury levels in fish from the western basin of Lake Erie and lower Detroit River with mercury levels in the preserved specimens taken prior to 1937 and during the interval from 1942-1964. The trends are the same as observed in Lake St. Clair. Prior to 1937, fish in categories II, III, and IV had mean mercury levels of 0.2 ppm or less. The one muskellunge taken in 1922, represented in Category I, had a mercury level of 0.37 ppm. In the 1943-64 period, the mercury levels increased slightly in the three categories of fish. By 1970, the mercury levels had risen until only fish in Category IV had mercury levels of less than 0.3 ppm.

Effect of Heat on the
Biological Magnification Problem

Temperature plays an important role in the life cycle of all aquatic organisms. And, with the nation's electric power requirements increasing yearly, it will become increasingly important to assess the impact of large masses of hot water on the inorganic and organic mercury dynamics of the lakes and streams of our nation. No data are now available with which to predict the effect of heat on the environmental mercury problem.

Because all biological processes are, to some extent, thermochemically controlled, introducing heated water into an aquatic ecosystem could have far-reaching effects on the environmental mercury problem. Although no data are available to predict the effect of an increase in the water temperature on the rate at which mercury or methylmercury are produced or biologically assimilated by aquatic organisms, the following observations are apparent.

1. The physiology of all aquatic organisms is directly affected by temperature, and this effect is usually manifested through increased rates of metabolism and the increased activity of each respective organism.[150] This higher rate of metabolism could also accelerate the rate at which aquatic organisms accumulate mercury and organomercury compounds.

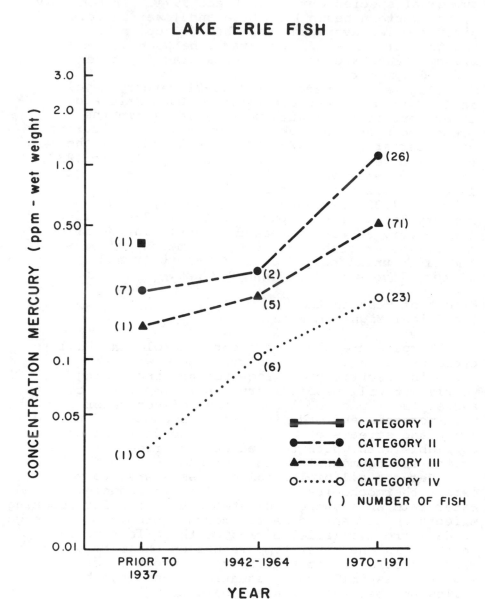

Figure 7.   *Comparison of mean mercury levels of preserved
museum fish specimens from Detroit River and
Lake Erie with mean mercury levels of fresh fish
taken in 1970 from the same area.*

2. The warmer water could not only stimulate the primary growth of fish food, but it could also extend the duration of feeding activities into the colder months.[150-151] Since the levels of methylmercury in fish bear a direct correlation with the size of the fish,[152] higher mercury levels must be expected in those fish.

3. Increased water temperatures stimulate the growth of bacteria. Therefore, if the biological methylation of mercury to methylmercury is a function of the concentrations of bacteria, increased levels of methylmercury can be expected in the benthic organisms that feed on these bacteria.

4. The chemical interconversions of various forms of mercury to mercuric ion (see Figure 4) are favored at elevated temperatures. Therefore, the mercuric ion required for the biological methylation process will be more readily available.

5. All forms of mercury, especially methylmercury, are more soluble at elevated temperatures and, therefore, are more readily available to the organisms of the aquatic ecosystem.

Mode of Mercury Toxicity
to Aquatic Organism

Various reports have summarized the toxicity of mercury compounds to aquatic organisms.[153-159] The acute toxic action of mercuric ions results from damage to the gill tissues and/or the formation of a film of coagulated mucus that fills the inter-lamellar spaces and prevents the normal movement of the gill filaments. Therefore, the necessary contact between the gill tissues and the water is interrupted, and the gaseous exchange is impeded to such an extent that the fish die from asphyxiation.[160] Lloyd[161] attributes this asphyxiation to a breakdown and swelling of the gill epithelium which obstructs the ability of the gills to exchange gases. Additionally, Meyer[162] demonstrated that mercuric ion inhibits the active uptake of sodium into the gills of goldfish and thereby causes an increased sodium loss from the fish. And Backstrom[163] has noted that because they are hypertonic, freshwater fish must continuously dispose of the water they absorb osmotically and replace the salts that are lost by diffusion as well as by excretion. Therefore, any impairment of the sodium uptake through the gills by the action of mercury may be very harmful to the fish.

There is evidence to indicate that because of the extremely low levels of mercury, the chronic and subchronic toxic action of mercury compounds, especially lipid soluble organomercurials, may not react destructively with the gill membranes. Therefore, more of the mercurial is absorbed into the blood and other internal tissues of the organism. Moreover, if the various forms of mercury react in a similar manner in aquatic organisms as in mammals and birds, similar distribution patterns and elimination rates for inorganic and organic mercurials could be expected. Because of their higher solubility in lipids, however, the alkylmercurials, especially methylmercury, are accumulated in the red blood cells and the central nervous system. Also, since alkylmercurials are inherently stable, the degradation of methylmercury to the mercuric form required for elimination from the aquatic organisms is minimal.

Although literature is abundant on the toxicity of mercury in the aquatic environment, these reports often differ widely. Many of the data relate to acute toxicity experiments of extremely short duration. Furthermore, the wide range of concentrations at which mercury is toxic to aquatic organisms indicates that the chemical and physical factors as well as the normal environmental variables affect the degree of toxicity of a particular mercurial. In general, mercury toxicities have been shown to be a function of several water quality parameters: temperature, pH, organic pollution loadings, hardness, alkalinity, heavy metal loadings, and dissolved oxygen. Furthermore, the relative toxicity of mercury to aquatic organisms varies widely with the species, its life stage, and the state of adaptation of the animals' acclimation to environmental conditions in the aquatic ecosystem. Therefore, any estimation of the toxicity of mercury or any of its compounds to aquatic organisms is extremely difficult.

Carpenter[164] stated that there is no theoretical lower limit for the toxicity of the salts of heavy metals, and even trace amounts of mercuric chloride would be toxic to fish if the exposure were long enough. Boetius[165] supported this conclusion in part, however, since the poisonous nature of mercury is accumulative, an absolute quantity of mercury must be accumulated in order to produce a lethal effect on any given organism. Boetius also

established that the product of the concentration
of mercuric ion and the survival time of a given
species is a constant which is a function of the
species, body weight, life stage, and the physical
and chemical properties of its aquatic ecosystem.

Mercury Distribution Studies in Aquatic Organisms with Radioactive Tracers

Radioactive tracers have been important aids
in studies of the uptake and elimination of mercury
from aquatic organisms. Hibiya and Oguri[166] used
radioactive mercury to study the rate of absorption
of mercury through the gills of the Japanese eel,
Anguilla japonica, and an absorption rate of
$(0.3 - 3.2) \times 10^{-5}$ mg per hour was found.
Backstrom[163] determined with detailed autoradio-
graphic techniques the distribution of radioactive
mercuric, methylmercuric, and phenylmercuric nitrate
in pike (*Esox lucius*), salmon (*Salmo salar*),
speckled trout (*Salvelinus fontinalis*), pike-perch
(*Lucioperca lucioperca*) and perch (*Perca
fluviatilis*). His data show that injections of
methylmercury caused a steady increase of mercury
in the muscles and brain of the fish while injections of mercuric and phenylmercuric nitrate
resulted in an increasing uptake primarily in
kidneys, spleen, and liver.
In a study of the distribution of mercury-203
labeled methylmercury and phenylmercury in pike,
Ohmono, *et al.*[167] established that stomach activity
remained high for three weeks, although it slowly
decreased after the first week when methylmercury
nitrate was administered orally by means of an
injection with plastic canal. Radioactivity in the
flesh reached a maximum after one week, and even
after 3 weeks nearly 50% of the administered methylmercury was located in the flesh. Moreover, the
radioactivity of all other organs continued to
increase for at least 3 weeks after administration.
Conversely, phenylmercuric nitrate concentrated
principally in the stomach, intestines, and liver.
Therefore, methylmercury nitrate spread rapidly and
widely throughout all organs and tissues of the
fish while the overall distribution of phenylmercuric
nitrate was much less extensive. Three weeks after
its administration, more than 70% of the phenylmercuric nitrate was still in the digestive organs
and kidneys while none was observed in the nervous

Table 6

Biological Half-Life of the Slow Component of Excretion for Ionic and Protein-Bound Mercurials

Aquatic Organism	Mercury Form	Method of Administration	Duration of Experiment (days)	Temp °C	Fast Excretion as % of Original Activity (days)	Slow Excretion Biological 1/2 Life (days)	Reference
1. Flounder (*Pleuronectes flesus*)	MMN	POP	100	13-19	33±4%	780±120	168
2. Flounder (*Pleuronectes flesus*)	MMN	POI	100	13-19	12±4%	700±50	168
3. Flounder (*Pleuronectes flesus*)	MMN	IMI	100	13-19	-- --	1200±400	168
4. Flounder (*Pleuronectes flesus*)	MMN	OBS	28	13-19	17±3	169±61	170
5. Flounder (*Pleuronectes flesus*)	MMN	OBS	77	13-19	23±2	430±115	170
6. Flounder (*Pleuronectes flesus*)	PMN	OBS	28	13-19	28±10	58±8	170
7. Flounder (*Pleuronectes flesus*)	PMN	OBS	77	3-19	40±7	164±16	170
1. Perch (*Perca fluviatilis*)	MMN	OBS	9-28	13-19	12±5	112±21	170
2. Perch (*Perca fluviatilis*)	MMN	OBS	77	3-19	8±8	470±37	170
3. Perch (*Perca fluviatilis*)	MMN	OBS	77	3-19	30	>>470	170
4. Perch (*Perca fluviatilis*)	PMN	OBS	34	3-19	32	157	170
1. Pike (*Esox lucius*)	MMN	POP	130	13-19	~10%	750±50	168
2. Pike (*Esox lucius*)	MMN	POI	130	13-19	~5%	640±120	168
3. Pike (*Esox lucius*)	MMN	IMI	130	13-19	-- --	780±80	168
4. Pike (*Esox lucius*)	MMN	OBS	20	16-19	10±5	140±40	170
5. Pike (*Esox lucius*)	MMN	OBS	84	3-19	11±1	490±30	170
6. Pike (*Esox lucius*)	PMN	OBS	21-53	11-18	35-41	64±4	170
7. Pike (*Esox lucius*)	PMN	OBS	84	3-18	30	190	170
1. Roach (*Leucisus rutilus*)	MMN	OBS	8	15-17	12±4	25±9	170

1. Sea Bass (*Serranus scriba*)	MMN	OBS	60	17-23	4-6	267±27	171
1. Eel (*Anguilla vulgaris*)	MMN	POP	130	13-19	~17	910±40	168
2. Eel (*Anguilla vulgaris*)	MMN	POI	130	13-19	~17	1030±70	168
3. Eel (*Anguilla vulgaris*)	MMN	IMI	130	13-19	-- --	1030±80	168
1. Mollusc (*Tapes decussata*)	MMN	IMI	99	17-23	20	481±40	171
2. Mollusc (*Tapes decussata*)	MC	IMI	--	-- --	-- --	~100	169
3. Mollusc (*Tapes decussata*)	MC	ASW	--	-- --	-- --	10 days	169
4. Mollusc (*Tapes decussata*)	MC	RSF	--	-- --	-- --	5 days	169
5. Mollusc (*Lymnaea stagnalis*)	MMN	OBS	14-29	18±2	12%	27±13	170
6. Mollusc (*Mytilus galloprovincialis*)	MMN	IMI	98	17-23	20	1000	171
1. Mussel (*Pseudanodonta complanata*)	MMN	IMI	275	13±6	7	435	170
2. Mussel (*Pseudanodonta complanata*)	MMN	IMI	63-106	17±3	10±3	129±22	170
3. Mussel (*Pseudanodonta complanata*)	MMN	IMI	54-106	17±3	9±5	86±10	170
4. Mussel (*Pseudanodonta complanata*)	PMN	IMI	6-40	8±2	3±3	43±7	170
5. Mussel (*Pseudanodonta complanata*)	MN	IMI	12-15	8±2	3±3	23±6	170
1. Crayfish (*Astacus fluviatilis*)	MMN	IMI	63	18±2	5	297	170
2. Crayfish (*Astacus fluviatilis*)	MMN	OBS	14-63	18±2	13±4	144±37	170
1. Crab (*Carcinus maenas*)	MMN	IMI	35	17-23	4-6	400±50	171
1. Seal (*Pusa hispida*)	MMN	POP	153	-- --	20	~500	172

*Error given as mean deviation from mean.

MMN - methylmercury nitrate   POI - per os, free ionic form
PMN - phenylmercuric nitrate   IMI - into muscle, ionic form
MC - mercuric chloride   ASW - absorption from sea water
MN - mercuric nitrate   RSF - resorption from food
POP - per os, proteinate-bound   OBS - orally administered into stomach of organism

system. Moreover, the toxicity and distribution patterns of methylmercury were the same for both "free" ionic methylmercury nitrate and "protein-bound" methylmercury nitrate which was produced by reacting ionic methylmercury nitrate with fresh bovine liver homogenate. These research findings were corroborated in a later study of pike and Rainbow trout.[137]

The length of the biological half life of some mercury compounds in aquatic organisms was demonstrated by Miettinen and co-workers.[168-172] They studied the distribution and excretion of mercuric, phenylmercuric, and methylmercuric nitrate in fish, mussels, molluscs, crayfish, crabs, and a seal. The biological half life of mercury in these water inhabitants was determined radiographically with mercury-203. The labeled mercurial was administered to the animals orally or by injection. The data are presented in Table 6. For the fish (flounder, roach pike, perch, and eels), the biological half life of methylmercury varied between 100 and 1030 days. For phenylmercuric nitrate it varied between 58 and 190 days. In each case a period of rapid excretion was followed by a period of very slow elimination of the mercury. Among the other aquatic organisms the biological half life of methylmercury nitrate varied between 27 ± 13 and 500 days. For phenylmercuric and mercuric nitrate these values ranged from 10 to 43 days.

Toxicity of Mercury in the Freshwater Ecosystem

With respect to the toxicity of inorganic mercury in the form of mercuric ion, short term studies indicate that concentrations in the range of 1 ppm are fatal to fish.[165,173-174] For long term exposures of 10 days or more, mercury levels as low as 10 to 20 ppb have been shown to be fatal to fish. Uspenskaya[175] established that concentrations of 10 and 20 ppb of mercury were fatal to *Phoxinus* in 80-92 and 19-32 days respectively.

Belding[176] established that the lowest concentration of mercuric chloride fatal to brook trout within 24 hours was 12.5 ppm. However, Rushton[177] reported that 10 ppm of mercuric-chloride killed rainbow trout (*Salmo gairdneri*) in only 15 minutes.

Since 1914 when organomercurials were first introduced into agriculture, increasingly larger quantities of them have been lost to the environment.

While the organomercurials are generally more toxic
to aquatic organisms than inorganic mercury compounds are, fish are able to survive relatively
higher concentrations with little ill effect for
short periods of time. Many of the early fish
toxicity studies with respect to organomercurials
resulted from the use of pyridylmercuric acetate
for the control of external parasites. It has been
used extensively for this purpose in freshwater
hatcheries.[144,178-192] For example, Rucker and
Whipple[182] report that steelhead trout (*Salmo
gairdneri*) fry and three-inch blueblack salmon
(*Oncorhynchus nerka*) are able to survive in 10 ppm
of pyridylmercuric acetate for one hour with no
toxic effects. Willford[191] reported on 24 and 48
hour LC_{50} values for pyridylmercuric acetate concentrations that range between 2.6 to 16 ppm and
2.5 to 12.8 ppm respectively for six different
species of freshwater fish. Clemmens and Sneed[188,192]
published data on an extensive series of bioassay
studies to determine the effect of pyridylmercuric
acetate on channel catfish fingerlings (*Ictalurus
punctatus*). They report that at LC_{50} values of 0.39
the fish lasted 72 hours, whereas at 3.2 ppm, they
endured only 24 hours.

Hamamoto[193] compared the tolerance of carp
(*Cyprinus carpio*) to phenylmercuric acetate,
propionate, butyrate, and valerate. He reported
LD_{50} values that range from 154 to 275 ppb for 24
hours and 76 to 208 ppb for 48 hours. The increased
toxicity of alkylmercurials over arylmercurials was
established in another study by Hamamoto.[194] The
24 and 48 hour LD_{50} values for the homologous series
of ethyl, propyl, butyl, and amylmercuric chloride
were found to range from 43-132 ppb and 8-75 ppb
respectively. Akiyama[195] reported on the 24-hour
median freshwater tolerance limit (TLm) values of
eggs, developing embryos, fry and adults of a
cypridont fish *Oryzias latipes* for phenylmercuric
acetate and methyoxyethylmercuric chloride. In
general, methoxyethylmercuric chloride was less
toxic than phenylmercuric chloride. With methoxyethylmercuric chloride the 24-hour TLm values for
eggs, fry and adults was 0.50-1.30, 1.48 and 1.74
ppm, respectively. The values with phenylmercuric
acetate were 0.18-0.26, 0.30 and 0.21 ppm.

On the basis of 120 hour bioassay tests conducted on minnows (*Notropis cornutus, Notropis
whipplii,* and *Notropis atherenoides*), Van Horn and
Balch[196] determined that the minimum lethal

concentrations of pyridylmercuric acetate, pyridylmercuric chloride, phenylmercuric acetate, and
ethylmercuric phosphate are 150, 40, 20, and 800 ppb
respectively. Rucker and Amend[144] established that
ethylmercuric phosphate has a greater rate of gill
absorption than pyridylmercuric acetate, and they
demonstrated that a different distribution of each
mercurial is formed in the various tissues of the
fish. In a later paper Amend, *et al.*[197] explained
that these biological absorption differences are
primarily due to the different chemical properties
as well as to the physiological degradation products
of each mercurial. Moreover, the poisonous effects
on rainbow trout (*Salmo gairdneri*) of a single large
exposure to ethylmercuric phosphate (125 ppb for one
hour) increased with an increase in the water temperature or chloride ion concentration or a decrease
in the dissolved oxygen content of the water.[197]

The distribution of methylmercury in pike (*Esox
lucius*) was investigated by Backstrom[163] and
Miettinen, *et al.*[137] Miettinen estimated the LD_{50}
value of methylmercury for northern pike to be 15±3
mg per kilogram of the fish's live body weight when
the methylmercury was administered as a single dose,
while 12 ppm was toxic to rainbow trout in 1.5 to 2
hours. However, when the same amount was fed to the
fish as two portions with an interval of from one
to two days, only 25 per cent of the fish died after
94 days. These experiments also established that
the cause of death for these fish was kidney failure.
Miettinen[137] also reported that the chemical form of
methylmercury administered to the fish made little
difference in its toxic effect. Moreover, the
methylmercury bound to sulfhydryl groups of proteins,
as it would exist in nature, is just as toxic as the
free unbound ionic form. Therefore, through gill
absorption and long term ingestion of contaminated
food, fish accumulate much higher levels of methylmercury than they could tolerate in a single large
dose.

Fish have also been studied to explain how
mercurial diuretics are assimilated into the urine.
Cafruny and Gussin[198] investigated the urinary excretion of 1-[3-(chloromercuri)-2-methoxypropyl]-
urea (Chlormerodrin*) and o-{[3-(hydroxmercuri)-2-
methoxypropyl] carbamoyl} phenoxyacetic acid sodium
salt (Mersalyl*) tagged with mercury-203 from the
aglomerular fish *Lophis americanus*. After the mercurials were injected into the muscles of the fish
at two separate sites rostral to the kidneys, mercury

---

*Registered name.

levels in the urine and blood plasma were determined.
The ratios of urine mercury to plasma mercury for the
acidic Mersalyl were 19.6 whereas the value was only
0.51 for the non-acidic Chlormerodrin. The authors
concluded that the mercurial transport mechanism of
fish is affected by the acidity of the mercurial.

Jackim, *et al.*[199] investigated the effects of
mercury poisoning on five liver enzymes in the
killifish (*Fundulus heteroclitus*); acid and alkaline
phosphatase, catalase, xanthine oxidase and ribo-
nuclease. After these enzymes were treated with
mercury, the effects were determined by 96 hour
bioassays. The results differed significantly from
those reported for unexposed fish. Therefore, the
authors suggested that observing the changes in
liver enzyme activity would be a useful biochemical
autopsy method to diagnose sublethal mercury
poisoning in fish.

While there appeared to be no appreciable uptake
of mercury in aquatic plant tissues, Hannerz[135] established that the mercury levels were 10 to 20 times
higher in the submerged portions than the emergant
portions because of surface adsorption. However,
levels as high as 140 µg/Kg have been reported in
the freshwater algae (*Nitella*) by Johnels, *et al.*[134]
Therefore, while little mercury is translocated from
the water into the plants, animals are able to filter
and accumulate any mercury that is present in the
aquatic ecosystem.

Recently, Dolar, *et al.*[200] investigated the
magnitude and rate of mercury accumulation by a
rooted aquatic plant, *Myriophyllus spicatum*
(watermilfoil) from both solution culture and from
mercury contaminated lake sediments. Each of the
mercury compounds affected plant growth adversely,
and most resulted in a loss of plant tissue during
the experiment. Plants grown in solution cultures
accumulated inorganic mercury more readily than
organic mercury. The concentration factors for
inorganic mercury varied between 100 and 500 times
over the initial 0.5 ppm present in the water. For
organic mercury these concentration factors varied
between 45 and 132. When watermilfoil was rooted
in sediment treated with various mercury compounds
considerable mercury accumulated in the plant
tissue which was in direct relation to the amount
of mercury in the sediment. Moreover, in contrast
to the results obtained in the solution culture
experiment, a much greater uptake occurred from the
organic rather than from the inorganic mercury forms.

These data are explained through the supposition by the authors that the inorganic forms of mercury were precipitated largely as unavailable mercuric sulfide. Nevertheless, the results of these experiments show that some rooted aquatic plants can be an important factor in the translocation of mercury from contaminated sediments.

The toxic effect of mercuric chloride on the freshwater algae *Chlorella vulgaris* was investigated by den Dooren de Jong.[201] His data show that mercuric ion was so toxic to *Chlorella vulgaris* that only 27 ppb as mercury was required to establish an inhibitory environment while the highest concentration that these organisms could tolerate was 13 ppb as mercury. Harriss, *et al.*[202] disclosed that levels of diphenylmercury, phenylmercuric acetate, methylmercury dicyandiamide and N-methyl-mercuric-1,2,3,6-tetrahydro-3,6-methano-3,4,5,6,7, 7-hexachlorophthalimide in the range of 1 to 50 ppb caused a drastic decrease in the rate of photosynthesis of the following freshwater phytoplankton: *Merismopedia sp.* (Agmenellum), *Navicula sp.*, *Crucigenia sp.*, *Straurastrum sp.*, and *Ankistrodesmus sp.*

Johnels, *et al.*[134] examined various invertebrates and determined their accumulation of mercury. The midge (*Chironomus plumosus*) accumulated 230 µg/Kg; the stonefly (*Isoperla*) had 72 µg/Kg; 65 µg/Kg in the sowbug (*Asellus aquaticus*); 49 µg/Kg in the burrowing alderfly (*Sialis*); 140 µg/Kg in zooplankton and 440 µg/Kg in the tail muscle of a crayfish. In an area receiving mercury pollution these investigators found that the sowbug, burrowing alderfly, and caddisfly had mercury levels of 2400, 5500, and 1700 µg/Kg respectively. Water fleas (*Daphnia magna*) are more susceptible to mercuric chloride than fish are. For example, six ppb could immobilize the *Daphnia* in less than 64 hours.[203] Bringham and Kuhn[204] reported 30 ppb as the threshold toxicity level of mercuric chloride for *Daphnia*. In other studies of invertebrates Hannerz[135] reported that freshwater snails (*Planorbis sp.* and *Lymnaea stagnalis*) accumulate higher levels of phenyl, methoxyethyl, and methylmercury than of mercuric chloride.

Nolan, *et al.*[205] tested 6 mercurated phenols and found them highly effective against the freshwater snail, *Australorbis glabratus* with concentrations of 3 to 10 ppm killing 100% of these organisms in 24 hours. In a later study, Bond and Nolan[206] tested 32 mercury compounds including 6 mercurated phenols studied previously[205] and found them to be highly toxic against *Australorbis glabratus*. Various inorganic mercuric salts were 100% effective in the range of 3-10 ppm as the respective compound, while most of the organic mercurial compounds induced 100% mortality of the test organisms at the 1-3 ppm level. It was noted that on the whole there is a high moluscicidal potency above a given concentration followed by a rapid loss of efficiency upon dilution. Furthermore, there appeared to be no direct correlation between the mercury content and the activity of the respective mercury compound.

McMullen[207] found that diphenylmercury and *p*-tolylmercuric chloride are moderately toxic to the amphibious snail *Oncomelania nosophora*. The 96 hour LD_{50} values of these compounds was determined to be about 40 ppm. Jones[208] reported that the flatworm (*Polycelis nigra*) has a 48-hour toxicity threshold of 200 ppb for mercuric chloride as mercury.

Hendrick and Everett[209] studied the effects of methylmercury dicyandiamide (MMD) on juvenile Louisiana red crawfish, *Procambarus clarkii*, and established the 5 day median tolerance limit (TLm) to be 1.83 ppb based on the active ingredient MMD.

Getsova and Volkova[210] investigated the ability of various insects to accumulate radioactive mercury-203. They found biological magnification ratios relative to the water concentration of mercury of 4000 for the dragonfly (*Aeschna grandis*), of 5240 for the midge (*Glyphotolaelius punctatolineatus*), and of 216 for the fly (*Eristalis tenax*). In a later study Warnick and Bell[211] established 96-hour TLm values of 2.0 mg/l as mercuric ion for the stonefly (*Acroneuria lycorias*), mayfly (*Ephemerella subvaria*), and the caddisfly (*Hydropsyche betteni*). The researchers concluded that these insects appear to be much less sensitive to the toxic effects of mercury than many fish are.

MERCURY IN THE MARINE ECOSYSTEM

Toxicity of Mercuric Ion

Very little is known about mercury in sea water. It occurs mainly as the anionic complex ions $HgCl_3^-$ and $HgCl_4^=$ which do not appear to adsorb to particulate matter as readily as the cationic forms of mercury found in fresh water. The average mercury concentration of sea water has been given as 0.03

ppb. However, much variation has been noted depending on the location of the sample (see Table

Several investigators have reported that the toxic effects of mercuric chloride are decreased in the marine environment, and this effect is primarily due to chemical changes in the mercuric ion species. Binet and Nicolle[212] studied the toxic effect of 7.4 ppm of mercuric chloride on *Gasterosteus leiurus* as the chloride ion was increased. The data show that the toxicity of mercuric chloride increased with the rising level of sodium chloride to approximately 2000 ppm. However, as the level of sodium chloride was raised further, the toxicity of mercuric chloride decreased. The minimal toxicity effects were realized at sodium chloride concentrations of about 16,000 ppm. At that point the survival time of the *Gasterosteus leiurus* was nearly doubled. Moreover, at the level of sodium chloride corresponding to sea water, *i.e.* 22,000 ppm, the survival time was still approximately 50% greater than similar fish experienced in a fresh water solution. Boetius[165] reported essentially the same results for *Gasterosteus aculeatus*. However, in similar studies, Jones[213] reported that the addition of sodium chloride did not change the toxicity of mercuric chloride to *Phoxinus phoxinus*. These divergent results may be due, at least in part, to the physiological differences between the two euryhaline *Gasterosteus* species in comparison with the freshwater *Phoxinus* species.

Mercuric chloride is a covalent compound which is normally classified as a Lewis Acid. In a fresh or distilled water environment, mercuric chloride is only slightly dissociated. Thus, molecules of $HgCl_2$ predominate while the cationic species $HgCl^+$ and Hg^{++} are present in very small quantities. Moreover, the chloride ions are so firmly bound that only anionic species $HgCl_3^-$ and $HgCl_4^=$ occur. The simultaneous equations for distilled and freshwater are:

$$HgCl_2 \rightleftharpoons HgCl^+ + Cl^- \quad K = 3.3 \times 10^{-7} \qquad (1)$$

$$HgCl^+ \rightleftharpoons Hg^{++} + Cl^- \quad K = 1.8 \times 10^{-7} \qquad (2)$$

When additional chloride ions are available in the marine environment, the equilibrium conditions described by these equations are shifted to the left. This decreases the overall concentration of $HgCl^+$ and Hg^{++} species in solution, and the formation of undissociated mercuric chloride is favored as well as the anionic complex mercury ion species of $HgCl_3^-$

and $HgCl_4^=$. These reactions are represented by the following equations:

$$HgCl_2 + Cl^- \rightleftharpoons HgCl_3^- \qquad Kf_1 = 7.1 \qquad (3)$$

$$HgCl_3^- + Cl^- \rightleftharpoons HgCl_4^= \qquad Kf_2 = 10 \qquad (4)$$

Therefore, the overall stoichiometry involved in the marine environment appears to be:

$$HgCl_2 + M^+Cl^- \rightleftharpoons MHgCl_3 \rightleftharpoons M^+ + HgCl_3^- \qquad (5)$$

$$HgCl_2 + 2M^+Cl^- \rightleftharpoons M_2HgCl_4 \rightleftharpoons 2M^+ + HgCl_4^= \qquad (6)$$

where M^+ represents a cationic metal species, most commonly sodium ion.

Boetius[145] reports that, in general, the undissociated mercuric chloride and the anionic complex species of $HgCl_3^-$ and $HgCl_4^=$ are less toxic forms to marine organisms than the cationic species of $HgCl^+$ and Hg^{++}. Moreover, he postulates that the weakened toxic effects of mercuric chloride in sea water may be directly related to the decreased concentrations of the toxic cationic species $HgCl^+$ and Hg^{++} that result from the increased availability of chloride ion in that environment.

If $HgCl^+$ and Hg^{++} are, in fact, the toxic species of mercuric chloride, its toxicity to aquatic organisms is a function of its degree of dissociation. These data are more significant for the freshwater environment because increasing dilution causes the mercuric chloride to dissociate further. Moreover, because of its very small dissociation constant, some investigators have erroneously assumed that mercuric chloride exists predominately in its undissociated form as $HgCl_2$ in extremely dilute solutions. With electrolytic conductance experiments, Boetius[165] demonstrated the relationship between concentrations of $HgCl_2$ and the dissociated toxic ion $HgCl^+$ as a function of dilution. Boetius found that a thousand fold dilution increased the dissociation one hundredfold while the concentrations of toxic ions were only decreased by a factor of 10. Thus, the dissociation of the weak electrolytic mercuric chloride is greatly accelerated in dilute solutions which, in part, explains the great "toxic stability" of mercuric chloride in the freshwater environment. Finally, Boetius noted that phenylmercuric acetate appeared to be slightly more toxic in sea water than in freshwater. This difference was attributed to either physiological causes or the possibility that the toxic substance undergoes a chemical change in sea water.

Besides noting the chemical changes in the mercuric ion species, Doudoroff[214] has suggested that the decreased toxicity of the mercuric ion to marine fish could be related to their sodium ion uptake mechanisms (see Figure 4). Doudoroff reasons that because marine fish normally ingest much sea water in connection with their osmoregulation, they have little need to absorb added salts through their gills.

Toxicity of Mercury in the Marine Ecosystem

A series of studies showed a wide variation in toxicity levels among marine organisms. These differences depend on both the species of marine organism and the chemical form of the mercurial. North and Clendenning[215] and Clendenning and North[216] reported that levels of 500 and 1000 ppb of mercury as mercuric chloride caused respective decreases of 50 and 100% in photosynthesis activity of the giant kelp (*Macrocystis pyrifera*) after 96 hours. Moreover, Harriss, *et al.*[202] established that concentrations as low as 0.1 ppb of selected organomercurial fungicides decreased both the photosynthesis and the growth of laboratory cultures of marine *Nitzschia delicatissum* as well as of some freshwater phytoplankton species. And Ukeles found that ethylmercuric phosphate (Lignasan*) was lethal to marine phytoplankton at 60 ppb, and levels of less than 0.6 ppb drastically limited the growth of these organisms.

Glooschenko[218] established that the neritic marine diatom *Chaetoceros costatum* was able to accumulate carrier free mercury-203 both in the light and dark, although slightly more in the dark. This indicates that the uptake of mercury is not dependent on photosynthesis. Moreover, cells killed with formalin accumulated more mercury per cell than any other population, showing that accumulation was due to adsorption or increased membrane permeability to mercury in the dead cells. These findings agree with the work of Krauskopf[219] and Vinogradov.[220] Krauskopf established that sea water was undersaturated with mercury and adsorption of mercury onto dead plankton could account for up to 98% of the mercury removed from sea water, while Vinogradov reported that some marine algae contain 30 ppb of mercury on a dry weight basis.

*Registered name.

Boney, *et al.*[221] tested relative toxicities of a series of alkylmercurial chlorides as well as phenylmercuric chloride, phenylmercuric iodide, mercuric chloride, and mercuric iodide on the growth and viability of sporeling of the marine algae *Plumaria elegans* and established that the organic mercurials were more toxic than the inorganic mercurials. In a later study Boney and Corner[222] compared the toxicities of mercuric and propylmercuric chloride with 6 species of marine algae as test organisms. They found that propylmercuric chloride was more toxic than mercuric chloride to sporelings of *Plumaria elegans, Polysiphonia lanosa,* and *Spermothamnion repens.* However, when *Antithamnion plumula, Ceramium pedicellatum,* and *Ceramium flabelligerum* were used as the test organisms, the toxicities of the mercuric compounds were much closer in value. These findings are consistent with the fact that the species of algae which are more sensitive to the propylmercuric chloride possess a considerable amount of lipid material in their cell membranes.

Hoffmann[223] reported on the minimum lethal concentrations of a series of organic and inorganic mercury compounds for phyto and zooplankton. Zooplankton collected in the fall were shown to be more resistant to the mercurials than similar organisms collected in the spring. Moreover, since a log to log plot of exposure time *vs* concentration of the poison was linear, Hoffmann suggested that the adsorption process plays a part in the poisoning action.

Crustaceans are also readily poisoned by various inorganic and organic mercury compounds when the mercury is accumulated primarily in the gills, antennary glands, and central nervous system in addition to its penetration into the crustacean's tissues. Corner and Sparrow[224] studied the toxicities of mercury as mercuric chloride, mercuric iodide, and ethylmercuric chloride to three marine crustaceans *Artemia saline* (Brine shrimp), *Elminius modestus* (Barnacle larvae), and *Acartia clausi* (adult copepod). Compared with *Artemia*, the other two species were found to be much more sensitive to each of the mercuric salts tested when mercuric iodide or ethylmercuric chloride were used. The 215 hour LD_{50} values for *Acartia, Elminius,* and *Artemia* were 0.05, 0.30, and 20 ppm respectively. The most resistant species, *Artemia*, required a mercurial chloride concentration of approximately 800 ppm before a 2.5 hour LD_{50} level was reached.

In a subsequent study Corner and Sparrow[225] investigated the toxicities of mercuric chloride, mercuric iodide, and methyl, ethyl, *n*-propyl, *n*-butyl, *n*-amyl, *iso*-propyl, *iso*-amyl, and phenyl mercuric chlorides to larvae of *Artemia salina* and *Elminius modestus*. All of the organic mercurials and mercuric iodide were found to be more toxic to these species than mercuric chloride. In addition, the primary alkylmercurials were more toxic than the corresponding secondary compounds. As the homologous series of primary compounds ascends, the toxicities increase. Furthermore, the toxicities of these poisons for animals are closely related to their corresponding lipoid solublites. These results support the theory that organic mercury poisons with high lipoid solublites act by penetrating the test animals more rapidly than inorganic mercuric chloride can act.

Later, Corner and Rigler[226] used radioactive mercury-203 to investigate further modes of action by mercuric chloride and *n*-amylmercuric chloride as poisons on *Artemia salina, Elminius modestus,* and *Leander serratus* (prawn). From previous studies[225] it was established that the toxicity of *n*-amylmercuric chloride relative to mercuric chloride is 20 times to *Elminius* and 1000 times to *Artemia*. They concluded that the different toxicities of these two mercurials can be explained to a considerable extent by the differences between the rates at which the compounds are taken up by both species of test animals.

Knapik[227] investigated the effect of mercuric nitrate on the survival of certain crustacean species and found a variability in sensitivity from one species to another. *Neomysis vulgaris* was the most sensitive to the toxic action of mercuric nitrate followed by *Palaemonetes varians* and *Gammarus locusta. Rhithropanopeus harrisi tredentatus* had the longest survival time of the four. The survival times for these species in an aqueous medium containing 7.65 ppm mercury as mercuric nitrate were 1.5, 2.5, 4.4, and 10.8 hours. Knapik related the toxic effects of mercuric nitrate to the structure of the animal's outer covering and, to a lesser extent, to the speed with which the particular species moved in the water. Barnes and Stanbury[228] studied the effect of mercuric chloride on the copepod *Nitocra spinipes* and reported a 24-hour LD_{50} value of approximately 0.70 ppm mercury.

In another study, Corner[229] reported on the toxic effects of mercuric and *n*-amylmercuric chloride

on the spider crab, *Maia squinado*. Mercuric chloride
mercury was found in the blood and antennary glands
in concentrations above the levels introduced into
the sea water. The animals excrete small but increasing amounts of mercury in the urine. Moreover,
most of the mercury in the blood is associated with
the protein, and the levels in the blood remain
constant for several weeks after the animal is returned to sea water. Mercury also concentrates in
other body tissues, and relatively large amounts can
be found in the gills. When *n*-amylmercuric chloride
was introduced, mercury was again concentrated in
the gills and in various internal organs. However,
the levels found in the blood are very small while
no traces of mercury are found in the urine.

Weiss[230] conducted a study in the Biscayne Bay
area of Florida to determine the comparative tolerance
of some fouling organisms to paint containing 14-90%
mercury based on the dry weight of the paints. His
data indicate the following order of decreasing
tolerance of some fouling organisms: *Polysiphonia
sp., Balanus amphitrite, Bugula neritina, Balanus
improvisus, Watersipora cucullata, Anomia sp.,
Enteromorpha sp., Hydroides parvus,* hydroids and
tunicates.

Wisely and Blick[231] reported LD_{50} values for
some species of bryozoans (*Watersipora cucullata* and
Bugula neritina), tubeworms (*Spirorbis lamellosa* and
Galeolaria caespitosa), bivalve molluscs (*Mytilus
edulis planulatus* and *Crassostrea commercialis*) and
the brine shrimp (*Artemia salina*). The 2-hour LD_{50}
concentrations (as mercuric ion) required to kill the
larvae of these species were 0.10, 0.20, 0.14, 1.2,
13.0, 180, and 1800 ppm respectively.

Hunter[232] studied the toxicity of mercuric
chloride under various conditions to the marine
amphipod *Marinogammarus marinus*. These data show
that the LD_{50} values for *Marinogammarus marinus*
immersed in sea water contained mercury concentrations ranging from 10 to 50 ppm in time spans that
ranged from 28 to 5 hours. Furthermore, since the
toxic effect of mercury was found to be independent
of the salinity or dissolved oxygen concentrations
of the sea water, Hunter concluded that mercury
acts directly on *Marinogammorus marinus* by poisoning
the organism's protoplasm whereas copper was suspected of indirectly upsetting some vital metabolic
process.

Pyefinch and Mott[233] studied the sensitivity
to mercuric chloride of the barnacles *Balanus
balanoides* and *Balanus crenatus* as well as their

larvae. Over the period following metamorphosis,
the six-hour LD_{50} values for these two species
varied with the age of the organism. In general,
it was found that *Balanus crenatus* (LD_{50} ∿ 3.0 ppm
as mercury) were less sensitive to the toxic effects
of mercury than *Balanus balanoides* (LD_{50} ∿ 0.70 ppm
as mercury). Furthermore, mercury appears to have
more pronounced toxic effects on the nauplii of
these two species of barnacles. In another study,
Clark[234] reported that 90% of the adult *Balanus
balanoides* and *Balanus eburneus* were killed at the
1 ppm level of mercuric chloride. However, the
metamorphosis of cyprids was affected only when
the levels of mercuric chloride reached about 10
ppm.

Craig[235] reported on mercury levels in two
mollusks, *Venus mercenaria* and *Spisula sp.*, from
Delaware Bay and Woods Hole. The levels of mercury
(0.82 and 0.84 ppm) in *Spisula sp.* taken at these
locations were almost identical. However, the
mercury levels reported for *Venus mercenaria* were
1.11 and 2.71 ppm for the clams taken from Delaware
Bay and Woods Hole respectively. Unfortunately,
the reliability of these data are questionable
because the mercury levels for sea water taken in the
Delaware Bay and Woods Hole areas were reported to
be 10.5 and 0.13 ppm respectively.

SUMMARY AND CONCLUSIONS

The impact of mercury on the aquatic environment
is clearly a function of the sources and levels of
mercury in that ecosystem. However, the aspect of
the problem which makes mercury more of an environmental threat than the other heavy metals is the
biological methylation and subsequent biological
magnification which it undergoes in the aquatic
environment.

Since the discovery of mercury levels as high
as 7 ppm in fish taken from Lake St. Clair during
early 1970, various state and federal agencies
throughout the United States have moved quickly to
identify and halt all known sources of direct mercury
pollution into the aquatic ecosystem. Since that
time the major direct sources of mercury pollution
have either been removed or drastically curtailed.
Nevertheless, indirectly, large quantities of mercury
are still being lost to the environment and greatly
contribute to the environmental mercury problem.
These indirect sources fall into two major categories:

the first source is the inadvertent or accidental release of mercury resulting from the use, misuse, and disposal of mercury and mercury containing industrial and consumer products. Thus, the mercury compounds incorporated in a wide variety of consumer products such as water-based paints, paper products, cosmetics, pharmaceuticals, batteries, household and laundry disinfectants, soaps, and the run-off of mercury lawn and garden fungicides ultimately find their way into the aquatic environment. The second major source of mercury in the environment occurs as a result of the exploitation of other mineral resources. Thus, mercury associated with coal, sulfide ores, and other natural minerals are released as these resources are utilized by industry.

Therefore, it is apparent that even though the direct sources of mercury pollution are being brought under control, enough mercury still reaches the aquatic environment to continue the problem through the mechanism of biological methylation which changes this mercury to methylmercury that can be biologically accumulated by the various trophic levels of aquatic organisms.

The biological conversion of inorganic mercury into methylmercury within the aquatic ecosystem is the single most important factor contributing to the environmental mercury problem. Since the conversion of inorganic mercury into methylmercury has been shown to be a reaction catalyzed by microorganisms, it is apparent that urban and industrial areas contributing both organic wastes and inorganic mercury provide an ideal environment for the biological mercury methylation process to occur.

Very little is known about the biological mechanisms in aquatic organisms that are responsible for the concentration of the various forms of mercury in their bodies. Nevertheless, it has been shown that aquatic organisms are able to concentrate alkylmercury compounds in their bodies either directly from the water or through the food chain much more effectively than either inorganic or aryl mercuricals. Moreover, the magnitude of mercury which is biologically accumulated appears to be determined by the species of the organism, its exposure interval, its feeding habits, metabolic rate, age, and size.

Finally, the literature contains many data on the lethal concentrations of the various mercury-containing compounds toward a wide variety of aquatic organisms; but the information has lead to

many conflicting interpretations. Additionally, very
few data now exist on lethal mercury levels in the
tissues of aquatic organisms. Also, at the present
time, very few experimental data are available on
which to base aquatic mercury levels which would
not be considered dangerous for aquatic organisms
under conditions of chronic exposure. Thus, it is
apparent that many questions need to be answered
before levels of safety can be established for
mercury in aquatic organisms.

ACKNOWLEDGMENTS

*The author acknowledges the support of this study by
the United States Department of the Interior, Office of Water
Resources Research.*
*The author also acknowledges and thanks Richard A. Cole
of the Michigan State University Institute of Water Research,
I. R. Jonasson and R. W. Boyle of the Geological Survey of
Canada, Ottawa, Ontario and David H. Klein of Hope College,
Holland, Michigan, for permission to reproduce illustrations,
tables or copyrighted material.*

REFERENCES

1. D'Itri, F. M. *The Environmental Mercury Problem.* A
   Report to the Michigan House of Representatives resulting
   from House Resolution 424, Great Lakes Contamination
   (MERCURY) Committee, Joseph M. Snyder, Chairman, 289 pp.
   (1971).
2. D'Itri, F. M. Unpublished Data, Michigan State University,
   East Lansing, Michigan. (1971).
3. Klein, D. H. Unpublished data, Hope College, Holland,
   Michigan (1971).
4. Goldschmidt, V. M. and Cl. Peters. *Nachr. Ges. Wiss.
   Gottingen, Math.-Physik, Klasse* (1933), in German.
   Chem. Abstr., 27:5690.
5. Goldschmidt, V. M. *Ind. Eng. Chem., 27*:1100 (1935).
6. Fuchs, W. *Ind. Eng. Chem., 27*:1099 (1935).
7. Gibson, F. H. and W. A. Selvig. *Bureau of Mines Technical
   Paper 669* (Washington, D.C.: U.S. Government Printing
   Office, 1944), 23 pp., *Chem. Abstr., 39*:1272.
8. Goldschmidt, V. M. *J. Chem. Soc., 655* (1937).
9. Inagaki, M. *J. Coal Res. Inst., 2*:229 (1951), in Janapese,
   Chem. Abstr., 49:7221a.
10. Headlee, A. J. W. and R. G. Hunter. *Ind. Eng. Chem., 45*:
    548 (1953), also *West Virginia Geol. Survey 13A,* 36
    (1955), *Chem. Abstr., 50*:551h.
11. Ruch, R. R., H. J. Gluskota, and E. J. Kennedy. *Environ-
    mental Geology Notes No. 43* (Urbana, Illinois: Illinois
    State Geological Survey, February 1971), 14 pp.

12. Joensuu, O. I. *Science, 172*:1027 (1971).
13. Karasik, M. A., A. G. Dvornikov, and V. Ya. Petrov. *Mineralog, i. Geokhim. Pivdenno-Skhidnoi. Chastini Ukr. RSR, Akad. Nauk Ukr. RSR* 53 (1963), in Russian, *Chem. Abstr., 61*:5399d.
14. Karasik, M. A., A. E. Vasilevs'ka, V. Ya. Petrov., and E. A. Ratekhin. *Geol. Zh., Akad. Nauk Ukr. RSR 22*:53 (1962), in Russian, *Chem. Abstr., 57*:2513f.
15. Dvornikov, A. G. *Dokl. Akad. Nauk SSSR, 172*:199 (1967), in Russian, *Chem. Abstr. 66*:57653g.
16. Dvornikov, A. G. *Dopov. Akad. Nauk Ukr. RSR, Ser. B, 30*: 732 (1968), in Ukranian, *Chem. Abstr., 70*:13512p.
17. Vasilevskaya, A. E., and V. P. Schcherbakov. *Dopovodi Akad. Nauk Ukr. RSR* 1494 (1963), in Russian, *Chem. Abstr. 60*:10433b.
18. Karasik, M. A., A. G. Dvornikov, G. K. Talalaev, and K. K. Zemblevskii. *Koks Khim.* 14 (1967), in Russian, *Chem. Abstr., 67*:110361j.
19. Bol'shakov, A. P. *Geochemistry International No. 3*:459 (1964). Transl. from *Geokhimiya No. 5*:477 (1964).
20. Kashkai, M. A. and T. N. Nasibov. *Geokhimiya No. 9*:1132 (1968), in Russian, *Chem. Abstr., 69*:108745k.
21. Kvornikov, A. B. *Dopov. Akad. Nauk, Ukr, RSR, Ser. B, 29*:828 (1967), in Ukranian, *Chem. Abstr., 67*:110396z.
22. Pakter, M. K., D. P. Dubrovskaya, A. V. Pershin, and G. K. Talalaev. *Koks Khim., 11*:43 (1968), in Russian, *Chem. Abstr., 70*:70038n.
23. Pakter, M. K., D. P. Dubrovskaya, A. V. Pershin, and G. K. Talalaev. *Khim. Tverd. Topl.* 145 (1967), in Russian, *Chem. Abstr., 68*:71081s.
24. Holdsworth, E. C. *Chemical Age* (London), *100*:9 (1970).
25. Aston, F. W. *Nature, 119*:489 (1927).
26. Kirby, W. *J. Soc. Chem. Ind. 46*:422R (1927).
27. White, D. E. *Geol. Soc. Am. Bull., 68*:1659 (1957).
28. Bailey, E. H., P. D. Snavely, and D. E. White. *U.S. Geol. Surv. Prof. Paper 424-D,* D306-D309 (1961).
29. Trost, P. B., Geochemical Consultant, Lakewood, Colorado. Statement made at the International Conference on Environmental Mercury Contamination, Ann Arbor, Michigan, September 30, 1970.
30. White, D. E. *Geochemistry of Hydrothermal Ore Deposits*, Barnes, H. L., ed. (New York: Holt, Rinehart, Winston, Inc., 1967), Chapter 13, pp. 575-671.
31. Institute of Geological Sciences, *Natural Environmental Research Council* (London: Her Majesty's Stationery Office, 1970), p. 240.
32. Young, W. H., and J. J. Gallagher. *1968 Minerals Yearbook* (Washington, D.C.: U.S. Government Printing Office, 1969), Volume II, p. 301.

33. Lorenz, W. C. *1968 Minerals Yearbook* (Washington, D.C.: U.S. Government Printing Office, 1969), Volume II, p. 379.

34. Klein, D. H. Reprint of a paper presented at the Mercury in the Western Environment Conference, Portland, Oregon, February 25-26, 1971.

35. Hanson, A. *Mercury in Man's Environment.* Proceedings of the Royal Society of Canada Symposium, Ottawa, Canada, February 15-16, 1971, pp. 22-33.

36. Komlev, G. A., T. N. Kleandrov, V. S. Chakhotin, L. K. Udalov, and V. F. Makarov. *Izv. Akad. Nauk, Uz. SSR, Ser. Takhn, Nauk. 8:66* (1964), in Russian, *Chem. Abstr., 62:8727h.*

37. *Bureau of Mines Information Circular 8252* (Washington, D.C.: U.S. Government Printing Office, 1965).

38. Irukayama, K. *Advan. Water Pollut. Res., Proc. 3rd Int. Conf., Munich, Germany, September, 1966, 3:153* (1967).

39. Kurland, L. T., S. N. Faro, and H. Siedler. *World Neurology, 1:370* (1960).

40. Takeuchi, T. Reprint of a paper presented at the International Conference on Environmental Mercury Contamination, Ann Arbor, Michigan, September, 1970.

41. Saukov, A. A. *Geochemistry of Mercury.* (Moscow: Academy of Sciences, 1946), also *Tr. Inst. Geol. Nauk, Akad. Nauk SSSR No. 78, Miner-Geokhim. Ser. No. 17,* (1946), 129 p. in Russian.

42. Ardin'yan, N. Kh. *Petrography, Mineralogy and Geochemistry, No. 46* (Moscow: Institute of Geology of Mineral Deposits, 1964) (translation OTS, 64-21933), 118 pp.

43. Brown, A. S. *Symposium on Geochemical Prospecting,* Ottawa, 1966, Proc., Canada Geol. Survey Paper 66-54, (1967), pp. 72-83.

44. Fredrick, G. H., and H. E. Hawkes. *Econ. Geology, 61:* 744 (1966).

45. Jonasson, I. R., and R. W. Boyle. *Mercury in Man's Environment,* Proceedings of the Royal Society of Canada Symposium, Ottawa, Canada, February 15-16, 1971, pp. 5-21.

46. Williston, S. H. *J. Geophy. Res. 73:7051* (1968).

47. Ardin'yan, N. Kh., A. I. Trortskii, and G. A. Bela. *Geochemistry International No. 4:670* (1964). Translated from *Geokhimiya, No. 7:654* (1964).

48. Stock, A., and F. Cucuel. *Naturwissenschaften, 22:* 390 (1934).

49. Saukov, A. A., and N. Kh. Ardin'yan. Compt. rend. *Acad. Sci. U.R.S.S. 32:358* (1941), *Chem. Abstr., 37:19603.*

50. Andersson, A. *Grundforbattring, 20:95* (1967), *Chem. Abstr., 69:51225j,* in Swedish.

51. Ardin'yan, N. Kh., N. A. Ozerova, and S. K. Gipp. *Tr. Inst. Geol. Rudn. Mesterozhds, Petrogr., Mineralog. i Geokhim,* 5 (1963), *Chem. Abstr., 59:7262a.*

52. Vinogradov, A. P. *Geochem.*, 641 (1962).
53. Hawkes, H. E., and J. S. Webb. *Geochemistry in Mineral Explorations*. Harper's Geoscience Series (New York: Harper and Row, 1962), 415 pp.
54. Martin, J. T. *Analyst, 88*:413 (1963).
55. Andersson, A. "The Mercury Problem," *Oikos Supplementum, 9*:13 (1967).
56. Kimura, Y., and V. L. Miller. *Anal. Chim. Acta, 27*: 325 (1962).
57. Goldwater, L. J. *J. Royal Inst. of Public Health and Hyg., 27*:279 (1964).
58. Fujimura, Y. *Jap. J. Hyg., 18*:10 (1964).
59. Sergeyev, Ye. A. *Proceeding of the First Al-Union Conference on Geochemical Methods of Prospecting for Ore Deposits*, Krasnikov, V. I., ed. (Moscow, 1957), in Russian.
60. Cholak, J. *Proc. Natl. Air Pollution Symp., 2nd* (Pasadena, Calif., 1952), p. 13.
61. McCarthy, J. H. *Mining Eng., 20*:46 (1968).
62. D'Itri, F. M., and D. H. Klein. Unpublished data, Michigan State University and Hope College, Michigan, respectively.
63. "Threshold Limit Values for 1967," Adopted at the 29th Annual Meeting of the American Conference of Governmental Industrial Hygienists, Chicago, Ill., May 1-2, 1967.
64. Erikson, E. "The Mercury Problem," *Oikos Supplementum, 9*:13 (1967).
65. Brune, D. *Anal. Chem. Acta., 44*:15 (1969).
66. Proust, J. L. *J. Phys., 49*:153 (1799).
67. Garrigou, F. *Comp. rend., 84*:963 (1877).
68. Willm, E. *Comp. rend., 88*:1032 (1879).
69. Bardet, J. *Comp. rend., 157*:224 (1913).
70. Stock, A. *Svensk Kem. Tid., 50*:342 (1938), *Chem. Abstr. 33*:20693.
71. Wiklander, L. *Grundforbattring, 21*:151 (1968), also *Geoderma, 3*:75 (1969), *Chem. Abstr., 71*:6391n.
72. Dall'Aglio, M. *Origin and Distribution of the Elements*, Ahrens, L. H., ed. (Oxford: Pergamon Press, 1968), pp. 1065-1081.
73. Brune, D. and D. Landstrom. *Radiochemica Acta, 5*: 228 (1966).
74. Hinkle, M. E., and R. E. Learned. *U.S. Geol. Survey Prof. Paper 650-D* (1969), pp. D251-D254.
75. Wershaw, R. L. "Sources and Behavior of Mercury in Surface Water," In *Mercury in the Environment*, Geol. Survey Prof. Paper 713 (Washington, D.C.: U.S. Government Printing Office, 1970), p. 29.
76. Chau, Y. K., and H. Saitoh. *Environ. Sci. Technol., 4*: 839 (1970).
77. Shcherbina, V. V. *Geochem* (1956), p. 486.
78. Mekhomina, G. I. *Pochvovedenia, no. 11*:116, *Chem. Abstr., 72*:110364r.

79. Saxby, J. D. *Rev. Pure Appl. Chem., 19*:131 (1969).

80. Kraynov, S. R., G. A. Volhov, and M. Kh. Korol'kova. *Geochem. Int., 3*:108 (1966).

81. Bayev, V. G. *Dokl. Akad. Nauk. SSSR, 181*:1249 (1968), Eng. translation in *Geochem.*, 211 (1968).

82. Ardin'yan, N. Kh. *Tr. Inst. Geol. Rudn. Mestorozhd., Petrogr. Mineralog. i Geokhim.*, 9 (1962), *Chem. Abstr., 57*:16336d.

83. Ardin'yan, N. Kh. *Izv. Akad. Nauk, Arm. SSR., Geol. i Geog. Nauki, 16*:73 (1963), *Chem. Abstr. 58*:7237a.

84. Mason, B. *Principles of Geochemistry*, 2nd ed. (New York: John Wiley and Sons, Inc., 1966), p. 187.

85. Hosohara, K., H. Kozuma, K. Kawasaki, and Tsuruta. *Nippon Kagaku Zasshi, 82*:1479 (1961), *Chem. Abstr., 56*: 5766g.

86. Hosohara, K. *Nippon Kagaku Zasshi, 82*:1107 (1961), *Chem. Abstr., 56*:4535d.

87. Saukov, A. A. *Geochemic* (Berlin, VEB-Berlag Technik, 1953).

88. Takeuchi, T. Department of Pathology, Kumamoto University, School of Medicine, Kumamoto, Japan, Personal Communication.

89. Leatherland, T. M., J. D. Burton, M. J. McCartney, F. Culkin. *Nature, 232*:112 (1971).

90. Harriss, R. C. *Nature, 219*:54 (1968).

91. Turekian, K. K. and H. Wedepohl. *Bull. Geol. Soc. Amer., 72*:178 (1961).

92. Bostrom, K. and D. E. Fisher. *Geochim. Cosmochim. Acta 33*:743 (1969).

93. Burton, J. D. and T. M. Leatherland. *Nature, 233*:23 (1971).

94. D'Itri, F. M., C. S. Annett, and A. W. Fast. *Mar. Tech. Soc. J., 5*:10 (1971).

95. *Investigation of Mercury in the St. Clair River - Lake Erie Systems*, FWQA, Great Lakes Regional Office, National Field Investigation Center, May, 1970. 108 pp.

96. Annett, C. S., M. P. Fadow, F. M. D'Itri, and M. E. Stephenson. *Mich. Acad., 4*:325 (1972).

97. Kennedy, E. J., R. R. Ruch, and N. F. Shimp. *Ill. State Geol. Sur., Environ. Geol. Notes No. 44*, March, 1971. 18 pp.

98. Thomas, R. L. "The Distribution of Mercury in the Sediments of Lake Ontario," Paper presented at the 14th Conference on Great Lakes Research, University of Toronto, April 19-21, 1971.

99. Konrad, J. G. *Wis. Dept. of Nat. Res. Research Rep. No. 74* (1971), 17 pp.

100. Klein, D. H., and E. D. Goldberg. *Environ. Sci. Tech., 4*: 765 (1970).

101. Jernelov, A. and S. Freyschuss. *Present Possibilities for Restoring Mercury Polluted Lakes* (Stockholm: Swedish Institute of Water and Air Safety, February 1968).

102. Ogner, G., and M. Schnitzer. *Science, 170*:317 (1970).
103. Fagerstrom, T. and A. Jernelov. *Water Res., 5*:121 (1971).
104. Cranston, R. E. and D. E. Buckley. *Environ. Sci. Tech., 6*:274 (1972).
105. Landner, L. *Report No. B76* (Stockholm: Swedish Water and Air Pollution Research Laboratory, August 1970), 11 pp.
106. Jensen, S. and A. Jernelov. *Nordforsk Biocindenformation, No. 10* (March, 1967), and *No. 14* (February, 1968).
107. Jensen, S. and A. Jernelov. *Nature, 223*:753 (1969).
108. Wood, J. M., C. G. Rosen, and S. F. Kennedy. *Nature, 220*:173 (1968).
109. Dunlap, L. *Chem. and Eng. News, 49*:22 (1971).
110. Fujiki, M. J. *Kumamoto Med. Soc., 39*:494 (1963), in Japanese.
111. Bertilsson, L. and H. Y. Neujahr. *Biochem. 10*:2805 (1971).
112. Imura, N., E. Sukegawa, S. K. Pan, K. Nagao, J. Y. Kim, T. Kwan, and T. Ukita. *Science, 172*:1248 (1971).
113. Yamaguchi, S., M. Matsumoto, M. Hoshide, S. Matsuo, and S. Kaku. *Arch. Environ. Health, 23*:196 (1971).
114. Landner, L. *Swedish Water and Air Pollution Research Laboratory Report No. B76*, August, 1970.
115. Werner, J. In *Chemical Fallout*, Miller, M. W., and G. G. Berg, eds. (Springfield, Ill.: Charles C. Thomas Publ., 1969), Chapter 4, pp. 68-74.
116. Hutchinson, G. E. *A Treatise on Limnology*, Vol. 1 (New York: John Wiley and Sons, 1957), Chapter 11, pp. 691-726.
117. Bouveng, H. *Modern Kemi, No. 3*:45 (1968), in Swedish.
118. Larsson, J. E. *Swedish Environmental Protection Board*, Stockholm, 44 pp. (1970).
119. Magos, L., A. H. Tuffery, and T. W. Clarkson. *Brit. J. Ind. Med., 21*:294 (1964).
120. Furukawa, K., T. Suzuki, and K. Tonomura. *Agr. Biol. Chem.* (Tokyo), *33*:128 (1969), in English.
121. Yamada, M., M. Dazai, and K. Tonomura, *Hakko Kogaku Zasshi, 47*:155 (1969). In Japanese, *Chem. Abstr., 70*: 90584r.
122. Suzuki, T., K. Furukawa, and K. Tonomura. *Rep. Ferment Res. Inst. 36*:1 (1969), *Biol. Abstr., 51*:103357.
123. Suzuki, T., K. Furukawa, and K. Tonomura. *Hakko Kogaku Zasshi, 46*:1048 (1968), in English, *Chem. Abstr. 70*: 50290k.
124. Matsumura, F., Y. Gotoh, G. M. Boush. *Science, 173* (1971).
125. Tonomura, K., K. Maeda, F. Futai, T. Nakagami, and M. Yamada. *Nature, 217*:644 (1968).
126. Tonomura, K., and F. Kanzaki. *Biochem. Biophys. Acta, 184*:227 (1969).

127. Tonomura, K., M. Maeda, F. Futai, T. Nakagami, and M. Yamada. *Nature, 217*:644 (1968).

128. Tonomura, K., T. Nakagami, F. Futai, K. Maeda, and O. Tanabe. *Rep. Ferment. Tes. Inst., 32*:25 (1967), in Japanese with English summary, *Biol. Abstr., 51*: 97053.

129. Tonomura, K., K. Maeda, and F. Futai. *J. Ferment. Technol., 46*:685 (1968).

130. Tonomura, K. *Biochem. Biophys. Acta, 182*:227 (1969).

131. Tonomura, K., K. Maeda, and F. Futai. *Hakko Kogaku Zasshi, 46*:685 (1968), in English, *Chem. Abstr., 70*: 9485q.

132. Harris, J. O., A. Eisenstark, and R. D. Dragsdorf. *J. Bacteriol., 68*:745 (1954).

133. Corner, E. D. S. and F. H. Rigler. *J. Mar. Biol. Ass. 36*:449 (1957).

134. Johnels, A., T. Westermark, W. Berg, P. I. Persson, and B. Sjostrand. *Oikos, 18*:323 (1967).

135. Hannerz, L. *Report No. 48*, Fisheries Board of Sweden, Institute of Freshwater Research, Drottningholm, (1968), p. 120.

136. Hasselrot, T. B. *Report No. 48*, Fisheries Board of Sweden, Institute of Freshwater Research, Drottningholm, (1968), p. 102.

137. Miettinen, V., E. Blankenstein, K. Rissanen, M. Tillander, J. K. Miettinen, and M. Valtonen. "FAO Technical Conference on Marine Pollution and Its Effects on Living Resources and Fishing," Rome, Italy, December 9-18 (1970).

138. Hughes, W. L. *Ann. N. Y. Acad. Sci. 65*:454 (1951).

139. Westoo, G. *Acta Chem. Scand., 20*:2131 (1966).

140. Noren, K. and G. Westoo. *Var Foeda, 19*:13 (1967).

141. Jernelov, A. *Limn. and Oceanography, 15*:958 (1970).

142. Chapman, W. M., H. L. Fisher, and M. W. Pratt. UCRL-50564 Lawrence Radiation Laboratory, University of California, Livermore, Calif. (1968), 50 pp.

143. Underdal, B. and T. Hastein. *Oikos, 22*:101 (1971).

144. Rucker, R. R. and D. F. Amend. *Prog. Fish Cult., 31*: 197 (1969), *Biol. Abstr., 51*:29800.

145. Hamilton, A. In the paper by E. G. Bligh in *Mercury in Man's Environment* (Ottawa, Ontario: Royal Society of Canada, 1971), p. 87.

146. Wobeser, G., N. O. Nielsen, and R. H. Dunlop. *J. Fish. Res. Bd. of Can., 27*:830 (1970).

147. Bails, J. D. Fish Division, Michigan Dept. of Natural Resources, Lansing, Michigan. Unpublished data, 1970.

148. Berg, W., A. G. Johnels, B. Sjostrand, and T. Westermark. *Oikos, 17*:71 (1966).

149. Evans, R. J., J. D. Bails, and F. M. D'Itri. *Environ. Sci. Tech., 6*:901 (1972).

150. MacNamara, E. E. State of New Jersey, Dept. of Conservation and Economic Development, Trenton, New Jersey, October, 1966.

151. Wurtz, C. B., and C. E. Renn. John Hopkins University Cooling Water Studies for Edison Electric Institute, RP-49, June, 1965.

152. Johnels, A. G., P. I. Persson, T. Westermark, B. Sjostrand, and W. Berg. *Oikos* (Supplement), *9*:39 (1967).

153. Swedish National Institute of Public Health. *Methyl Mercury in Fish - A Toxicological-Epidemeologic Evaluation of Risks* (Stockholm, Sweden: Nordisk Hygienisk Tiedskrift, 1971), Supplement 4, 289 pp.

154. U.S. Department of the Interior, *Mercury in the Environment, Geological Survey Professional Paper 713* (Washington, D.C., 1970), 67 pp.

155. Doudoroff, P., and M. Katz. *Sewage and Industrial Wastes, 25*:802 (1953).

156. Ellis, M.M. *Detection and Measurement of Stream Pollution*, Bulletin No. 22, U.S. Department of Commerce, Bureau of Fisheries, 1937.

157. Nelson, N. *Environ. Res., 4*:1 (1971).

158. Wallace, R. A., W. Fulkerson, W. D. Shults, and S. W. Lyon. *Mercury in the Environment - The Human Element.* (Oak Ridge, Tennessee: Oak Ridge National Laboratory, 1971), ORNL-NSF-EP-1, 61 pp.

159. McKee, J. E. and H. W. Wolf. *Water Quality Criteria.* 2nd ed. (The Resources Agency of California, State Water Quality Control Board, 1963), pp. 216-219.

160. Carpenter, K. E. *Ann. Appl. Biol., 12*:1 (1925).

161. Lloyd, R. *Ann. Appl. Biol., 48*:84 (1960).

162. Meyer, D. K. *Federation Proc., 11*:107 (1962).

163. Backstrom, J. *Acta Pharmacol. Toxicol., 27*:1 (1969).

164. Carpenter, K. E. *J. Exp. Biol., 4*:378 (1927).

165. Boetius, J. *Meddelelser fra Danmarks Fiskeri-og Havundersogelser, 3*:93 (1960), *Biol. Abstr., 37*:16971.

166. Hibiya, T., and M. Oguri. *Bull. Jap. Soc. Sci. Fish., 27*:996 (1961).

167. Ohmomo, Y., V. Miettinen, E. Blankenstein, M. Tillander, K. Rissanen, and J. K. Miettinen. Unpublished report on Studies on the Distribution of [203]Hg-Labelled Methyl Mercury and Phenyl Mercury in Pike, Department of Radiochemistry, University of Helsinki, Finland, Presented at the Fifth RIS-Symposium, Helsinki, May 19-20, 1969. 18 pp.

168. Jarvenpaa, T., M. Tillander, and J. K. Miettinen. *Suomen Kemi. B43*:439 (1970).

169. Unlu, M., J. K. Miettinen, and S. Keches. FAO Technical Conference on Marine Pollution and Its Effect on Living Resources and Fishing, Rome, December 9-18, 1970.

170. Miettinen, J. K., M. Tillander, K. Rissanen, V. Miettinen, and Y. Ohmomo. Paper presented at Northern Mercury Symposium of Nordforsk, Stockholm, October 10-11, 1968; also Miettinen, J. K., M. Tillander, K. Rissanen, V. Miettinen, and Y. Ohmomo. *Proc. 9th Conference on Radioisotopes* (Tokyo: Atomic Energy Society, Japan, 1969), pp. 474-478.

171. Miettinen, J. K., M. Heyraud, and S. Keckes. FAO Technical Conference on Marine Pollution and its Effects on Living Resources and Fishing, Rome, December 9-18 1970, 15 pp.

172. Tillander, M., and J. K. Miettinen. FAO Technical Conference on Marine Pollution and its Effects on Living Resources and Fishing, Rome, December 9-18, 1970, 9pp.

173. Jones, J. R. E. *J. Exp. Biol., 16*:425 (1939).

174. Weir, P. A., and C. H. Hine. *Arch. Envir. Health, 20*: 45 (1970).

175. Uspenskaya, V. I. *Gig. Sanit., 11*:1 (1946).

176. Belding, D. L. *Trans. Amer. Fisheries Soc., 57*:100 (1927).

177. Rushton, W. *Salmon and Trout Mag. 23*:42 (1920).

178. Van Horn, W. M., and M. Katz. *Science, 104*:557 (1946).

179. Rucker, R. R. *Prog. Fish. Cult., 10*:19 (1948).

180. Burrows, R. E. and D. D. Palmer. *Prog. Fish. Cult. 11*: 147 (1949).

181. Snieszko, S. F. *Prog. Fish. Cult., 11*:153 (1949).

182. Rucker, R. R. and W. J. Whipple. *Prog. Fish. Cult., 13*: 43 (1951).

183. Rogers, E. D., B. H. Hazen, S. B. Friddle, and S. F. Snieszko. *Prog. Fish. Cult., 13*:71 (1951).

184. Dequine, J. F. *Prog. Fish. Cult., 13*:103 (1951).

185. Foster, R. F., and P. A. Olson. *Prog. Fish. Cult., 13*: 129 (1951).

186. Allison, R. *Prog. Fish. Cult., 19*:58 (1957).

187. Allison, R. *Prog. Fish. Cult., 19*:108 (1957).

188. Clemmens, H. P. and K. E. Sneed. *Prog. Fish Cult., 20*: 8 (1958).

189. Clemmens, H. P. and K. E. Sneed. *Prog. Fish Cult., 20*: 147 (1958).

190. Hammer, G. L. *Prog. Fish Cult., 22*:14 (1960).

191. Willford, W. A. In *Investigations in Fish Control*, Bureau of Sport Fisheries and Wildlife, Fish and Wildlife Service (Washington, D.C., Government Printing Office, April, 1967).

192. Clemmens, H. P. and K. E. Sneed. *Special Scientific Report - Fisheries No. 316*, Fish and Wildlife Service (Washington, D.C.: Government Printing Office, 1959), 10 pp.

193. Hamamoto, Y. *Nippon Nogei Kagaku Kaishi, 34*:994 (1960).

194. Hamamoto, Y. *Nippon Nogei Kagaku Kaishi, 34*:997 (1960).

195. Akiyama, A. *Bull. Jap. Soc. Sci. Fish.*, *36*:563 (1970).
196. Van Horn, W. M. and R. Balch. *Tappi, 38*:151 (1955).
197. Amend, D. F., W. T. Yasutake, and R. Morgan. *Trans. Amer. Fish. Soc., 98*:419 (1969).
198. Cafruny, E. J. and R. Z. Gussin. *J. Pharmacol. Exp. Therap., 155*:111 (1967).
199. Jackim, E., J. M. Hamlin, and S. Sonis. *J. Fish. Res. Bd. Can. 27*:383 (1970).
200. Dolar, S. G., D. R. Keeney, and G. Chesters. *Environ. Lett., 1*:191 (1971).
201. den Dooren de Jong, L. E. *Antonie van Leeuwenhoek, 31*: 301 (1965).
202. Harriss, R. C., D. B. White, and R. B. Macfarlane. *Science, 170*:736 (1970).
203. Anderson, B. G. *Trans. Am. Fish. Soc., 78*:96 (1948).
204. Bringmann, G. and R. Kuhn. *Gesundheits-Ing., 80*:115 (1959).
205. Nolan, M. O., H. W. Bond, and E. R. Mann. *Am. J. Trop. Med. and Hyg., 2*:716 (1953).
206. Bond, H. W. and M. O. Nolan. *Am. J. Trop. Med. and Hyg. 3*:187 (1954).
207. McMullen, D. B. *Am. J. Trop. Med. and Hyg. 1*:671 (1952).
208. Jones, J. R. E. *J. Exp. Biol., 17*:408 (1940).
209. Hendrick, R. D. and T. R. Everett. *J. Econ. Entom., 58*:958 (1965).
210. Getsova, A. B. and G. A. Volkova. *Proc. Fed. Amer. Soc. Exptl. Biol., 24*:683 (1964).
211. Warnick, S. L. and H. L. Bell. *J. Wat. Poll. Control Fed., 41*:280 (1969).
212. Benet, L. and P. Nicolle. *Compt. Rend. Soc. Biol., 134*:563 (1940).
213. Jones, J. R. E. *J. Exp. Biol., 17*:325 (1940).
214. Doudoroff, P. In *The Physiology of Fishes, Volume II, Behavior*, Brown, M. E., ed. (New York: Academic Press, 1957), pp. 403-430.
215. Clendenning, K. A., and W. J. North. In *Proc. 1st Int. Conf. on Waste Disposal in the Marine Environment* (New York: Pergamon Press, 1960), p. 82.
216. North, W. J. and K. A. Clendenning. *Ann. Prog. Rep. Inst. Marine Resources,* Univ. Calif. LaJolla IMR Ref. 58-11 (1958).
217. Wheles, R. *Appl. Microbiol. 10*:532 (1962).
218. Glooschenko, W. A. *J. Phycol., 5*:224 (1969).
219. Krauskopf, K. B. *Geochim. Cosmochin. Acta, 9*:1 (1956).
220. Vinogradov, A. P. *The Elementary Chemical Composition of Marine Organisms* (New Haven, Conn.: Sears Foundation 325 pp.
221. Boney, A. D., E. D. S. Corner, and B. W. Sparrow. *(Bonnem) Schm. Biochem. Pharm., 2*:37 (1959).
222. Boney, A. D. and E. D. S. Corner. *J. Mar. Biol. Ass. U.K., 38*:267 (1959).

223. Hoffmann, C. *Kiel. Meeresforsch.*, 7:38 (1950).
224. Corner, E. D. S., and B. W. Sparrow. *J. Mar. Biol. Ass. U.K.*, 35:531 (1956).
225. Corner, E. D. S., and B. W. Sparrow. *J. Mar. Biol. Ass. U.K.*, 36:459 (1957).
226. Corner, E. D. S., and F. H. Rigler. *J. Mar. Biol. Ass. U.K.*, 37:85 (1958).
227. Knapik, M. *Acta Biologica. Cracoviensia*, 12:17 (1969).
228. Barnes, H. and F. A. Stanbury. *J. Exp. Biol.* 25:270 (1948).
229. Corner, E. D. S. *Biochem. Pharmacol.*, 2:121 (1959).
230. Weiss, C. M. *Biol. Bull. Mar. Biol. Lab., Woods Hole,* 93:56 (1947).
231. Wisely, B., and R. A. P. Blick. *Aust. J. Mar. Freshwat. Res.,* 18:63 (1967).
232. Hunter, W. R. *J. Exp. Biol.*, 26:113 (1949).
233. Pyefinch, K. A. and J. C. Mott. *J. Exp. Biol.* 25:276 (1948).
234. Clark, G. *Biol. Bull. Mar. Biol. Lab., Woods Hole,* 92: 73 (1947).
235. Craig, S. *JAOAC,* 66:1000 (1967).

SECTION II

USES OF BIOASSAY,
GENERAL PRINCIPLES EMPHASIZED

CONTRIBUTORS TO SECTION II

V. M. Brown. Water Pollution Research Laboratory of the
     Department of the Environment, Stevenage, England.
John G. Eaton. United States Environmental Protection
     Agency, National Water Quality Laboratory, Duluth,
     Minnesota.
Charles E. Stephan. Newton Fish Toxicology Laboratory,
     United States Environmental Protection Agency,
     Cincinnati, Ohio.

2. CONCEPTS AND OUTLOOK IN TESTING THE TOXICITY OF SUBSTANCES TO FISH

V. M. Brown. Water Pollution Research Laboratory of the Department of the Environment, Stevenage, England

INTRODUCTION

Pollution biologists seem to have developed, by and large, in a state of unawareness of the concepts of toxicological studies in other fields. Some of these primary concepts, and the terminology applied, in preliminary toxicity studies with fish, are therefore discussed here, albeit at an elementary level, in the hope of eliminating some of the confusion which exists.

TESTING: TERMINOLOGY

Measuring the toxicity of substances to fish is sometimes described as "fish toxicology." Not only is this presumptuous, toxicology being a subject greater than the small part that toxicity testing usually represents, but worse, it is not even correct. In parallel with the correct use of "toxicology" in, for example, "heavy-metal toxicology," "fish toxicology" is the subject of the toxic properties of fish and should therefore only be used to describe such studies.

Frequently the toxicity test is referred to as a "bioassay." In its widest interpretation biological assay might be considered to include the measurement of the toxicity of substances. However, the term "bioassay" has been largely associated with the measurement of drugs, where the fundamental interest is not in measuring the degree of response associated with a particular level of stimulus (as it is in toxicity studies), but with determining

73

from the degree of response, the strength of the stimulus. Laurence and Bacharach[1] stated that "it is today accepted that bioassay cannot be legitimately so described unless a standard preparation is used both for a comparison of activity with the test material and as a means of defining the unit in which the activity is to be expressed." (Standard preparations are exemplified by those established internationally by WHO for the assay of pharmacologically active compounds and in arbitrary international units of which the potency of test preparations is expressed.) This formally excludes the use of the term to describe toxicity tests, and for this reason and to avoid confusion which arises in interpretation, it seems advisable that toxicity tests should not be described as "bioassays." Distinction between the two is made in this paper.

Next there is the problem of describing the material being tested. Toxicity tests with fish are probably most frequently carried out to determine the effects of polluting materials, and for the purposes of this paper pollution can be considered as being a change in the quality of a water caused by the addition, either deliberate or accidental, of wastes typically originating in the activities of man, which makes the water unfavorable to fish. Broadly speaking, such pollutants can conveniently be classified on the basis of the way in which they exert their effects on fish, and, in particular, whether such reactions are, in conventional terms, essentially chemical (as those "classically" associated with poisons such as metal salts, hydrogen cyanide) physical (such as those of mineral particles, radionuclides, hot water, hypertonic solutions) or biological (as those of viruses, bacteria), even though such a classification admits of some scientific and semantic ambiguities.

As this paper is about toxicity it is necessary here to define those substances known as "poisons" (or more typically in the USA as "toxicants"). A typical dictionary definition of poisons is that they are substances which, when taken into or formed in the body, destroy life or impair health.[2] However, as most substances, including those not generally considered to be poisons, have this capacity if absorbed in sufficiently large doses the description is inadequate without some quantitative limitation. This makes any definition of a

poison arbitrary, but on a *pro rata* basis from
limits quoted for man the definition of a poison
for fish would be limited to those substances
causing harm at doses of about 7×10^{-4} (or less)
of the body weight. Unfortunately the dose absorbed
by fish is very rarely measured, although we could
well do with much more information in this area and
on the elimination rate of poisons. It does not,
however, seem possible to relate in any simple way
the dose which would be received by a fish from a
knowledge of the concentration of poison present in
a water and, therefore, the attribute used here to
define poisons must be their ability to produce
harmful effects by their specific chemical properties. Consequently, substances such as chemically
"inert" suspended solids, for example, should not
be described as poisons or toxicants as they sometimes are, not only for the above reasons, but
because they are not taken into the body. Although
they may bring about chemical changes indirectly
in the affected tissues, these are not specific to
the chemical properties of the waste.

In a major contribution to toxicity studies in
recent years "toxicant" was defined as "a general
term for any environmental variable." This definition is uncritical, and so at variance with both
the etymological origin and customary usage of the
word as to seem unacceptable. The increasing habit
of describing poisons as "toxins" is also to be
deplored. Toxin is not a synonym for poison.
Toxins are a particular class of poisons, usually
albuminous substances, of high molecular weight,
produced by animals or plants. to which the body
can respond by the production of antitoxins, and
while all toxins are poisons the reverse is not
true.

THE RESPONSE CURVE: TIME AND
CONCENTRATION

Next I should like to examine what we observe
when we carry out toxicity tests and how the data
are treated. There appears to be some confusion in
this area in the literature with regard to the
relevance of the statistical methods of bioassay.
In measuring toxicity the initial objective is to
define the concentrations (and their distribution)
at which the poison is capable of producing some
selected harmful response in a population of animals
under controlled conditions (duration and nature)

of exposure. The appropriate way to do this is by use of the quantal response[3] from which the relation between concentration and percentage effect can be defined, or as Bliss[4] put it "the dosage-mortality curve is the better standard of reference for most toxicological investigations." Pollution biologists, however, have tended to use the quantitative response of survival time, although this is quite incapable of yielding anything other than a subjective estimate of effective concentrations. Nevertheless, the procedure continues to be recommended for toxicity studies with fish, often with the suggestion that it has some inherent superiority and gives a more precise definition of toxicity than do quantal data. This appears to be based on a misconception. In *bioassays* in which response time is being observed as the effect parameter, comparison of the potency of two substances solely in terms of the period of time after which 50 percent of the population has responded to each is far less informative than are data on the response time for each animal. From these latter data not only can median response times be compared but also their variances, and from these the reality of any observed differences in survival times can be assessed statistically. This condition hardly applies to toxicity testing. In toxicity testing the quantitative response and the quantal response yield different information. While both allow construction of a curve (the so-called "toxicity curve"), which relates concentration and time necessary for some particular effect to be observed, when time is the effect parameter, only the position and slope of, for example, the time-mortality curve at each test concentration can be statistically defined, with the result that the error in the observation can only be expressed in units of time (Figure 8). Quantal data, on the other hand, allow statistical definition of the position and slope of, for example, the concentration-mortality curve for each selected period of exposure and allow the error at any time to be expressed in the units in which the information is required, that is, concentration (Figure 9). The two types of information permit construction of the curves shown in Figure 10.

In making a toxicity test a decision is required as to the duration of the test and this obviously should be made on a rational basis and not primarily on one of convenience as is often the case. If risk of exposure to a poison is likely to be for

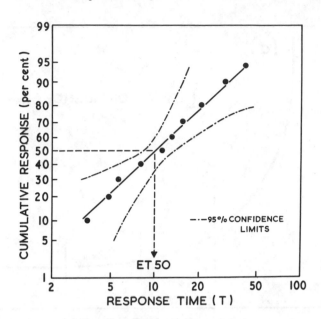

Figure 8.   *Quantitative response of a single population in concentration C_1 (distribution of individual threshold exposures C_1T_1, C_1T_2---C_1T_{10})*

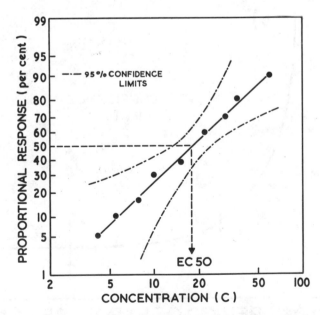

Figure 9.   *Quantal response of nine test populations after time T_1 (distribution of percentage responded in each test population under nine conditions of exposure T_1C_1, T_1C_2---T_1C_9)*

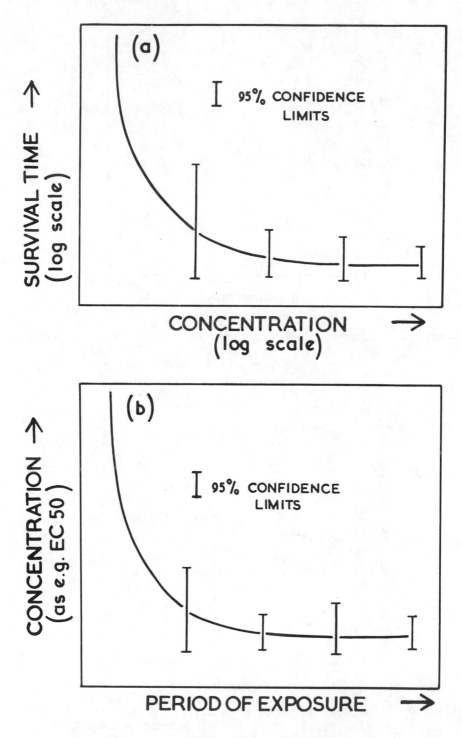

Figure 10.   *Toxicity curves using (a) quantitative data and*
(b) quantal data

the lifetime of the fish, then a test for one hour,
for example, is not enough, but then neither necessarily is any other "brief" period. Nevertheless,
irrespective of the situation to which the information is to be applied, arbitrarily selected short
periods of time are advocated as the "suitable," or
"correct" ones. Ideally, such practices should have
no place in environmental management, as they take
no cognizance of the way the fish is responding.
Fundamentally, the duration of the test should be
determined from a knowledge of the rate of onset and
of elimination of the effect selected as evidence
of poisoning in the test population. This in its
turn means that a "toxicity curve" must frequently
be described for each response of interest. The
duration of the test can, therefore, only be as-
certained empirically and, as a general principle,
recommendations for arbitrary or "standard" periods
of exposure are best ignored.

The "toxicity curve" represents another area
where problems arise with regard to the shape and
meaning of the curve. The validity and usefulness
of attempting mathematical description is questionable here. Basically, when the origin of the
properties of toxicity curves is considered, it
becomes apparent that all the various curves are
essentially one. A fundamental tenet of toxicology
is that for every substance (except perhaps carcino-
genic ones) there is some dose, no matter how small,
below which the substance will fail to exert any
harmful effect on the exposed animal. However, in
the present context, poisoning is a product of
concentration and time, and for a statement about
dose to have a precise meaning a period of exposure
must be stipulated. It may be that the statement
of effect is required for application to a lifetime's
exposure of the animal (perhaps the ideal in most
pollution studies) rather than to some lesser, or
particular, period of its life, but the difference
that the time context makes to the interpretation
of such a statement does not seem to be generally
recognized. The toxicity curve, and the outcome of
the exposure, is as much a result of what the animal
does to the poison as what the poison does to the
animal, and, as the animal is not a single unvarying
physiological entity, but is less able to tolerate
a particular poison in some states than it is in
others, this should be recognized in describing the
outcome of a test and in making any interpretation
thereof. Within these limitations a concentration

of poison will exist below which the selected
response will cease to be demonstrated and at which
the observed response remains relatively unchanged
for "prolonged" periods of exposure. Similarly,
because some finite interval of time must elapse
before a poison reaches the critical site and produces a harmful reaction, there is a concentration
range over which the exposure period required to
produce the response is relatively constant. Because of these two factors, we get the typical
exponential curves known for so many poisons, as
illustrated for three examples, A, B, C, in Figure
11.

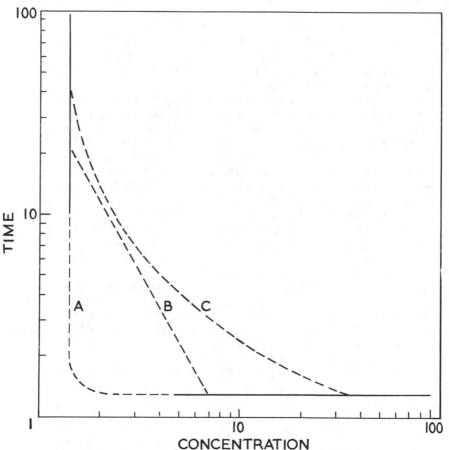

Figure 11. Toxicity curves plotted against logarithmic axes.

It is not only customary, but has been recommended, that for graphical presentation both median
response time and concentration should be

logarithmically transformed.   (Curves relating median
effective concentrations to period of exposure have
usually been presented without transformation, or
with only the time data logarithmically transformed.)
Some confusion with the reasons for making such
transformations in bioassay is evident here.   In the
initial analysis of experimental data from toxicity
tests, use of the logarithmic values of either time
(in time-effect curves) or concentration (in concentration effect curves) usually transforms the
"lognormal" sigmoid cumulative per cent effect curve
to the normal sigmoid curve with consequent analytical
advantage.   Broadly speaking, however, no such advantage is gained by logarithmic transformation of
the data of the "toxicity curve."   To call such a
transformed curve "logarithmic" is incorrect.
(Similarly it is wrong, for example, to call a
curve "harmonic" when data are plotted against the
reciprocal of time if the resultant curve is not
rectilinear, and typically such a transformation
produces a sigmoid curve.)   Plotting the logarithms
of the values is obviously of considerable advantage
where wide limits of time or concentration need to
be shown on readable scales and on a reasonable
sized piece of paper, and I would suggest that such
practicalities as these are the main reasons for
presentation in this way.   A more important consideration, however, which does not appear to have
been recognized, is that by plotting the logarithms
of the values it is possible readily to detect
whether or not a curve which on arithmetic axes
appears to have become "asymptotic" with the time
axis really has done so.   In the case of curves B
and C in Figure 12, premature termination of a test
(after say 10 h) could lead to wrong conclusions
being drawn as to the type of response and the
asymptotic concentrations.
    "Unusual" toxicity curves have been described;
these include not only those which approximate to a
logarithmic curve but also those showing more than
one inflection.   However, as long as the intensity
of response never decreases as the concentration
increases, these curves are not in any way unusual.
(Truly unusual curves which have been reported tend
to result from unwanted or unknown changes occurring
accidentally in the quality or intensity of the
stimulus and are probably better described as curves
of "unusual" tests.)   At the simplest level then,
the typical toxicity curve shows three things.   It
shows firstly, where the curve is asymptotic to the

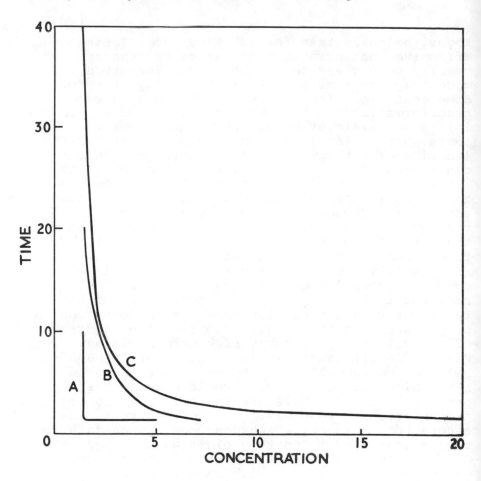

Figure 12. Toxicity curves plotted against arithmetic axes (broken portions only of curves in Figure 4).

concentration axis, the range of concentrations over which the selected proportion of the fish population which is being described has no ability to cope with the poison, and secondly, where the curve is asymptotic to the time axis, the concentration above which it cannot achieve homeostasis with the poison. It is this latter part of the curve which is of greatest interest in defining the direct effects of a poison on a fish in pollution control. Because in any response of this sort a number of physiological compartments is typically involved, the transition from one slope to the other is not abrupt. Thirdly, therefore, the inflected portion of the curve shows the progressive achievement of homeostasis in these compartments as concentration decreases, until all (and therefore the whole fish) are in this state for some extended period of time. The limits of concentration and time between which this transition occurs depend on the number of compartments and the rates of the processes involved, and will differ with the poison and the fish species. If these limits are very wide, it is then likely that only this region of the curve will be "discovered" in a typical toxicity test. The results may then be, or may seem to be, suitably described by logarithmic transformation of the data, and the rectilinear curve so produced then be described as an "abnormal" curve (as for example the broken curve B in Figure 11).

Some of the "problems" of the effects of temperature also originate from confusion in this area. It has been considered for many years that the sensitivity of fish to poisons increases as temperature increases. This conclusion was based on results from toxicity tests of very brief duration, in which toxicologically high concentrations of poisons (within the range at which the curve is asymptotic with the concentration axis) were used. In this range, however, as already indicated, the fish has little or no ability to modify the effects of the poisons, these being completely overwhelming (typically the whole population responding) and the test, in which time is used as the effect parameter, simply determines the effect of temperature on reaction rate, rate and temperature proving to be positively correlated. The only part the animal plays in this test is in supplying the reacting system. In tests made at concentrations in the region of the curve asymptotic with the time axis, however, and the area of greatest interest to us, the fish has, and is demonstrating, a capacity to "detoxify" the poison. Here, in poikilotherms, high temperatures (within the physiological range) increase metabolic rates, and in general should favor "detoxification" processes so that in this concentration region fish are likely to be most sensitive at low temperatures. This has recently been found to be the case for some poisons (phenol, cyanide). Consequently, statements that poisons are "less toxic" and "more toxic" at high temperatures do not conflict. Both statements are true, depending largely on whether or not the outcome of the poisoning is essentially determined by "biological" or by "chemical" processes (and whether time or concentration is the dependent variable).

It is important in interpreting the typical "toxicity curve" to recognize that this curve is an "artifact" in the sense that it is incomplete and is only observed because of the limits of time and concentration arbitrarily selected for investigation. Correctly, any response should be described in relation to its position on the curve as, for example, "occurring within the concentration-asymptotic exposure period" or "within the time-asymptotic (or "time-independent," "time-invariant") concentration." Beyond these ranges further inflections will occur as shown in Figure 13b and c. Because the fish must eventually die there must be some period of increasing time, after which the typical curve shows a further inflection and terminates on the time axis. Similarly, as the concentration increases the curve is likely to inflect toward the concentration axis as, for example, the major site of damage in the fish changes, or as the intensity and nature of the damage are no longer due solely to the chemical properties of the poison but become attributable to hypertonicity of the solution. The type of information being plotted in studies of this sort is similar to that shown in Figure 13a (but differs in that it does not usually start at true zero for the existence time of the population). In this part of Figure 13 hypothetical population survival curves are demonstrated for a variety of environments ranging from the extremely adverse (e.g., poison concentration of 200 mg/l) in which survival is very brief, to extremely favorable (the absence of the poison) and in which the population demonstrates its "normal" survival pattern. The survival times for 50 percent of each population in Figure 13a are plotted in Figure 13b, where the effect discussed above can be seen. The increased degree of inflection which would occur in this curve, were the broken curves for 10 and 5 mg/l in Figure 13a to be the true survival curves rather than those shown by the solid lines for these concentrations of poison, is illustrated in Figure 13c. From data acquired at the asymptotic concentration of the curve about, for example, the death or survival of fish, it does not seem possible to determine the period of time which will elapse before the inflection toward the time axis occurs (at least without much more fundamental toxicological information),

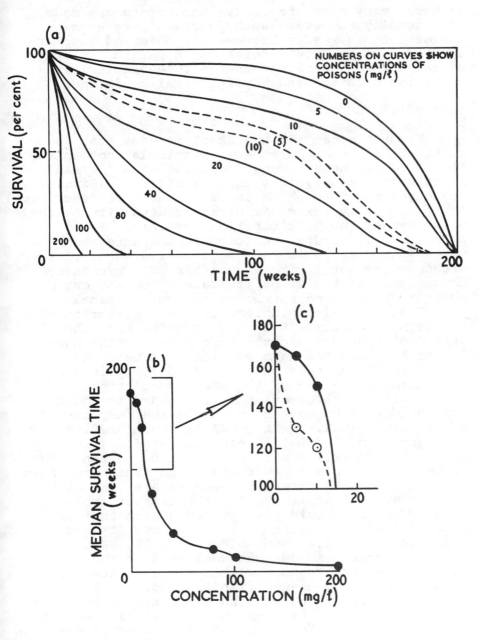

Figure 13. Population survival curves.

although many such predictive statements are made. Additionally, however, such "guesses" are even poorer than they seem because the observed data are derived from, and only refer to, a particular state of animal and ignore the change in susceptibility occurring with age and physiological condition.

INTERPRETATION

Far too often the statement is made that "a substance is 'toxic'" at some particular concentration, "toxic" being wrongly equated with "lethal"; if lethal is meant, it should be stated. (Similarly any other response which is used should be described.) Again, "acutely toxic" is often equated with "lethal" when there is no justification for this. As in pathology, an "acute" response is characterized perhaps as much by its nature (break-down of tissue/ organ systems at a rate at which repair processes cannot compensate) as by its rate of onset, but such breakdown need not be extensive or lethal.

There are also problems in the application of the term "sublethal." To have meaning both the period of exposure and the percentage of the population to which the term applies must be stated, otherwise it can only be assumed to apply in some way to the life-time of the whole population.

A major problem exists with regard to the terminology to be used to describe the response of the median animal. In the USA the term "median tolerance limit" (TL_m) is commonly used. This term seems to have arisen not in studies on poisons, but in those made into the effects of physical stressors (agents producing stress[5]), such as heat, on fish and in which areas it still has pertinent application. That it should take precedence in the minor area of toxicity testing with fish, over the adequate and versatile terminology of LD50, ED50 (and their analogues and derivatives) used in the major fields of bioassay and toxicology seems, however, idiosyncratic and unjustifiable. Additionally, because TL_m refers only to survival and to 50 percent of the population, it lacks the versatility of the "standard" terminology and cannot be used to describe responses of different quality and proportion, or indeed the population variance, all of which are required in putting the information in context in pollution control. It also admits of some confusion in its use of the word "tolerance" in that this word has at least two other meanings in bioassay and toxicology.

"Tolerance" was recommended by Finney,[3] for example, as being equivalent, but preferable, to the word "threshold" (the minimum intensity of stimulus required to produce a selected response in a single test animal under controlled conditions) as used in other studies. With certain qualifications (discussed below) this parallels its use in TL_m, except that it applies to single animals and does not automatically refer to lethal effect. "Tolerance" is, however, also used to describe acquired resistance following repeated exposure to a poison.[6] Consequently, it is evident that TL_m is a rigid, inefficient, and potentially confusing term. The most efficient terminology is one which is self-defining, and in the case of the lethal response "lethal concentration" (LC) cannot be improved upon. As far as responses other than death are concerned "effective concentration" (EC), while not self-defining, is at least no less so than TL_m and is more versatile. Consequently it seems that pollution biologists would be better served by adopting the "standard" terminology of bioassay/toxicity testing.

There is then the problem of the terminology for the concentration at which the LC50 becomes effectively constant for some prolonged period of time. This concentration, for which I introduced the term "asymptotic LC50," has been described, but only in studies on the toxicity of substances to fish, in a number of ways, including "ultimate median tolerance limit," "incipient lethal level," "threshold," and "lethal threshold concentration," all of which demonstrate a certain amount of vagueness. It is probably convenient to consider first the term "threshold."

As applied to fish tests this word is frequently misinterpreted (with much justification) by both scientists and non-scientists alike, as being the concentration below which a substance is "safe." In fish tests the term "threshold" has generally been used, implicitly or explicitly, to describe the concentration "at which toxicity ceases" (for 50 percent of the population), *i.e.*, the concentration at which 50 percent of a test population should survive for some prolonged period. In this, an absence of any coherent or logical usage of the word in relation to its use elsewhere is shown. To have any meaning at all the threshold for any stimulus can only sensibly be given the definition attached earlier and fundamentally must be a

property of the individual animal. The word has
this connotation in physiology, psychology,
toxicology generally, and in bioassay.[3],[7] (In
industrial hygiene, however, it has been used to
refer to the effective minimum value for the population involved.) This does not mean that the
term cannot be applied to, or quantitatively defined
for, any proportion of a population. It is as apt
to speak of the threshold concentration for 2 percent,
or 80 percent of the population, for example, as
it is of that for 50 percent. The proportion of
the population for which the value applies must
therefore be stated. Where it refers to 50 percent
of the population it is the median threshold concentration, and when the response is death, it is
the median lethal threshold concentration. The
typical "toxicity curve" is a plot of the median
lethal threshold concentrations for different
periods of exposure and proper definition requires,
therefore, that the period of exposure be stated.
Thus, we can have the 2-h median lethal threshold
concentration, the 48-h median lethal threshold
concentration, etc., or, more succinctly, the 2-h
LC50, the 2-d LC50. (Similarly one can now define
the 4-h LC10, the 3-d LC70, the 3-d EC70, and so
on, each of these being the threshold concentration
for the stated percentage of the population after
the stated period of exposure.) When the curve
resulting from plotting the logarithm of the median
threshold concentration against logarithm of time
becomes "asymptotic" to the time axis, it must still
be defined with respect to time and either the
period of exposure required for this to be observed
stated or else the concentration be described in
some way such as the "asymptotic median threshold
concentration" or "the time-independent range of
the median threshold concentration." To call it
simply the "threshold" or the "threshold concentration" is indiscriminate and unsatisfactory. To
call it the "lethal threshold concentration" is
probably worse, as the definition already given by
many of the users of the word to "threshold" is
that it is lethal without further qualification.
In this wrong context "threshold" refers only to
the concentration at which the curve for the median
response becomes parallel to the time axis so that
it cannot be used to describe the same portion of
the curve for some other percentage response.
However, the term threshold can be used for this
purpose if required as in "the asymptotic threshold

concentration for e.g. death/overturn etc., of 10 percent of the population" (but more conveniently "the asymptotic EC10," or for death "the asymptotic LC10"). Other terms ("ultimate TL_m," "incipient lethal level," "incipient LC50") are unsatisfactory for a number of reasons. They do not, for example, simply describe the observed data, but contain a predictive element, and, as indicated earlier, such prediction is not in my opinion justified. As Mayers stated[8] "the end results of acute poisoning obviously cannot be extrapolated to exposures which characterize ... the environment; where multiple stress (not all chemical in origin) must be taken into account, as well as cumulative, delayed, and long-range effects." They are also unnecessarily restrictive, all referring only to death, which means they cannot be used to describe other responses. Even more importantly they contain "nonsense terms" within themselves or within their definition. In the one case, survival at the so-called "threshold" concentration is described as "indefinite." This either means that it is undefined, which is obvious and unnecessary to state, or that it is unlimited. If this latter is meant then the fish are described as immortal. The proposition should be rather that "survival is prolonged" or "survival is for some period of time exceeding the duration of the test." For the same reason "incipient lethal level" is unsatisfactory. Additionally, however, it includes the word "incipient." If incipient is given its proper meaning (*i.e.*, beginning) then the phrase means the level "beginning to be lethal." However, if a concentration can be recognized as having this quality then it can only properly be defined as a "lethal" level. The alternative is a "non-lethal" level. The word "ultimate" suffers from similar disadvantages in this context in that its usage is not consistent with its dictionary definition ("furthest, last, final, limiting") and yet no new definition is given to it. The ultimate median tolerance limit at some late stage in life (when the natural death rate becomes high) must be an extremely small concentration. There is no need for a predictive element in this definition and it is better avoided.

The basic initial requirements from a toxicity test (within some defined conditions) are then: illustration of the toxicity curve (EC50 versus time of exposure preferably) with the confidence limits for each observation shown so that the

validity and limits of a subjectively fitted curve
can be assessed; the value of the asymptotic EC50
and its 95 percent confidence limits (and these
latter in preference to only the former); the slope
(and its 95 percent confidence limits) describing
the distribution of EC values within the population
at this time (using, for example, the slope-function
of Litchfield and Wilcoxon[9]). Data not wanted are
virtually meaningless lists of EC50 values for
different poisons for arbitrary periods of exposure
or such things as statements of the relative
toxicities (based on EC50 values at such times) of
two poisons, as these may well be reversed with
prolongation of exposure. Such a situation is
shown in Figure 14a where after time T_1 the median
fish of the population in poison E is more sensitive
than that in poison D, whereas the reverse is the
case after time T_2.

The same comment applies to comparisons of
sensitivities of different species of fish after
arbitrary periods of exposure. Statements as to
the relative sensitivities of different species
should in any case be avoided as these only have
meaning in environmental control if the most sensitive states of each species present in the
environment in question are compared. Likewise,
comparison of the EC50 values of the median fish
of two populations is not sufficient to define a
particular effect of a pollutant on these populations and can be quite misleading in pollution
control. For example, for two populations F and G
(Figure 14b) it may be that the EC50 of G is
greater than that of F (that is the median fish of
population G is less sensitive than the median fish
of population F). However, the slopes of the
concentration-mortality curves may so differ that
the EC90 of F is greater than that of G. Consequently, instead of population F (the apparently
more sensitive species on the basis of the EC50)
being wiped out by a certain increase in the concentration of poison, population G could suffer
this fate. A similar situation is shown for populations F and H, except that in this case both
populations have the same EC50.

It cannot be sufficiently stressed in the face
of the plethora of EC50 and LC50 values which has
been, and is being, produced that, in pollution
studies directed toward environmental management,
the concentration producing a response in 50 percent
of a population is, in contrast to the usefulness

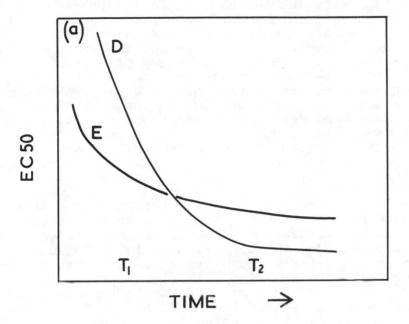

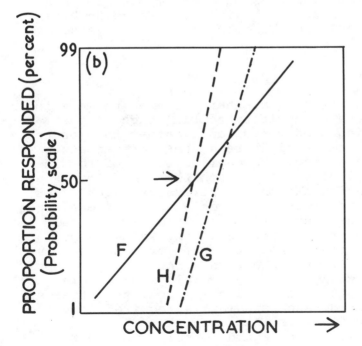

Figure 14. Comparisons of sensitivities.

of such data in bioassay, of little consequence and, without definition of the distribution of threshold values within a population, can have no worthwhile practical application. Even so, the fundamentally low value of such data, in terms of statistical limitations alone, should be recognized. Thus, for example, the average median effective concentration (EC50) of a poison, for some given period of exposure, determined from tests made with groups of twenty fish, could in an "infinitely" large population cause this same response in any proportion between 22 and 78 percent (at the 99 percent confidence level). Similarly, the concentration observed as being the EC20 could give between 4 and 51 percent response. Where the EC50 is determined using groups of only 5 fish then in a large population this concentration could affect between 0 and 100 percent. The detection of lower levels of response is therefore difficult. A harmful effect occurring for example, with a true probability of 0.02 (that is, in 1 in 50 fish) in a large population would only be detectable, at the 95 percent confidence limit, by using test groups of some 150 fish; an effect occurring at a probability level of 0.01 would need groups of 300 fish. That practical difficulties make it impossible in general to make tests with such large numbers of fish should not be allowed to obscure such facts.

GENERAL COMMENTS

It is not possible here to examine more important aspects of toxicity testing such as the overwhelming need for chronic rather than acute tests; or the possible relations between acute and chronic poisoning. However, because of the impossibility of simulating in the laboratory the complex interrelations of stressors which exist in the field, further appreciation of the effects of pollutants must rely heavily on ecological studies and this approach is one which is being made at the Water Pollution Research Laboratory, Stevenage. There is also a strong case for applying the techniques of epidemiological studies to fish populations in moderately polluted streams. Feed-back from such studies is required to put laboratory findings into context.

Finally, with regard to pollution control, I should like to comment on the search by some workers

for a "standard" test and a "standard" test animal.
This seems to show a basic lack of comprehension of
the purpose of toxicity testing for environmental
control purposes and how this differs from bioassay.
Any method of bioassay is essentially an analytical
"tool," in which it is important to try to eliminate
variability attributable to anything other than the
substance being assayed. It is a "tool" which is
used to measure the potency of drugs (any large
error in the assay of which could put a patient at
risk), and which must be available for use by
different personnel in different laboratories and
yet still give similar results. Thus standardiza-
tion of equipment, reference compounds, and procedure
(including pre-test treatment of the animals) is
imperative to control variability. A "standard"
animal is also required. This is not a "representative" animal but one which has high sensitivity
to the test material and produces a constant and
unequivocal response in the particular assay
involved. Selective inbreeding to produce
homozygous strains achieves this (by minimizing
genetic variability). There is no single universally
applicable "standard" animal--the one optimal to
the test is used; a strain known to be extremely
sensitive to one drug will not necessarily be more
sensitive than some other strain to a second
drug.[10,11] While there are circumstances in toxicity
testing in which it could be advantageous to use a
homozygous strain, in general, for pollution control
purposes, this is not the case. Such a practice
could in fact be misleading, for the range of sensitivities occurring over a number of strains can
be wider than that in the original outbred stock.

As there is nothing "standard" about any poison,
animal or exposure in the environment, no "standard"
test can be advocated. The only standards to be
applied in toxicity testing are those of good experimental technique and sound scientific practice.
Toxicity testing is an exploratory study and must
be as versatile and unlimited as the conditions and
situation to which the study is to apply. *The more
standardized the test or test animal, the less
applicable the information obtained is likely to
become.* This fact has been stressed again and
again by numerous eminent authorities: "because
of the differences in physical, chemical and biological properties of ... toxic substances ... it
is not possible, nor indeed desirable, to standardize
the methods of investigation used in the study of

toxicity";[12] "we have no system of tests to which we would all subscribe as being the most certain to reveal toxic effects."[13] It is most important for the pollution biologist to be aware of this.

Nevertheless, because tests of short duration, in which whole-body quantal responses are observed, still continue in many cases to be the only ones recommended for, or carried out in, studying the toxicity of solutions to fish, some of their impli- cations have been considered here. However, the complexities and difficulties involved in making a meaningful study of toxicity, and the dangers in- herent in simple approaches and interpretations, particularly when extrapolations and unwarranted assumptions are made, cannot be too strongly empha- sized. Tests of this type cannot adequately define the toxic properties of a substance and should in fact not have, as they often do at the present time, this as their presumed objective. Their role should be that of providing information for the development of comprehensive tests based on a proper identifica- tion and assessment of the source, nature and degree of hazard from a given pollutant, and these con- siderations should be the starting point for the design of all tests.

ACKNOWLEDGMENT

Crown copyright. Reproduced by permission of Her Britannic Majesty's Stationery Office.

REFERENCES

1. Laurence, D. R. and A. L. Bacharach. *Evaluation of Drug Activities: Pharmacometrics,* Vol. 1 (London: Academic Press, 1964).
2. Geddie, W. *Chambers Twentieth Century Dictionary.* (London: W. & R. Chambers Ltd., 1959).
3. Finney, D. J. *Probit Analysis.* (Cambridge: Cambridge University Press, 1952).
4. Bliss, C. I. *Ann. Appl. Biol. 24*:815 (1937).
5. Selye, H. *The Physiology and Pathology of Exposure to Stress.* Montreal: Acta, 1950).
6. Czaky, T. Z. *Introduction to General Pharmacology.* London: Butterworths, 1969).
7. Bliss, C. I., and M. A. Cattell. *Rev. Physiol. 5*:479 (1943).
8. Mayers, M. R. *Occupational Health.* Baltimore: The Williams & Wilkins Co., 1969).

9. Litchfield, J. T., and F. Wilcoxon. *J. Pharmac. exp. Ther. 96*:99 (1949).

10. Brown, A. M. *J. Pharm. Pharmac. 14*:406 (1962).

11. Brown, D. M., and B. O. Hughes. *J. Pharm. Pharmac. 14*: 399 (1962).

12. Truhaut, R. *Proceedings of the International Symposium on Maximum Allowable Concentrations of Toxic Substances in Industry*. (London: Butterworths, 1961).

13. Barnes, J. M. *Proc. Eur. Soc. Study Drug Toxicity 2*: 57 (1963).

3. CHEMISTRY AND FISH TOXICOLOGY

Charles E. Stephan. Newton Fish Toxicology Laboratory, U.S. Environmental Protection Agency, Cincinnati, Ohio.

ABSTRACT

Chemists should be an integral part of all toxicological activities involving chemical toxic agents. In experimental toxicology, chemists can aid in the selection of toxicants to be tested, help design toxicity tests, measure and characterize the level of the toxicant to which the subjects are actually exposed, determine the fate of the toxicant after it comes in contact with the subjects, help determine the mode of action of the toxicant, aid in detecting some of the effects of the toxicant on the subjects, and devise ways to prepare special materials and toxic agents designed by toxicologists. In applied toxicology, chemists can help improve the accuracy of predictions of what may happen or explanations of what has already happened and help prevent grossly inaccurate extrapolations. Fish toxicology usually involves exposures of the subjects to water containing a toxic agent rather than direct application of a toxic agent to the subjects. Understanding any interaction between the toxic agent and the water can be very important in terms of designing, conducting, interpreting and applying the results of toxicity tests with fish. Information on the accumulation of toxicants in various tissues can be used in several ways. Analytical chemists can identify materials that are finding their way into the environment and whose biological activities should be studied. However, since toxicology is fundamentally biology, chemists must contribute to toxicology but must not dominate it.

INTERDISCIPLINARY CHARACTER

Almost without exception, all activities in both experimental and applied toxicology have biological, chemical, and statistical aspects. Because of the interdisciplinary character of toxicological activities and because few people are really competent in more than one field, biologists, chemists, and statisticians should be involved in all such activities. Competence in chemistry cannot atone for incompetence in biology; nor can competence in biology make up for a lack of competence in chemistry. In addition, competence in chemistry and biology cannot be used to best advantage without sufficient experience in experimental design and data analysis.

Chemists should understand the interdisciplinary character of toxicology for two reasons. First, at least a part of what is called environmental chemistry is closely related to, if not actually a part of, one or more fields of toxicology, such as fish toxicology. Therefore at least this part of environmental chemistry should get some direction from toxicology. Undoubtedly, environmental chemistry can and will contribute much by itself, but it will contribute much more useful information if it considers its relationship to the appropriate fields of toxicology. Such consideration will also dispel any misconception that environmental chemistry is just one application of analytical chemistry. Toxicology, and therefore environmental chemistry, needs expertise in all areas of chemistry--organic, inorganic, physical, and nuclear, in addition to analytical chemistry.

The second reason for stressing an interdisciplinary point of view is that toxicology is primarily biology, not chemistry. Although chemistry must provide an understanding of the chemical aspects of toxicological activities, the primary interest is in the biological aspects, that is, the effects of toxic agents on living entities. Chemists must contribute to toxicology, but not dominate it.

TEAM APPROACH

Dominance, or the overemphasis of one component, can become a problem in any multidisciplinary effort, depending on the individuals involved. In addition, a toxicological team consisting of one or more biologists, chemists, and statisticians occasionally forgets the main reason for the existence of the team and conducts little more than detailed investigations of minor

factors. The existence of these two potential
problems is the major justification for toxicologists,
who themselves may not be experts in biology, chemistry, or statistics. Toxicologists are supposed
to maintain a broad overview of the basic toxicological questions and guide the team. Thus the team
should consist of at least four people rather than
three. Because the biological, chemical, statistical,
and toxicological aspects are all vital, each
toxicological activity should be planned and reviewed
periodically by a team which includes a biologist,
a chemist, a statistician, and a toxicologist.

The concept of a team approach is faced with
several obstacles, especially when it is applied
to fish toxicology. First, many people feel that
teams in general, and interdisciplinary teams in
particular, are more trouble than they are worth.
Second, many toxicologists are more interested in
mechanisms and modes of action than in the broad
overview. Third, even though much research to
determine water quality criteria is essentially
toxicology, very few trained toxicologists are
working in this field. On the other hand, the team
approach to toxicology can be extremely useful when
the team members recognize its advantages and want
to make it work. This is essential when a toxicologist is not available. The regrettable lack of
toxicologists makes it all the more important for
chemists working in the field of fish toxicology
to consider carefully what they should and should
not try to do, so they may avoid both dominance
and frustration.

EXPERIMENTAL TOXICOLOGY

Experimental toxicology is concerned with
developing, conducting, and interpreting the results
of toxicity tests. Poorly designed tests are often
inefficient, sometimes provide no information at
all, and even more important they may provide false
information and be misleading. Chemists can contribute to experimental toxicology in a variety
of ways. They can aid in the selection of toxicants that should be tested; help design toxicity
tests; measure and characterize the level of the
toxicant to which the subjects are actually exposed; determine the fate of the toxicant after
it comes in contact with the subjects; help
determine the mode of action of the toxicant; aid

in detecting some of the effects of the toxicant on the subjects; and devise ways to prepare special materials and toxic agents that have been designed by toxicologists.

Chemists are in a good position to identify actual or potential environmental contaminants because many of these are used or produced by the chemical industry. Therefore chemists should know the chemical and physical properties of the materials and the amounts used in various applications. In addition, analytical chemists can monitor the environment to determine the existing levels of particular contaminants. The major monitoring programs now underway are for pesticides and heavy metals, but they are not as extensive as they should be. In addition, many other kinds of potential environmental contaminants are not covered by any of the existing monitoring programs. Outside of the monitoring programs, significant levels of materials are occasionally found by accident in unsuspected places. It is the chemist's responsibility to develop screening techniques for the analysis of a wide variety of environmental samples for a wide variety of materials. Mercury and polychlorinated biphenyls were first identified as possibly significant environmental contaminants through the analysis of such samples. Aquatic organisms such as fish and oysters are becoming a very important source of samples for monitoring the environment. By selecting the right kinds of samples, chemists can use biology to their own advantage. Fish and other aquatic life will sometimes provide a significant amount of preconcentration for the chemist and can often provide both long-term composite and short-term instantaneous samples. The analysis of a variety of samples from one aquatic environment is important because it is not impossible for different kinds of samples from one aquatic environment to give apparently contradictory results. For example, analysis of the water may not indicate contamination, whereas analysis of fish or sediment may show significant contamination. Chemists must help insure the validity of the results of such analyses through the use of duplicate samples, good collection and storage techniques, spiked samples, confirmation by alternative analytical techniques, etc.

At all times, chemists must keep in mind the biological aspects of environmental contamination. For instance, the environmental impact of a substance may depend more on its biological properties than on its chemical and physical properties. Also, analytical methods used in monitoring programs must provide valid measurements at levels that are biologically important and at background levels so that trends can be detected.

Fish toxicology usually involves exposures of the subjects to a toxic agent in a water rather than the direct application of the toxic agent to the subjects. Thus, understanding the interaction of the toxic agent and the water can be very important in terms of designing, conducting, interpreting, and applying the results of toxicity tests with fish. In many cases the properties of the solution or suspension--including toxicity--will depend on how it is prepared and handled. Very often the toxicant will react chemically with the water or be biologically transformed by organisms in the water. In addition, there may be complications due to pH control, volatilization, sorption, or uneven distribution of insoluble toxicants due to floating or settling. Chemists can often foresee things that might occur and suggest ways they might be overcome. However, chemical solutions to such problems may introduce further complications. Each particular problem must be solved in light of the overview of the reasons for conducting the toxicity test in the first place. For instance, it is often suggested that a complication, such as insolubility, can be overcome by the use of an additive, such as an emulsifying agent or surfactant. Unfortunately, such additives may be toxic themselves, or they may affect the results of the tests through chemical or biological synergism or antagonism. Whether or not there is an interaction can be determined biologically, but this is rarely done. Chemists must make sure that the cure for the problem is not as bad as, or worse than, the problem itself. When possible, the chemist should suggest several alternative solutions so that the advantages and disadvantages of each can be examined.

Another contribution that chemists can make is in the measurement and characterization of the level of the toxicant to which the subjects are actually exposed. At a minimum, measurement of the level of the toxic agent will serve as a check on the design of the test, the operation of equipment, and the occurrence of human error. However, biologists and chemists are beginning to realize that many unsuspected but important things happen when toxicants are added to water. Thus it is often important to characterize the toxic agent, not just measure its level. Although predictions concerning what will happen when a toxic agent is added to water can be an aid in designing toxicity tests, it is always wise to have the predictions confirmed by chemical analyses. These may indicate that the test needs to be redesigned or may furnish information that is vital for the interpretation and application of the results of the test.

On the other hand, the results of chemical analyses may be misinterpreted. In toxicity tests with low levels of some toxicants, chemical analyses will show that the levels of the toxicant in the water are significantly--perhaps as much as fifty percent--less than they are supposed to be. This may have a chemical explanation, or it may have a biological one if the subjects or other aquatic organisms are removing such significant amounts from the water. Alternatively, analyses might be interpreted as indicating the presence of a toxicant when it is not present, if the analytical method cannot differentiate between the parent compound and a degradation product. Once a problem has been discovered and an explanation has been found, someone still must decide what should be done about it.

The determination of the fate of a toxicant after it comes in contact with the subject is usually the job of an analytical chemist. The toxicant may be taken up by the subject, in which case it may or may not be transformed, stored, or excreted. One of the most challenging tasks in environmental chemistry is to devise ways to analyze any specific tissue for any specific toxicant or degradation product at the levels that are biologically important. Some approaches to the determination of the fate of toxicants, such as the use of tracers, can provide misleading results if the possibility of degradation is not taken into account. Interest in degradation products is increasing rapidly because such products may be either more or less toxic and persistent than the parent compounds.

Data on toxicant concentrations in various tissues have three very practical uses: first, they may identify a possible human health hazard; second, they may be useful in identifying the causes of fish kills; and third, they may detect environmental contamination. Unfortunately, the concentrations of toxicants in fish tissues have not received much attention until recently, partly because techniques for chemical analyses were lacking and partly because laboratory biologists were not aware of the potential practical importance of such information. The high levels of mercury and pesticides recently found in fish gave impetus to the laboratory study of toxicant concentrations in tissues. In addition, interest in scientific methods for determining the causes of fish kills is increasing. In the past, most fish kills were solved on the basis of circumstantial evidence, but in many cases today, circumstantial evidence is not good enough for legal action.

Chemists can play a major role in determining the means by which a toxicant affects a subject by studying the route of uptake and the site of action. Often the detection of the effects of a toxic agent on the subject is considered to be the responsibility of the biochemist, the physiologist, or the toxicologist, but much of the actual work involves analytical chemistry. Although a chemist is not always needed to use an analytical method, an analytical chemist should be the most qualified person to develop such a method and might be helpful in interpreting the analytical data.

An interesting recent development in toxicology is the correlation of structure with biological activity. Although this development is not too important in fish toxicology at the present time, it may well become so in the future. Undoubtedly toxicologists will find it easy to design compounds that chemists will find difficult to prepare.

Applied Toxicology

In addition to the contributions that chemists can make to experimental fish toxicology, they can also make important contributions to applied fish toxicology. One responsibility of applied toxicology is to extrapolate from what happens in the laboratory to what happens in real life. This extrapolation may be either a prediction of future events or an explanation of past events. Because of the present state of the art, this extrapolation is sometimes little more than an educated guess. Chemists can often help improve the accuracy of the extrapolation or at least help prevent grossly inaccurate extrapolations by giving biologists and toxicologists a better understanding of what happens chemically in toxicity tests and in the environment.

Unfortunately, what takes place in a laboratory is not always a good indication of what will or will not take place in a field situation. In the most publicized recent case--that of mercury--chemists themselves probably contributed to the problem by predicting that elemental mercury would be inert and would not pose a hazard. Elemental mercury may be chemically inert, but it apparently is not biologically inert. This illustrates an important point. It is always inviting to predict that toxic agents will behave in a field situation the same as they do in a laboratory situation, but this is often not the case. Chemistry is only one part of environmental chemistry. The other equally important part is the environment. Although chemistry and toxicology may be more complex in the great outdoors, this is where the problems occur.

Another aspect of applied toxicology is the use of biological tests in practical applications outside the laboratory. In spite of all that chemists can do, they should be the first to admit that in most cases the results of chemical tests cannot accurately indicate biological properties. For example, chemical analyses of a body of water cannot reliably indicate whether or not that body of water can and will support a good crop of fish, except in extreme cases. Generally, biological tests will be better measures of biological properties than chemical tests, even if the biological tests are not as well developed as many of the chemical tests. Environmental protection needs toxicological accuracy much more than it needs statistical precision. Chemists must work with biologists so that the biological tests are chemically sound and are used to their best advantage.

The next step beyond applied toxicology is, of course, the formulation of laws. Here many political, economic, social, and legal considerations must be taken into account in addition to the toxicological ones. In the long run, protection of the environment may depend more on good laws than on good toxicology. However, biologists, chemists, statisticians, and

toxicologists must work together to furnish the
fundamental toxicological data necessary if man is
to have any possibility of protecting his environment and himself. A good working relationship
between environmental chemistry and toxicology will
benefit both fields of interest and mankind in
general.

4. RECENT DEVELOPMENTS IN THE USE OF LABORATORY
   BIOASSAYS TO DETERMINE "SAFE" LEVELS OF
   TOXICANTS FOR FISH

John G. Eaton. United States Environmental Protection
Agency, National Water Quality Laboratory, Duluth,
Minnesota.

ABSTRACT

Chronic (long-term) bioassays are of primary
value in determining "safe" levels of toxicants.
All such tests conducted at the National Water
Quality Laboratory involve exposures through the
reproductive period of the life cycle and subsequent
exposures of the eggs and young. Test procedures
considered adequate are available for bluegills,
fathead minnows, brook trout and *Daphnia magna*, and
procedures are being developed for several additional
fish and invertebrate species. Various short-term
tests have also been developed for use in conjunction
with chronic tests. The ratio between the concentration of some substance producing a selected
chronic response and that causing 50% mortality in
two days to two weeks is termed an application factor. This factor is then used in making a provisional
estimate of the chronically "safe" concentration of
a toxicant for species for which only acutely lethal
concentrations have been determined. Thirty- to
sixty-day tests have been used to examine the
sensitivity to certain pollutants of eggs or larvae
of many different species when previous chronic
exposures have shown these life stages to be consistently very sensitive. Various short-term
physiological and behavioral tests are also being
examined in conjunction with chronic exposures, as
possible means of estimating chronically toxic
concentrations.

INTRODUCTION

The objective of the work with water pollutants being carried on at the National Water Quality Laboratory, Duluth, Minnesota, is to establish the maximum concentrations at which these substances may be present in aquatic environments without damaging their biota. To obtain this end, all life stages and vital functions of a wide variety of the organisms must be protected, so that the prey-predator, food-web relationships necessary for the production of economically important species such as fish will be preserved. Even organisms which decompose other organisms must be preserved to maintain a healthy environment. Therefore, as well as determining safe concentrations of toxicants for fish, the effects on other types of organisms must also be examined. Although testing the sensitivity of these other organisms is an important part of our research program, we have sometimes found that some fish species are as much or more sensitive than organisms below them in the food chain, and that the integrity of the aquatic environment will be preserved if the most sensitive fishes are protected. Where this is known to be true, the job of determining safe concentrations of pollutants becomes simpler.

LABORATORY PROCEDURES

At first one might think that the best way to determine the effects of pollutants on aquatic environments is by making field observations where a particular toxicant is known to be present. In the field however, the investigator is confronted by such a barrage of variables causing or influencing the observed effect that it becomes impossible to determine the proportion of the effect for which the toxicant is responsible. These variables include climatic factors, natural water chemistry, and very possibly additional pollutants or other man-made changes. Field observations do however, offer excellent possibilities for determining whether or not an aquatic environment is being polluted.[1]

Toxicants are therefore first studied singly in the laboratory, where many of the variables encountered in the field can be controlled, and where responses of organisms can be observed within a range of concentrations of the toxic substances for various time periods. Such tests are called bioassays

and the most informative of these are chronic exposures. We consider it imperative that chronic
tests include continuous exposure of the organisms
through at least one reproductive period and subsequent exposure of the eggs and young. Only this
kind of exposure demonstrates the concentrations at
which most life processes are protected. This type
of bioassay is essential for water-quality standards
programs, which permit the continuous presence of
pollutants up to a specified level. (Most present
standards specify allowable levels based on maximum
concentrations, rather than on averages or other
measurements.) The lowest concentration at which
direct harmful effects occur is frequently associated with reproductive process changes, such as
the number of eggs produced or their hatchability.
"Safe" toxicant concentrations as determined by
reproduction bioassays are often 10-100 times lower,
and sometimes as much as 200-500 times lower, than
concentrations determined by acute or short-term
bioassays using 50% mortality as an end point.

Because we consider chronic tests so important,
a significant part of our research has been devoted
to developing the fish-culture procedures necessary
to carry out reproduction chronic exposures in the
laboratory. It should be emphasized that the test
animals reproduce, or spawn in the case of fish, of
their own volition on substrates that simulate those
they utilize in the wild. These chronic tests are
presently run as a matter of course with bluegills
(*Lepomis macrochirus* Rafinesque), fathead minnows
(*Pimephales promelas* Rafinesque), brook trout
(*Salvelinus fontinalis* (Mitchill)), and *Daphnia
magna* (water fleas). Sufficient information should
be available soon to allow us to use several additional fish and invertebrate species with confidence.

Chronic testing usually includes exposure of
animals in duplicate to five or six toxicant concentrations along with a control; consecutive
concentrations usually differ from one another by
a factor of 2 or 3. Fish tests often start with
40-50 individuals per tank, and numbers are reduced
at intervals for closer examination for toxicant
effects and to adjust sex ratios so that only 6-20
remain at the time of spawning. Fish chronic
exposures routinely take about 10 months to a year
to complete, whereas *Daphnia* are exposed for only
3 weeks, as they go through an entire life cycle
in that time.

APPLICATION FACTORS

The results of bioassays to determine life-cycle "safe" concentrations of toxicants have been extended from the test species to other species that have not yet been exposed chronically, and to water types different from those of the laboratory exposures by use of so-called "application factors." These factors are experimentally derived and are multiplied by acute (short-term) bioassay results in an attempt to obtain estimates of chronically "safe" levels. Arbitrarily derived factors have also been used when chronic test results with a toxicant have not been available, but their predictability is poorer and their use requires much greater caution.

Experimentally derived application factors are obtained with a given fish species and water quality by relating the concentration found "safe" under conditions of a chronic exposure, to the concentration found lethal in an acute or short-term toxicity test (see the following example). The end point of the acute test is determined upon the death of 50% of the individuals exposed and is expressed as the concentration lethal to half the individuals exposed in a given length of time (*e.g.*, 96-hr LC50), or as the concentration at which it begins to take much longer to kill 50% of the individuals exposed than it did at higher concentrations. The latter is termed a lethal threshold-concentration determination, which Brown[2] points out is more correctly called a *median* lethal threshold-concentration determination and is usually numerically similar to the 96-hr LC50 value.

The acute test value is then used as the denominator, and the "safe" and lowest "unsafe" values determined in a chronic exposure are used as the numerators, to determine an application-factor range within which the actual "just safe" application factor must lie.

Tests at the National Water Quality Laboratory have demonstrated that for several toxicants the application-factor ranges correspond or overlap for the two or three species of fish tested with each toxicant. In one case the application-factor ranges for two species exposed to malathion overlapped even though the acute and chronic toxicities differed by a factor of 50 times.[3,4] From these data it has been assumed as a working method that the acute value, obtained for species that cannot be or have

Example

Acute LC50: 100 mg/liter

Chronic: "safe" - 5 mg/liter
        lowest "unsafe" - 10 mg/liter
        (actual "just safe" somewhere between these
            two)

Application-factor range: 5/100 - 10/100 = 0.05 - 0.1

Water Y: different water type in which the result
            (estimated "safe" chronic level) will be
            applied.

Application-factor range for species A, water X	Application factor range for species B, water X
0.1 - 0.05	0.08 - 0.04

0.1 - 0.04

LC50 for species C
    in water X   = 200 mg/liter
LC50 for species D
    in water X   = 50 mg/liter
LC50 for species E
    in water X   = 10 mg/liter
LC50 for species C
    in water Y   = 25 mg/liter

Estimated "safe" chronic values for species C, D, and
E in water X, and C in water Y:

        C) 200 x 0.04 = 8 mg/liter
        D) 50 x 0.04 = 2 mg/liter
        E) 10 x 0.04 = 0.4 mg/liter
    C in water Y) 25 x 0.04 = 1 mg/liter

---

not been tested chronically, can be multiplied by
the appropriate application factor to make a provisional estimate of the chronically "safe" levels
(see above example).

Water hardness, as it affects the biological
response of the organism, is known to have an effect
on heavy metal toxicity. Application-factor ranges
determined for fish exposed chronically to a single
toxicant (copper) have been compared thus far in
only two widely different water hardnesses.[5,6]

In both cases the ranges were similar, indicating that there is some justification for using laboratory derived factors to estimate chronically "safe" levels for other species in different water types. In such cases the factors would be applied to acute test results obtained in the different water types (water type Y in the example). The testing of this use of an application factor was one of the purposes of a controlled stream study discussed below.

The potential effects of water hardness and pH on metal solubility must be recognized. In many cases the insoluble fractions of the total metal present are known to be less toxic than the dissolved fraction. Thus, two water types in which the solubilities of a metal differ might have different application-factor ranges if the results are stated in terms of the total amount present. The same might be true of application factors derived for two species in a given water type if the sensitivities of the species differ widely. Studies are underway to determine exactly what constituents of metals in water are toxic, and we hope these investigations will lead to the development of more precise application factors. Organic ligands in the water are also known to influence the amount of metal present in a toxic form.

The toxicities of many types of pollutants (most organic pesticides for example) are not altered as much by hardness or most other common water characteristics. Since only one chronic test has been completed for each of most of the toxicants we have studied, experimentally derived applicantion factors for these toxicants represent results for one species of fish in one water type. Much more work is necessary to define all the limitations, but even the single application factors obtained so far are widely used. The tremendous variation in the ratios of acute to chronic results, ranging from five times up to over 500 times, indicates that the use of even these few experimentally derived application factors is more sound than the use of arbitrarily derived factors. Very seldom has the estimated "safe" chronic level been raised by applying an experimentally derived factor where an arbitrarily derived factor had been used previously. In the case of many toxicants for which chronic test data is not available however, the use of arbitrary factors is still necessary.

The preceding discussion of application factors is restricted to work with fish. At the present

time we do not know how well these ideas might apply
to the estimation of "safe" levels for various
groups of aquatic invertebrates. But we can assume
that larger application factors will be obtained
with at least those invertebrates for which a test
for 48 to 96 hrs occupies a large percentage of
their life cycle time.

OTHER METHODS FOR
ESTIMATING SAFE CONCENTRATIONS

Various types of "short-cut" methods are being
examined in addition to the use of application factors to estimate "safe" concentrations of toxicants.
The difference between application factors and
short-cut methods is that the latter involve the
direct use of pollutant effects observed at
chronically toxic levels, but which appear in a
shorter period of time than that necessary to carry
out a reproductive chronic test. Chronic bioassays
must still be conducted with a toxicant first, but
during the tests effects might be observed that can
be directly related to the toxicant levels found
to be "just safe" or "just unsafe." Or shorter
term exposures can be carried out later at these
levels to observe effects which could not be
monitored in the original chronic test system.
For example, fish might be tested at chronically
"safe" concentrations while held in metabolism or
swimming-performance chambers.
Thus, during chronic exposures with the metal
cadmium, which is probably very toxic to all animals,
one of the effects observed at the lowest "unsafe"
concentrations tested was mortality of eggs or early
larvae of bluegills and fathead minnows. No other
readily observable effects were observed at lower
concentrations. From this we have generalized
that exposures of eggs and fry of fish to cadmium,
the latter for 30-60 days after hatching, will
probably provide us with a better estimate of
chronic toxicity than any other short term test.
As eggs are available from many species of fish
that we cannot test chronically, this is a better
tool than the standard acute tests for investigating
cadmium toxicity to species such as pike, salmon,
bass, and many others. Because cadmium is only
partly soluble at concentrations causing harmful
effects to several fish species, this short-cut
method is probably better than the use of a cadmium
application factor.

113

Another short-cut method being investigated is the assessment of changes caused by the toxicant in the composition of fish, particularly that of the blood. Usually two or three samples of fish are removed from the test tanks during a chronic exposure, and the blood examined for changes in hematocrit, red blood cell numbers, amount of hemoglobin, and concentrations of various blood ions, proteins, enzymes, and other constituents. The only tests completed at present demonstrate that blood changes occurring after a short exposure of brook trout to copper could be used to predict "safe" levels in a chronic exposure lasting 11 months.[7] As might be expected, it did not allow prediction of a twofold greater sensitivity among the offspring than among the parents.

Several studies are underway in which attempts are being made to relate physiological or behavioral changes to chronic toxicity, such as that of Drummond and Spoor.[8] For the most part investigations in these areas are just beginning, and results are not yet available.

Earlier I referred to the difficulty of obtaining specific, quantitative information on toxicity from field observations alone. Nevertheless field studies are important, especially when it becomes necessary to demonstrate the applicability of laboratory studies to field situations. The interpretation of field results is facilitated, however, if some control is also exercised over the field conditions without disqualifying the selected area as representative of a natural ecosystem subject to most of the usual environmental influences. A study has nearly been completed by the Newtown Fish Toxicology Laboratory, a field station of the National Water Quality Laboratory, in which a natural stream in southern Ohio was intentionally polluted with copper to compare previously and concurrently run laboratory chronic bioassays with field results, and partly to check the predictability of the copper application factor for several of the resident species. Some control was exercised by placing screens at both ends of the half-mile-long test section so that fish movements in and out of the area could be monitored, and by introducing the copper at a rate proportional to the water flow in order to maintain a constant metal concentration. Preliminary results from that project seem to demonstrate the validity of predictions made from data obtained from laboratory studies, possibly with an additional slight avoidance effect whose observation is precluded in laboratory chronic tests. Thus, the results reassure us that our laboratory results are meaningful, although more and different kinds of studies are also needed.

REFERENCES

1. Cairns, John Jr., and Kenneth F. Dickson. "A Simple Method for Determining the Effects of Pollution Upon Aquatic Communities," Paper presented 162nd National Meeting of American Chemical Society, Washington, D.C., September 12-17, 1971.

2. Brown, V. M. "Concepts and Outlook in Testing the Toxicity of Substances to Fish," Paper presented at 162nd National Meeting of American Chemical Society, Washington, D.C., September 12-17, 1971.

3. Mount, Donald I. and Charles E. Stephan. "A Method for Establishing Acceptable Toxicant Limits for Fish-- Malathion and the Butoxyethanol Ester of 2,4-D," *Trans. Amer. Fish. Soc. 96:*185 (1967).

4. Eaton, John G. "Chronic Malathion Toxicity to the Bluegill (*Lepomis macrochirus* Rafinesque)," *Water Research 4:*673 (1970).

5. Mount, Donald I. "Chronic Toxicity of Copper to Fathead Minnows (*Pimephales promelas*, Rafinesque)," *Water Research 2:*215 (1968).

6. Mount, Donald I. and Charles E. Stephan. "Chronic Toxicity of Copper to the Fathead Minnow (*Pimephales promelas*) in Soft Water," *J. Fish Res. Bd. Canada 26:* 2449 (1969).

7. McKin, James M. and Duane A. Benoit. "Effects of Long-term Exposures to Copper on Survival, Growth, and Reproduction of Brook Trout (*Salvelinus fontinalis*)," *J. Fish. Res. Bd. Canada 28:*655 (1971).

8. Drummond, Robert A. and William A. Spoor. "A Method for Recording the Responses of Freeswimming Animals to Toxicants and Deleterious Environmental Conditions," Paper presented at 162nd National Meeting of American Chemical Society, Washington, D.C., Sepgember 12-17, 1971.

SECTION III

EXAMPLES OF END POINTS

AND INDICATORS USED IN BIOASSAY

CONTRIBUTORS TO SECTION III

Jelle Atema. Woods Hole Oceanographic Institution, Falmouth, Massachusetts.

Thomas G. Bahr. Institute of Water Research, Michigan State University, East Lansing, Michigan.

David Boylan. Hawaii Institute of Marine Biology, University of Hawaii, Honolulu, Hawaii.

Ronald Eisler. National Marine Water Quality Laboratory, West Kingston, Rhode Island.

K. A. Fishbeck. Institute of Plant Development and Department of Botany, University of Wisconsin, Madison, Wisconsin.

Harold W. Fisher. Biophysics Laboratory, University of Rhode Island, Kingston, Rhode Island.

G. R. Gardner. National Marine Water Quality Laboratory, West Kingston, Rhode Island.

G. C. Gerloff. Institute of Plant Development and Department of Botany, University of Wisconsin, Madison, Wisconsin.

E. H. Jackim. National Marine Water Quality Laboratory, West Kingston, Rhode Island.

Stewart Jacobson. Woods Hole Oceanographic Institution, Falmouth, Massachusetts.

Gilles LaRoche. National Marine Water Quality Laboratory, West Kingston, Rhode Island.

Alexander R. Malcolm. Northeast Water Supply Research Laboratory, Environmental Protection Agency, Narragansett, Rhode Island.

Dee Mitchell. Monsanto Company, St. Louis, Missouri.

Ruth Patrick. Academy of Natural Sciences of Philadelphia, Philadelphia, Pennsylvania.

Benjamin H. Pringle. Northeast Water Supply Research Laboratory, Environmental Protection Agency, Narragansett, Rhode Island.

John Todd. Woods Hole Oceanographic Institution, Falmouth, Massachusetts.

Cornelius I. Weber. Analytical Quality Control Laboratory, Environmental Protection Agency, National Environmental Research Center, Cincinnati, Ohio.

P. P. Yevish. National Marine Water Quality Laboratory, West Kingston, Rhode Island.

G. E. Zaroogian. National Marine Water Quality Laboratory, West Kingston, Rhode Island.

5. RECENT DEVELOPMENTS IN THE MEASUREMENT OF THE
   RESPONSE OF PLANKTON AND PERIPHYTON TO CHANGES
   IN THEIR ENVIRONMENT

Cornelius I. Weber. Chief, Biological Methods,
Analytical Quality Control Laboratory, Environmental
Protection Agency, National Environmental Research
Center, Cincinnati, Ohio

INTRODUCTION

In the broadest sense, a bioassay is a determination of the biological effects of some substance or
environmental condition, and includes the use of
organisms to detect or to measure the concentration
of substances or to indicate the nature of physical
conditions in the environment. Due to the complexity
of most natural aquatic communities, and the multi-
plicity and variability of environmental factors
acting upon them, it is usually difficult to identify and quantitate the effect of any one environmental factor on the entire community or a given
species within it. For this reason it is desirable
to carry out laboratory bioassays employing a single
species which is subjected to changes in one environmental factor at a time under controlled
conditions. However, the ultimate proof that
laboratory-derived bioassay data are adequate to
protect fish and other aquatic life must come from
studies of natural aquatic communities. Field
studies of the responses of aquatic organisms to
their environment play a major role in the program
of the U.S. Environmental Protection Agency, and
are employed in short-term surveys conducted to
assess the immediate, gross effects of separate
sources of pollutants on short segments of rivers
and on lakes, and to monitor long-term changes in
surface water quality.

The abundance, species composition and condition of aquatic organisms in natural communities are directly related to water quality. Field studies of water quality conducted by state and federal environmental protection agencies generally involve the assessment of the species composition and diversity, numerical density and biomass of four communities of aquatic organisms, namely: plankton, periphyton, macroinvertebrates, and fish. Plankton consists of the weakly-swimming or drifting micro-organisms, whereas the periphyton are microorganisms that grow upon or become attached to submerged surfaces. Both communities consist of bacteria, protozoa, actinomycetes, algae, microcrustacea and rotifers. The biomass and condition of the plankton and periphyton may be defined in terms of cell or organism counts, cell volumes, dry and ash-free weight, chlorophyll and adenosine triphosphate content, primarily productivity, and rates of respiration and nitrogen fixation. This report discusses some of the more recently developed methods of assessing plankton and periphyton bio-mass and condition which use chlorophyll and ATP content, and nitrogen fixation rates.

Chlorophyll

Chlorophyll *a* constitutes approximately 1-2% of the dry weight of organic matter in all algae, and provides a useful estimate of algal biomass. Also, the relative abundance of other chlorophylls, such as chlorophyll *b* and chlorophyll *c*, are in-dicative of the taxonomic composition of the algae.[1] Recent improvements in chlorophyll analyses have provided simplified and sensitive techniques, and yielded new parameters for describing the condition of chlorophyllous organisms. The simultaneous spectrophotometric measurement of concentrations of chlorophylls *a*, *b* and *c* in acetone extracts, developed by Richards and Thompson,[2] has been im-proved by the revisions of Parsons and Strickland[3] and the UNESCO committee,[4] and a more sensitive method for the measurement of chlorophyll *c* was developed by Parsons[5] and Ricketts.[6] The avail-ability of inexpensive scanning spectrophotometers which have a resolution of better than one nanometer (nm) permit more accurate measurement of the chlorophylls than previously possible in the average laboratory.

Other recent developments in chlorophyll determinations include the fluorometric method for measuring chlorophyll *a*, first employed by Yentsch and Menzel.[7] This method is two or three orders of magnitude more sensitive than the spectrophotometric method usually employed, and is reliable for chlorophyll *a* concentrations as low as 5 µg/liter of surface water. The fluorometric method has also been adapted to *in vivo* measurement of phytoplankton chlorophyll *a*.[8]

During the past decade, oceanographers have used the ratio between the organic matter and chlorophyll *a* in marine seston as an index of the condition of marine plankton. This approach has great potential for use in the federal water pollution control program as a means of measuring changes in plankton and periphyton species composition which may be directly related to changes in water quality. Normally, plankton and periphyton communities are dominanted by algae. Discharge of degradable non-toxic organic wastes into surface water usually results in a greater biomass of non-chlorophyllous, heterotrophic organisms than of algae which is reflected in a rise in the total biomass:chlorophyll *a* ratio. This ratio has been termed the Autotrophic Index (AI), and as originally defined, was associated with a particular type of instrumentation.[9-11] We have modified the formula for this index, making it independent of the instrumentation employed in the analysis.

$$\text{Autotrophic Index (AI)} = \frac{\text{Biomass (dry wgt organic matter)}}{\text{Chlorophyll } a}$$

where: (a) plankton biomass and chlorophyll *a* are expressed in mg/liter of surface water,
(b) periphyton biomass and chlorophyll *a* are expressed in mg/m^2 of surface from which the sample was removed.

AI values generally fall in the range of 50 to 100 for plankton and periphyton communities in unpolluted waters. Values reported for algal cultures, phytoplankton and seston are listed in Table 7. Waters containing a large amount of detritus,[22,23] as well as heterotrophic organisms, show large values for this index.

The applicability of this parameter to water quality studies is demonstrated by results obtained from periphyton communities which developed on glass slides suspended just below the surface of

Table 7

Autotrophic Indices

Sample	AI	Reference
Algal culture	55	Myers and Kratz[12]
Algal culture	70	Parsons[13]
Algal culture	96	Parsons et al.[14]
Algal culture	45	Steemann Nielsen[15]
Algal culture	48	Cobb and Myers[16]
Algal culture	40	Weber (Unpublished)
Algal culture	63	Ricketts[17]
Pond water	66	Copeland et al.[18]
Marine phytoplankton	200	Holm-Hansen[19]
Marine phytoplankton	80	Parsons[20]
Marine phytoplankton	76	Zeitzschel[21]
Lake seston	457	Weber[22]
Marine seston	146	Steele and Baird[23]
Marine seston	70	Ryther and Menzel[24]
Marine seston	81	Lorenzen[25]
Pond water	44	Moss[26]
Pond water	121	Moss[26]

the Ohio River for two-week intervals at two sites, one upstream and the other downstream of the outfall from a 180,000,000 gal/day sewage treatment plant at Cincinnati, Ohio. The sampling device used (Figure 15) was similar to the Catherwood Diatometer described by Patrick et al.[27]

The autotrophic indices at the two stations were significantly different for each sampling period (Figure 16). The average values obtained above and below the waste outfall were 177 and 1019, respectively. The values obtained for samples from the upstream station indicated the presence of a predominantly autotrophic (algal) periphyton, a diagnosis substantiated by microscopic analysis (Table 8). The organisms on the slides included the filamentous blue-green alga, *Schizothrix calcicola*, the green algae, *Closterium moniliferum*, *Stigeoclonium* spp. and *Mougeotia* spp., and the diatoms *Melosira varians*, *Navicula tripunctata* var. *schizonemoides*, *Gomphonema parvulum*, *Bacillaria paradoxa*, *Cymbella* spp., and *Synedra* spp.

In contrast to the upstream station, the higher indices obtained at the downstream station suggested the presence of a community containing a large number of heterotrophic, non-chlorophyllous organisms.

122

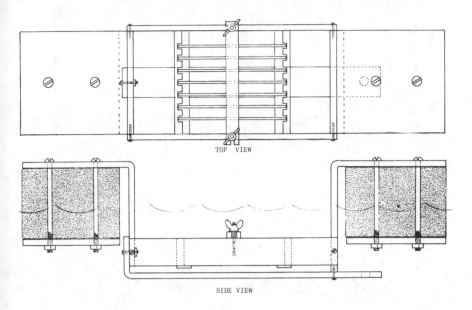

TOP VIEW

SIDE VIEW

Figure 15.   *Periphyton sampler. One-fourth inch plexiglass*
frame, 4"W X 17"L X 5"H, supported by two 4" X
4" X 2" styrofoam floats. Rack holds eight
1" X 3" glass microscope slides.

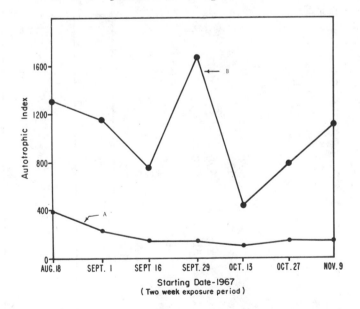

Figure 16.   *Autotrophic indices above (A) and below (B) a*
sewage treatment plant outfall on the Ohio River
at Cincinnati, Ohio.

123

Table 8

Composition of the Periphyton Growing on Glass Slides in the Ohio River Near Cincinnati, 1967

Organism		8/18-9/1	9/1-9/16	9/16-9/29	9/29-10/13	10/13-10/27	10/17-11/9	11/9-11/24
					Exposure Period			
Algae								
Filamentous	(a)*	17,301	13,461	1,472	820	0	0	0
blue-green	(b)	5	1,240	0	700	0	0	0
Coccoid	(a)	110	119	0	106	0	0	0
green	(b)	30	109	253	230	317	217	130
Filamentous	(a)	50	120	342	106	248	50	0
green	(b)	127	79	0	0	82	0	0
Pigmented	(a)	20	70	0	0	0	0	0
flagellates	(b)	0	20	60	20	40	33	81
Diatoms-	(a)	863	2,430	223	2,419	377	114	116
centric	(b)	425	526	1,065	2,100	1,055	3,489	1,957
Diatoms-	(a)	1,359	2,668	2,007	2,142	843	144	286
pennate	(b)	191	228	416	320	164	489	603
Algal cells	(a)	19,703	18,868	4,044	5,593	1,468	313	402
per mm²	(b)	778	2,202	1,794	3,370	1,658	4,228	2,771
Sphaerotilus	(a)							
	(b)	++		+	+++	+++	+++	+++
Stalked	(a)							
ciliates	(b)	++	++	++	+++	++		

* (a) = above outfall, (b) = below outfall, + = few, ++ = common, +++ = very abundant.

The periphyton on the slides at this station was dominated by the stalked ciliate protozoa, *Epistylis* and *Vorticella*, and the filamentous bacterium *Sphaerotilus natans*. Filamentous green algae (*Spirogyra* spp. and *Ulothrix* spp.), and the diatom *Melosira varians* were also present. The abundance of *Epistylis*, *Vorticella* and *Sphaerotilus* is considered by Klein[28] and Hynes[29] to be characteristic of moderate organic pollution.

The physiological condition of the algae can be determined by comparing the concentrations of chlorophyll *a* and its degradation product, pheophytin *a*, in the samples. Pheophytin *a* is formed when magnesium is lost from the porphyrin ring of chlorophyll *a*. This pigment is photosynthetically inactive, but has an absorption spectrum similar to that of chlorophyll *a*. If it is present in acetone extracts of pigments from periphyton and phytoplankton, pheophytin *a* may be erroneously measured as chlorophyll *a*. Recently developed fluorometric[7] and spectrophotometric methods[30] permit the simultaneous measurement of the concentrations of these two pigments. This is accomplished by measuring the optical density of the pigment solution at 663 nm (OD663) before and after acidification. Acidification of the solution quantitatively converts the chlorophyll *a* to pheophytin *a* and results in a decrease in the OD663 because pheophytin *a* has a lower specific absorption at that wavelength than has chlorophyll *a* (see Figure 17). In the absence of pheophytin *a*, a pigment solution containing chlorophyll *a* will yield a before:after acidification OD663 ratio of 1.70. In the absence of chlorophyll *a*, acidification of a pigment extract containing pheophytin *a* will result in no reduction of the OD663 (the before:after acidification OD663 ratio will be 1.0). In practice, the OD663 or the fluorescence of each pigment extract is measured before and after acidification, and the chlorophyll *a* and pheophytin *a* concentrations are calculated as described by Lorenzen.[30]

Autotrophic indices and the physiological condition of the algae (as described by the chlorophyll *a*:pheophytin *a* ratios) are two of several parameters we are measuring in a study of the effects of copper on the periphyton developing on glass slides in a small eutrophic calcareous midwestern stream.

125

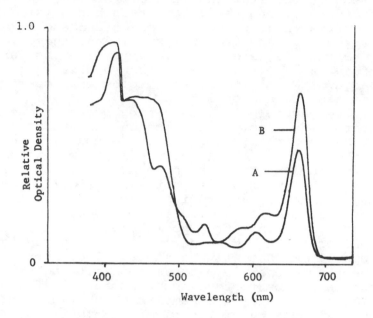

Wavelength (nm)

Figure 17. *Absorption spectra of a solution of chlorophyll in*
90% aqueous acetone before (B) and after (A)
acidification.

Adenosine Triphosphate

Until recently, total plankton biomass could
be estimated only by measuring the ash-free weight
of the particulate matter (seston) or by optically
determining the volume of the various components
of the plankton and applying empirically-derived
factors to convert volumes to dry weight of organic
matter.[14,31,32] Both methods have serious disadvantages. The error associated with biomass estimates based on ash-free weights is unknown, but
probably is large because the seston usually includes
much non-living organic matter.[33] Measurements of
plankton volume obtained using a microscope are
tedious and the factors employed in converting
volume to dry weight organic matter are little more
than coarse approximations. In addition, bacteria,
because of their size, are usually disregarded even
if very numerous.

Adenosine triphosphate (ATP) shows considerable
promise of providing a more accurate index of living

plankton and periphyton biomass because of its uni-
versality as an energy-mediating compound, and
because it is not found in significant amounts in
senescent, moribund and dead organisms. Current
methodology for measuring the concentration of ATP
in plankton employs the enzyme system responsible
for firefly luminescence. In this reaction, reduced
luciferin (the light-emitting substance involved in
theproduction of luminescence in the firefly) combines with the enzyme luciferase in the presence
of ATP to form a complex which is oxidized, yielding
one photon for each molecule of ATP expended.

LUCIFERIN-LUCIFERASE + ATP $\rightarrow$ LUCIFERIN-LUCIFERASE-AMP + PP

LUCIFERIN-LUCIFERASE-AMP + O_2 $\rightarrow$ LUCIFERIN-LUCIFERASE-AMP +
$$H_2O + LIGHT$$

The ATP is extracted from a concentrated plankton or periphyton sample by boiling it for 10 minutes
in Tris buffer adjusted to pH 7.75. An aliquot of
the ATP extract is injected into a luciferin-luciferase
preparation (available from biochemical supply houses),
and the emitted light is measured using either a
scintillation counter or a commercially-available
instrument specifically designed to measure ATP by
luminescence.
ATP was first used to assess the microbial biomass in water samples by Holm-Hansen and Booth,[34] who
applied a factor of 1.5 to convert the concentration
(in femtograms) of ATP to numbers of bacterial cells/
liter. They found that estimates of bacterial numbers
based on ATP concentrations were higher than those
obtained by standard bacteriological plating methods.
Hamilton, Holm-Hansen and Strickland[35] also found that
estimates of cellular carbon (living microbial biomass)
obtained from ATP measurements were higher than those
obtained using conventional methods (direct cell
counts). Holm-Hansen[19] compared ATP, chlorophyll *a* and
desoxyribose nucleic acid concentrations, cell numbers and volumes, and particulate (sestonic) carbon
and nitrogen in waters off the coast of southern California. Biomass estimates (expressed as micrograms
of carbon/liter) based on ATP and chlorophyll *a* concentrations, as well as cell numbers and cell volumes,
agreed well in samples from the euphotic zone.
Although the firefly method of measuring ATP
has been available for several years, information on
the concentration of ATP in surface waters is still
very scarce. Values reported for ocean waters range
from 10 to 800 nanograms ATP/liter (Table 9). Higher
concentrations were observed in the Ohio River below
the principal municipal waste treatment plant outfall

Table 9

ATP Content of Surface Water

Sample	ATP (ng/liter)	Reference
Pacific Ocean	210-360	Holm-Hansen[19]
Pacific Ocean	10	Hamilton *et al*[35]
Marseilles Bay	20-800	Daumas & Fiala[36]
Ohio River	300 *	Weber (Unpublished)
Ohio River	1700 +	Weber (Unpublished)
Little Miami River	2000	Weber (Unpublished)

*Above sewage treatment plant outfall.
+Below sewage treatment plant outfall.

and in the Little Miami River, a highly eutrophic tributary of the Ohio River.

The potential value of measurements of the concentration of ATP in surface waters as an index of the living plankton or periphyton biomass lies in the similarity in the ATP concentrations in bacteria, algae and other microorganisms for which data are available (Table 10). Although the concentrations range from 0.03 to 12.0 g ATP/mg dry weight, most values fall between 1.0 and 4.0 g/mg dry weight (average = 2.4). Preliminary work carried out with this technique at the Analytical Quality Control Laboratory indicates that the method is relatively simply and inexpensive, and the instrumentation is stable and reliable. The ultimate usefulness of the method in the water pollution control program will depend upon its acceptance and widespread application, and whether the relationship between ATP and dry weight organic matter evident in current data is substantiated in future work.

In addition to the above use of ATP, the method has many potential applications in laboratory and field bioassays of the effects of toxic substances and harmful physical conditions, because it permits immediate detection of mortality among populations of microorganisms. For example, the method has been used by the U.S. Environmental Protection Agency to measure plankton mortality resulting from entrainment in the cooling systems of thermal power generating plants.

Nitrogen Fixation

Nitrogen is a major cell constituent, and its abundance is an important factor determining the productivity of aquatic ecosystems. Accurate data

Table 10

Concentration of ATP in Various Organisms

Organism	ATP (μg/mg dry wgt)	Reference
Algae		
Amphidinium carteri	0.15	Holm-Hansen & Booth[34]
Chlorella vulgaris	2.0	Syrett[37]
Cyclotella nana	0.84	Holm-Hansen & Booth[34]
Ditylum brightwellii	1.2	Holm-Hansen & Booth[34]
Dunaliella tertiolectra	1.5	Holm-Hansen & Booth[34]
Monochrysis lutheri	0.28	Holm-Hansen & Booth[34]
Rhizosolenia sp.	1.6	Holm-Hansen & Booth[34]
Selenastrum capricornutum	1.4-3.4	Lee *et al.*[38]
Skeletonema costatum	0.31	Holm-Hansen & Booth[34]
Syracosphaera elongata	0.03	Holm-Hansen & Booth[34]
Bacteria		
Chromobacterium marinum	3.2	Hamilton & Holm-Hansen[39]
Escherichia coli	3.0	Lehninger[40]
Escherichia coli	3.0-9.0	Cole *et al.*[41]
Micrococcus sp.	1.2	Hamilton & Holm-Hansen[39]
Pseudomonas sp. (C-6)	4.4	Hamilton & Holm-Hansen[39]
Pseudomonas sp. (GL-7)	2.4	Hamilton & Holm-Hansen[39]
Pseudomonas sp. (GU-1)	2.4	Hamilton & Holm-Hansen[39]
Serratia sp.	2.0	Hamilton & Holm-Hansen[39]
Streptococcus faecalis	2.0-12.0	Forrest[42]
Vibrio sp.	2.4	Hamilton & Holm-Hansen[39]
Other		
Yeast	0.4	Grylls[43]
Microorganisms in activated sludge	1.4-2.0	Patterson *et al.*[44]
Phaseolus vulgaris	1.0-2.0	Jones[45]

on nitrogen budgets of surface waters, therefore, are essential for effective water resource management. Elemental nitrogen is known to be fixed by more than twenty species of blue-green algae and fifteen genera of bacteria.[46],[47] Biological fixation of elemental nitrogen, therefore, may be an important source of combined nitrogen in surface waters. The ability of an organism to fix nitrogen is also a great competitive advantage and plays an important role in plankton population dynamics.

Although the need for data on the rates of nitrogen fixation in aquatic ecosystems has increased greatly in recent years, the data are sparse because the methods were imprecise, time consuming, or required expensive instruments and reagents.[48] Nitrogen fixation in lakes, for example, has been estimated from Kjeldahl measurements of the gain in bound nitrogen in the seston,[49,50] from cell counts of nitrogen-fixing bacteria and blue-green algae,[51] from radioactive nitrogen (^{13}N) or the stable nitrogen isotope, ^{15}N, and from changes in the argon:nitrogen ratio in the gas above cultures ventilated with a 1:2 mixture of argon and nitrogen.[52] Until recently, the use of ^{15}N was considered the most convenient, sensitive and reliable method of estimating rates of nitrogen fixation in laboratory or *in situ* flask tests.[53,54]

Elemental nitrogen is reduced to ammonia by the nitrogenase enzyme system in the presence of ATP and magnesium ions.

$$N \equiv N + 6e \xrightarrow[\text{ATP} + \text{Mg}^{++}]{\text{Nitrogenase}} 2 NH_3$$

During a study of the inhibition of N_2 reduction caused by molecules of analogous structure (carbon monoxide and acetylene), Dilworth[55] and Schollhorn and Burris[56] discovered that acetylene was reduced to ethylene. Use of acetylene reduction to estimate rates of nitrogen fixation was first suggested by Hardy and Knight.[57] The reaction was subsequently adapted to *in situ* measurement of N_2 fixation rates by Stewart *et al*,[48] who quantitated the evolved ethylene by gas chromatography. This method is less time-consuming than the ^{15}N method, utilizes far less expensive reagents and instrumentation, and permits the use of incubation periods of five seconds to 30 minutes, compared to 24 hours required for the ^{15}N method. As little as one picomole of C_2H_4 can be detected, making the method 10^3 times more sensitive than the ^{15}N method and 10^6 times more sensitive than the Kjeldahl method.

The reduction of acetylene to ethylene requires only two electrons, whereas the reduction of N_2 to ammonia requires six electrons.

$$H - C \equiv C - H + 2e \rightarrow H_2C = CH_2$$

On the basis of the energy requirements, the theoretical yield of this reaction would be a ratio of three moles of ethylene produced to one mole of N_2 reduced. Reported ratios range from 2.4 to 8.6, but most values fall between three and five (Table 11). The rate of acetylene reduction is usually reported in nanomoles of ethylene produced per milligram of protein per minute. Rates for bacteria range from 20 to 40 nanomoles C_2H_4/mg protein/min, and are one or two

Table 11

Relationship Between Nitrogen Fixed and
Ethylene Produced (Mole Ratio)

Organism	C_2H_4/N_2*	Reference
Bacteria		
Azotobacter vinelandii	3.95	Hardy *et al.*[59]
Clostridium pasteurianum	4.22	Dilworth[55]
Clostridium pasteurianum	2.5	Schollhorn & Burris[56]
Blue-green algae		
Nostoc muscorum	3.73	Stewart *et al.*[60]
Anabaena cylindrica	2.99	Stewart *et al.*[60]
Anabaena flos-aquae	3.17	Stewart *et al.*[60]
Soil Microorganisms	5.2	Steyn & Delwiche[58]

*Mean value.

orders of magnitude greater than those for blue-green algae, which range from 0.1 to 4.7 nanomoles C_2H_4/mg protein/min (Table 12).[48,55,59,61,62]

Although ethylene is also produced during normal metabolism in plants from the beta cleavage of methionine, and is known to occur in fruits of vascular plants[67,68] and in the fungi *Pencillium digitatum*, *Alternaria citri*, and *Blastomyces dermatitus*,[69] it has not been reported in aquatic microorganisms, and is not likely, therefore, to interfere with the use of the acetylene method.

Cells of nitrogen fixing microorganisms grown in the presence of combined nitrogen such as nitrate or ammonia do not reduce nitrogen (combined nitrogen does not inhibit nitrogen fixation directly, but its presence prevents the development of the nitrogenase enzyme system). Also, cells of such organisms grown in the absence of combined nitrogen will gradually lose their ability to fix nitrogen if combined nitrogen is added to the culture. Photosynthesis is essential for nitrogen fixation by algae.[48] If nitrogen-fixing algae are placed in the dark, they lose their ability to fix nitrogen within approximately one hour. However, if returned to the light, they quickly resume N_2 reduction. Similarly, natural plankton populations taken from the surface of a lake and incubated in the aphotic zone fix less nitrogen than if incubated near the surface.[66] The rate of N_2 fixation in phytoplankton incubated *in situ* near the surface follows a diel pattern similar to that of photosynthesis, which is reduced as a result of photoinhibition during midday.

Table 12

Rate of Ethylene Production (Nanomoles C_2H_4/mg Protein/min)

Organism	*C_2H_4/mg protein/min*	*Reference*
Bacteria		
Azotobacter vinelandii	56.3	Hardy et al.[59]
Azotobacter chroococcum	31	Drozd & Postgate[61]
Chloropseudomonas ethylicum	5.8	Evans & Smith[63]
Clostridium pasteurianum	32.1	Dilworth[55]
Clostridium pasteurianum	12	Schollhorn & Burris[56]
Clostridium pasteurianum	46	Hardy et al.[59]
Blue-green Algae (cultures)		
Anabaena cylindrica	1.9	Stewart et al.[60]
Anabaena cylindrica	4.15	Stewart[64]
Anabaena flos-aquae	3.6	Stewart et al.[60]
Anabaena flos-aquae	1.3	Stewart[64]
Mastigocladius laminosus	1.0	Stewart et al.[60]
Nostoc entophytum	0.9	Stewart et al.[60]
Nostoc muscorum	3.6	Stewart et al.[60]
Nostoc sp.	0.4	Stewart et al.[60]
Tolypothrix sp	1.8	Stewart et al.[60]
Tolypothrix tenuis	0.91	Stewart[64]
Phytoplankton (concentrated)		
Anabaena sp.	0.10	Stewart et al.[48]
Aphanizomenon flos-aquae	0.77	Stewart et al.[48]
Calothrix sp.	0.01	Stewart et al.[48]
Gleotrichia echinulata	0.36	Stewart et al.[48]
Nostoc sp.	1.62	Stewart et al.[48]
Phytoplankton (unconcentrated)		
Lake Mendota	1.68	Stewart et al.[48]
Lake Erie	3.2	Howard et al.[65]
Little Arbor Vitae Lake	2.1	Rusness & Burris[66]
Lake Mendota	1.5	Rusness & Burris[66]

Quantitative measurements of acetylene reduction have been published for only a few lakes--Lake Mendota, Lake Mary, and Trout, Crystal and Little Arbor Lakes in Wisconsin, and Lake Mize, Florida.[48,66,70] The highest reported rate of nitrogen fixation was 840 mg N/m^2/day in Lake Mendota.[48] Most values, however, are four orders of magnitude less (Table 13).

The acetylene reduction technique has also been used to determine phosphorus availability in surface waters.[71] In this test, the rate of ethylene

Table 13

In Situ Nitrogen Fixation Rates in Surface Samples and in a Sediment as Determined by the Acetylene Reduction Method

Lake	(Note)	Nitrogen fixation rate		Reference
Surface Water				
Lake Mendota (Wisc)	(a)	93	µg N$_2$/1/hr	Stewart et al.[48]
Lake Mendota (Wisc)	(b)	42	µg N$_2$/1/hr	Stewart et al.[48]
Lake Mendota (Wisc)	(c)	1000	µg N$_2$/1/hr	Stewart et al.[48]
Lake Mendota (Wisc)	(d)	0.50	µg N$_2$/1/hr	Rusness & Burris[66]
Lake Mary (Wisc)	(e)	0.006	µg N$_2$/1/hr	Brezonik & Harper[70]
Lake Mize (Fla)	(f)	0.17	µg N$_2$/1/hr	Brezonik & Harper[70]
Little Arbor Vitae Lake (Wisc)	(g)	2.24	µg N$_2$/1/hr	Rusness & Burris[66]
Sediment				
Lake Erie		0.29	µg N$_2$/g/day	Howard et al.[65]

[a]Mean of hourly measurements taken during photoperiod (12 hours).

[b]Rate at a depth of one meter over a period of 30 minutes.

[c]Mean of 10 successive one-minute measurements of C$_2$H$_2$ reduction by a bloom of *Gleotrichia echinulata*.

[d]Maximum.

[e]Mean of single water samples from 4 levels in the epilimnion.

[f]Rate nearly uniform from surface to bottom.

[g]Maximum.

production was measured in phosphorus-starved *Anabaena flos-aquae* incubated in test water and in culture media containing several levels of orthophosphate (0.00, 0.010, 0.025, 0.050, and 0.100 mg P/liter) and was found to be directly proportional to the amount of phosphorus added, up to a concentration of 0.100 mg P/liter.

Because the rate of nitrogen fixation varies greatly with different types of organisms and with the concentration of combined nitrogen in the water it will not be possible to use nitrogen fixation rates to estimate the biomass of N_2-fixing microorganisms in surface waters. However, the acetylene reduction method will be very useful in measuring nitrogen budgets and in bioassays.

ACKNOWLEDGMENT

The author gratefully acknowledges the laboratory assistance of Donald Moore and Ben McFarland in the analysis of the periphyton samples discussed in this report.

REFERENCES

1. Round, F. E. *The Biology of the Algae.* (New York: St. Martins Press, 1965).
2. Richards, F. A., with T. G. Thompson. "The Estimation and Characterization of Plankton Populations by Pigment Analyses. II. A Spectrophotometric Method for the Estimation of Plankton Pigments," *J. Mar. Res. 11(2):156* (1952).
3. Parsons, T. R., and J. D. H. Strickland. "Discussion of Spectrophotometric Determination of Marine-plant Pigments, With Revised Equations for Ascertaining Chlorophylls and Carotenoids," *J. Mar. Res. 21(3):155* (1963).
4. UNESCO. *Monograph on Oceanographic Methodology. I. Determination of Photosynthetic Pigments in Sea-water.* (Paris, 1966).
5. Parsons, T. R. "A New Method for the Microdetermination of Chlorophyll *c* in Sea Water," *J. Mar. Res. 21(3):164* (1963).
6. Ricketts, T. R. "A Note on the Estimation of Chlorophyll *c*," *Phytochem. 6:1353* (1967).
7. Yentsch, C. S., and D. W. Menzel. "A Method for the Determination of Phytoplankton Chlorophyll and Pheophytin by Fluoroescence," *Deep-Sea Res 10:221* (1963).
8. Lorenzen, C. J. "A Method for the Continuous Measurement of *in vivo* Chlorophyll Concentration," *Deep-Sea Res. 13:223* (1966).
9. Taylor, M. P. *Thermal Effects on the Periphyton Community in the Green River* (Chattannoga, Tn.: TVA, Div. Hlth. & Safty, Water Qual. Br., 1967).

10. Tennessee Valley Authority. *Thermal and Biological Studies in the Vicinity of the Colbert Steam Plant (Tennessee River).* (Chattannoga, Tn.: TVA, Div. Hlth. & Safty, Water Qual. Br., 1967).
11. Tennessee Valley Authority. *Thermal and Biological Studies in the Vicinity of Widows Creek Steam Plant (Tennessee River).* (Chattannoga, Tn.: TVA, Div. Hlth. & Safty, Water Qual. Br., 1967).
12. Myers, J., and W. A. Kratz. "Relationship Between Pigment Content and Photosynthetic Characteristics in a Blue-Green Alga," *J. Gen. Physiol. 39*:11 (1955).
13. Parsons, T. R. "On the Pigment Composition of Eleven Species of Marine Phytoplankton," *J. Fish. Res. Bd. Can. 18(6)*:1017 (1961).
14. Parsons, T. R., K. Stephens, and J. D. H. Strickland. "On the Chemical Composition of Eleven Species of Marine Phytoplankters," *J. Fish. Res. Bd. Can. 18(6)*:1001 (1961).
15. Steemann Nielsen, E. "Chlorophyll Concentration and Rate of Photosynthesis in *Chlorella vulgaris*," *Physiol. Plant. 14*:868 (1961).
16. Cobb, H. D., and J. Myers. "Comparative Studies of Nitrogen Fixation and Photosynthesis in *Anabaena cylindrica*," *Amer. J. Bot. 51(7)*:753 (1964).
17. Ricketts, T. R. "On the Chemical Composition of Some Unicellular Algae," *Phytochem. 5*:67 (1966).
18. Copeland, B. J., K. W. Minter, and T. C. Dorris. "Chlorophyll *a* and Suspended Organic Matter in Oil Refinery Effluent Holding Ponds," *Limnol. Oceanogr. 9(4)*: 500 (1964).
19. Holm-Hansen, O. "Determination of Microbial Biomass in Ocean Profiles," *Limnol. Oceanogr. 14(5)*:740 (1969).
20. Parsons, T. R. "The Use of Particle Size Spectra in Determining the Structure of a Plankton Community," *J. Oceanogr. Soc. Japan 25*:172 (1969).
21. Zeitzschel, B. "The Quantity, Composition and Distribution of Suspended Particulate Matter in the Gulf of California," *Mar. Biol. 7(4)*:305 (1970).
22. Weber, C. I. "The Measurement of Carbon Fixation in Clear Lake, Iowa, Using Carbon-14," Ph.D. Thesis, Iowa State University, Ames, (1960).
23. Steele, J. H., and I. E. Baird. "The Chlorophyll *a* Content of Particulate Organic Matter in the Northern North Sea," *Limnol. Oceanogr. 10(2)*;261 (1965).
24. Ryther, J. H., and D. W. Menzel. "On the Production, Composition, and Distribution of Organic Matter in the Western Arabian Sea," *Deep Sea Res. 12*:199 (1965).
25. Lorenzen, C. J. "Carbon/Chlorophyll Relationships in an Upwelling Area," *Limnol. Oceanogr. 13(1)*:212 (1968).
26. Moss, B. "Seston Composition in Two Freshwater Pools," *Limnol. Oceanogr. 15(4)*:504 (1970).

27. Patrick, R., M. H. Hohn, and J. H. Wallace. "A New Method for Determining the Pattern of the Diatom Flora," *Not. Nat. 259*:1 (1954).
28. Klein, L. *River Pollution. II. Causes and Effects.* (London: Butterworths, 1962).
29. Hynes, H. B. N. *The Biology of Water Pollution.* (Liverpool Univ. Press, 1963).
30. Lorenzen, C. J. "Determination of Chlorophyll and Pheo-Pigments: Spectrophotometric Equations," *Limnol. Oceanogr. 12(2)*:343 (1967).
31. Mullin, M. M., P. R. Sloan, and R. W. Eppley. "Relationship Between Carbon Content, Cell Volume, and Area in Phytoplankton," *Limnol. Oceanogr. 11*:307 (1966).
32. Strathmann, R. R. "Estimating the Organic Carbon Content of Phytoplankton From Cell Volume or Plasma Volume," *Limnol. Oceanogr. 12(3)*:411 (1967).
33. Weber, C. I., and D. R. Moore. "Phytoplankton, Seston, and Dissolved Organic Carbon in the Little Miami River at Cincinnati, Ohio," *Limnol. Oceanogr. 12*:311 (1967).
34. Holm-Hansen, O., and C. R. Booth. "The Measurement of Adenosine Triphosphate in the Ocean and Its Ecological Significance," *Limnol. Oceanogr. 11(4)*:510 (1966).
35. Hamilton, R. D., O. Holm-Hansen, and J. D. H. Strickland. "Notes on the Occurrence of Living Microscopic Organisms in Deep Water," *Deep-Sea Res. 15*:651 (1968).
36. Daumas, R., and M. Fiala. "Evaluation de la matiere organique vivante dans les eaux marines par la mesure de l'adenosine triphosphate," *Mar. Biol. 3*:243 (1969).
37. Syrett, P. J. "Respiration Rate and Internal Adenosine Triphosphate Concentration in *Chlorella*," *Arch. Biochem. Biophys. 75*:117 (1958).
38. Lee, C. C., R. F. Harris, J. K. Syers, and D. E. Armstrong. "Adenosine Triphosphate Content of *Selenastrum capricornutum*," *Appl. Microbiol. 21(5)*:957 (1971).
39. Hamilton, R. D., and O. Holm-Hansen. "Adenosine Triphosphate Content of Marine Bacteria," *Limnol. Oceanogr. 12(2)*:319 (1967).
40. Lehninger, A. L. *Bioenergetics.* (Benjamin, N.Y., 1965).
41. Cole, H. A., J. W. T. Wimpenney, and D. E. Hughes. "The ATP Pool in *Escherichia coli*. I. Measurement of the Pool Using a Modified Luciferase Assay," *Biochim. Biophys. Acta 143*:445 (1967).
42. Forrest, W. W. "Adenosine Triphosphate Pool During the Growth Cycle in *Streptococcus faecalis*," *J. Bacteriol. 90*:1013 (1965).
43. Grylls, F. S. M. "The Chemical Composition of Yeasts," In *Biochemist's Handbook*, Long, C., ed. (New York: Van Nostrand, 1961), pp 1050-1053.
44. Patterson, J. W., P. L. Brezonik, and H. D. Putnam. "Measurement and Significance of Adenosine Triphosphate in Activated Sludge," *Environ. Sci. Technol. 4(7)*:569 (197

45. Jones, P. C. T. "The Effect of Light, Temperature, and Anaesthetics on ATP Levels in the Leaves of *Chenopodium rubrum* and *Phaseolus vulgaris*," *J. Exp. Bot. 21(66)*:58 (1970).

46. Fogg, G. E. "Nitrogen Fixation," In *Physiology and Biochemistry of Algae*, Lewin, R. A., ed. (Academic Press, 1962).

47. Stewart, W. D. P. "Nitrogen-Fixing Plants," *Science 158*:1426 (1967).

48. Stewart, W. D. P., G. P. Fitzgerald, and R. H. Burris. "*In situ* Studies on N_2 Fixation Using the Acetylene Reduction Technique," *Proc. Natl. Acad. Sci. 58*:2071 (1967).

49. Hutchinson, G. E. "Limnological Studies in Connecticut. IV. The Mechanism of Intermediary Metabolism in Stratified Lakes," *Ecol. Monogr. 11*:21 (1941).

50. Mortimer, C. H. "The Nitrogen Balance of Large Bodies of Water," *Off. Circ. Brit. Waterw. Assoc. 21*:1 (1939).

51. Kusnezow, S. E. "Die Rolle der Mikroorganismen in Stoffkreislauf der Seen," *Deut. Ver. der Wissensch. Ber.* (1959).

52. Fay, P., and G. E. Fogg. "Studies on Nitrogen Fixation by Blue-Green Algae. III. Growth and Nitrogen Fixation in *Chlorogloea fritschii*," *Mitra. Arch. Mikrobiol. 42*: 310 (1962).

53. Neess, J. C., R. C. Dugdale, V. A. Dugdale, and J. J. Goering. "Nitrogen Metabolism in Lakes. I. Measurement of Nitrogen Fixation with N^{15}," *Limnol. Oceanogr. 7*: 163 (1962).

54. Fogg, G. E., and A. J. Horne. "The Determination of Nitrogen Fixation in Aquatic Environments," In *Chemical Environment in the Aquatic Habitat*, Proc. IBP Symp., Amsterdam and Nieuwersluis, 1966, Golterman, H. L., and R. S. Clymo, eds., (Amsterdam: Noord-Hollandsche Uitgevers Maatschappij, 1967).

55. Dilworth, M. J. "Acetylene Reduction by Nitrogen-Fixing Preparation from *Clostridium Pasteurianum*," *Biochim. Biophys. Acta 127*:285 (1966).

56. Schollhorn, R., and R. H. Burris. "Study of Intermediates in Nitrogen Fixation," *Federation Proc. 25*:710 (1966).

57. Hardy, R. W. F., and E. Knight, Jr. "ATP-Dependent Reduction of Azide and HCN by N_2-Fixing Enzymes of *Azotobacter vinelandii* and *Clostridium pasteurianum*," *Biochim. Biophys. Acta 139*:69 (1967).

58. Steyn, P. L., and C. C. Delwiche. "Nitrogen Fixation by Nonsymbiotic Microorganisms in Some California Soils," *Environ. Sci. Technol. 4(12)*:1122 (1970).

59. Hardy, R. W. F., R. D. Holsten, E. K. Jackson, and R. C. Burris. "The Acetylene-Ethylene Assay for N_2 Fixation: Laboratory and Field Evaluation," *Plant Physiol. 43*: 1185 (1968).

60. Stewart, W. D. P., G. P. Fitzgerald, and R. H. Burris. "Acetylene Reduction by Nitrogen-Fixing Blue-Green Algae," *Arch. f. Mikrobiol. 62*:336 (1968).

61. Drozd, J., and J. R. Postgate. "Interference by Oxygen in the Acetylene-Reduction Test for Aerobic Nitrogen-Fixing Bacteria," *J. Gen. Microbiol. 60(3)*:427 (1970).

62. Jewell, W. J., and S. A. Kulasooriya. "The Relation of Acetylene Reduction to Heterocyst Frequency in Blue-Green Algae," *J. Exp. Bot. 21(69)*:874 (1970).

63. Evans, M. C. W., and R. V. Smith. "Nitrogen Fixation by the Green Photo-Synthetic Bacterium *Chloropseudomonas ethylicum*," *J. Gen. Microbiol. 65*:95 (1971).

64. Stewart, W. D. P. "Nitrogen Input into Aquatic Eco-systems," In *Algae, Man and the Environment*, Jackson, D. F., ed. (Syracuse Univ. Press, 1968).

65. Howard, D. L., J. I. Frea, R. M. Pfister, and P. R. Dugan. "Biological Nitrogen Fixation in Lake Erie," *Science 169*:61 (1970).

66. Rusness, D., and R. H. Burris. "Acetylene Reduction (Nitrogen Fixation) in Wisconsin Lakes," *Limnol. Oceanogr. 15(5)*:808 (1970).

67. Owens, L. D., M. Lieberman, and A. Kunishi. "Inhibition of Ethylene by Rhizobitoxine," *Plant Physiol. 48*:1 (1971).

68. Gane, R. "Production of Ethylene by Some Ripening Fruits," *Nature 134*:1008 (1934).

69. Jansen, E. F. "Ethylene and Polyacetylenes," In *Plant Biochemistry*, Bonner, J., and J. E. Varner, eds. (Academic Press, 1965).

70. Brezonik, P. L., and C. L. Harper. "Nitrogen Fixation in Some Anoxic Lucustrine Environments," *Science 164*:1277 (1969).

71. Stewart, W. D. P., G. P. Fitzgerald, and R. H. Burris. "Acetylene Reduction Assay for Determination of Phosphorus Availability in Wisconsin Lakes," *Proc. Natl. Acad. Sci. 66(4)*:1104 (1970).

6. DIATOMS AS BIOASSAY ORGANISMS

Ruth Patrick. Academy of Natural Sciences of
Philadelphia, Philadelphia, Pennsylvania.

SUMMARY

Diatoms are very useful as bioassay organisms.
They can be used in the laboratory and studies in
unialgal cultures, or whole communities can be
studied under semi-natural conditions. Bioassay
methods for batch and continuous flow laboratory
experiments are described. Also the method for
bioassaying diatom communities is set forth.
Furthermore diatoms can be used to monitor or
bioassay the ability of water to support algal
growth in rivers, estuaries, and lakes. This
method is also described.

Diatoms, which are unicellular plants belonging
to the large group of plants known as algae, are
very useful in monitoring pollution in lakes, rivers,
and estuaries. They would probably also be useful
in monitoring pollution in the open sea. They are
a very important food source for most forms of
aquatic life that feed upon plants, and they carry
out the process of photosynthesis which is so important in the generation of oxygen needed by all
organisms in order to carry out metabolic processes.
There are a great many species of diatoms and
these species have very different ranges of tolerance
to ecological conditions. As a result the diatoms
as a group live in a wide variety of conditions.
Because of the large number of species of diatoms,
one will find many species present in almost all
natural conditions. These species may be represented
by moderately sized populations, relatively small

populations, or relatively large populations depending upon the ecological conditions under which
they live. Because as a group they consist of many
species that have populations composed of varying
numbers of specimens, they are an excellent group
to treat statistically in analyzing their reaction
to varying ecological conditions.

Most of the characteristics used in identification are present on the diatom wall, which is
siliceous. Therefore they do not need to be treated
in special ways in order to preserve them. Furthermore, one can collect diatoms and retain them for
many years before studying them without losing their
characteristics for identification.

Uses of Diatoms as Indicators
of Pollution

Studies of diatom communities and variation in
the structure of these communities can be used in
many different types of studies to determine the
presence of pollution. One efficient way to obtain
a representative sample of diatoms living in an
area is by the use of a diatometer[1] (Figure 18),
which can be easily placed in a body of water.

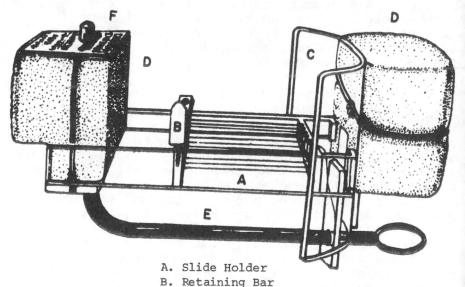

A. Slide Holder
B. Retaining Bar
C. Deflector
D. Styrofoam Float
E. Brass Rod
F. Identification Tag

Figure 18.   Catherwood Diatometer.

These instruments can be distributed over an entire
watershed in a relatively short period of time.
They can then be collected approximately two weeks
later, and the numbers of species, kinds of species,
and sizes of diatom populations growing on slides
positioned in the diatometers provide a reliable
index of the degree of degradation of the river,
lake, or estuary affected by pollution. This method
can also be used to determine the pattern and extent
of the effect of a waste. In a study on the Delaware
River about 150 diatometers were placed in concentric
circles around an outfall. In this way the limits
of the effect of the waste could be determined. The
results of these studies were checked by model
studies of flow patterns and were found to be very
reliable.

In using diatoms to continually monitor the
effect of pollution, one or more diatometers are
placed upstream above any possible effect of an
effluent and also placed at varying distances downstream from the outfall of an effluent in order to
continually monitor the effect of this effluent on
the receiving body of water.

Besides the studies of the kinds and numbers
of species and relative sizes of their populations--
which are useful in determining the amount, general
type, and severity of pollution--diatometers can
also be used to determine the nutrient levels in a
body of water. By placing the diatometers in a body
of water one knows exactly the type of substrate
that is being exposed and the length of time it is
exposed. The diatoms and other algae that collect
and grow on the glass slides can then be scraped
off these slides and the relative abundance of the
more common species, and residue, and volatile
weights determined. In this manner the relative
abilities of bodies of water to support a standing
crop of algae can be determined. If it is desirable,
chlorophyll and other pigment analyses as well as
^{14}C uptake studies can be made.

The diatom communities can also be used to
indicate the amount of radioactivity present. It
has been shown that diatoms concentrate radioactivity to levels many times that found in the
natural water or solution to which they are exposed.
This occurs by the adsorption of radioactivity onto
the jelly-like film that coats all diatoms and by
absorption into the cell. Diatoms have been used
at the Savannah River Atomic Energy Plant to determine the presence of radioactivity. It should be

noted that it has been found that once the exposure to radioactivity is eliminated the diatoms lose their activity rather rapidly.

Recent studies indicate diatoms may also be used for monitoring insecticides. When one considers that diatoms store fats and oils, it would be expected that they would concentrate insecticides.

The calibration of the diatometer to date has been carried out in fresh water and in estuaries of rivers and bays. We have not carried out the necessary studies to interpret the effect of pollution in the open sea, although this could be done fairly easily, and the interpretation would be based upon the principles set forth below for fresh water rivers and estuaries. Because the plankton of a sea may be less diverse than the plankton plus benthic forms in estuaries, I would expect the height of the mode and structure of the curve to be somewhat different.

METHODS AND PROCEDURES
FOR DETERMINING DIVERSITY

The method for determining the structure of the diatom community has been outlined by Patrick, *et al.*[1] Samples of natural diatom communities fit the model of a truncated normal curve (Figure 19). A shorter method for estimating numbers of species present, percent of populations composed of dominant species, and biomass has been outlined by Hohn.[2]

Interpretation

In most natural streams when one studies a large enough sample to construct a truncated normal curve, the number of observed species will vary from 100 to 175 depending on ecological conditions. A reduction of 50 percent is considered as definitely indicating unfavorable conditions. If possible, a diatometer should be maintained in a natural area of the river in order to determine the number of species one should expect, rather than using the figures 100 to 175. The percent of dominance should be less than 50% if the stream is not significantly perturbed by pollution.

The following is a guide to interpreting what type of pollution is causing perturbation in a river. It should be remembered that pollution is usually a mixture of various pollutants and there is a gradient of effects.

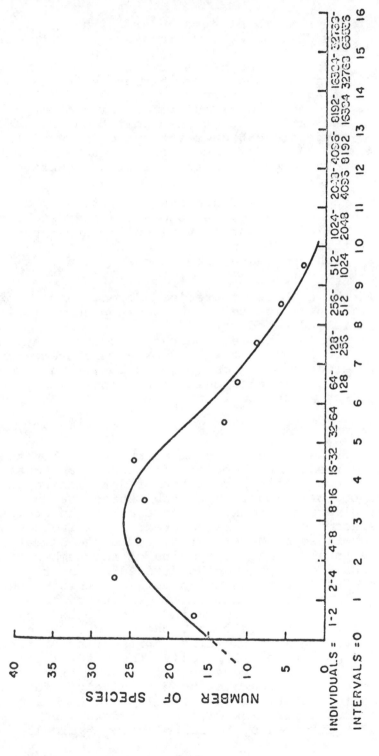

Figure 19.

If the numbers of species are high:
1. low dominance, low biomass - indicate certain types of toxicants such as low pH or moderately high temperature.
2. high dominance, high biomass - indicate organic pollution.
3. high dominance, moderately high biomass - indicate light organic pollution or temperature higher than normal.
4. moderate dominance, low biomass - indicate cold temperature or ecologically unfavorable conditions.

If the numbers of species are low:
1. low dominance, low biomass - indicate high toxicity. Rarely, very cold weather produces this condition.
2. high dominance, low biomass - indicate toxicity.

If the numbers of species are moderately low:
1. high dominance, high biomass - organic load is indicated.
2. high dominance, moderate biomass - may be a low level of toxicity or adverse temperature effects, or light organic load. This condition is difficult to diagnose except to say pollution is present.

Diversity Index

If it is not desired to carry out either the detailed reading or the semi-detailed reading as outlined above, another method of determining the effect of pollution is by establishing the Shannon-Weiner diversity index. For these indices at least 1000 specimens should be counted. This index has been described by MacArthur.[3] It is the summation for the total number of species of the number of individuals of each species divided by the total number of specimens in the sample and this number is multiplied by the natural logarithm of this number. If this diversity index is plotted against the number of species found, one can establish different slopes of the line representing different kinds of pollution. Typically, the natural community has a high diversity index and a high number of species. One that has been adversely affected by organic pollution has a fairly high number of species but a very low diversity index, whereas a community affected by toxic pollution typically has a low diversity index and a low number of species. There are, however, exceptions where invasion rates are high and the effect of toxic pollution is to inhibit reproductive rates. In such cases one may have a fairly high diversity and a high number of species but the biomass in this case will be extremely low. The biomass under organically enriched conditions is higher than that found under natural conditions of the same body of water at the same season of the year. We must, of course, realize that pollution as found in our surface waters is usually not due to a single factor, so combinations of toxic and organic effects are usually present. In such cases, dominance of a few species is generally evident, one may or may not have a large biomass; but typically the numbers of species are greatly reduced.

McIntosh[4] and McIntire[5] have discussed other methods for establishing the amount of diversity. Patton,[6] MacArthur,[3] McIntosh,[4] and McIntire[5] have described various ways of comparing diversity between two or more communities.

METHODS FOR USING DIATOMS AS BIOASSAY ORGANISMS IN THE LABORATORY

In these tests various concentrations of wastes are compared to a control to determine what concentrations will produce significant deleterious effects. The length of time of the batch tests is usually five to seven days. During this period of time one can feel fairly confident that all nutrients are present in sufficient amounts so that they will not inhibit growth. After such a length of time some of the minor elements may be deficient and changes may occur in the medium. If it is desirable to run a longer test, continuous flow tests are recommended. Both types of tests are described in these procedures.

Continuous flow tests, besides being desirable to study the effect of a waste over a long period of time (one to several months), are also desirable when wastes are unstable and therefore will either be decomposed or volatilized in a short period of time. They are also recommended when the waste substances interact with the dilution water and cause precipitates to form.

In some cases diatoms will not grow in dilution water which has high enough nutrient concentrations

145

to support growth for seven days. In these special cases continuous flow tests are recommended. It has been our experience that diatoms usually grow better under continuous flow conditions than in batch tests. Therefore when high productivity in a test is desired, the continuous flow is preferable even for short-term tests.

Methods for Performing Tests

The batch test method has been described by Patrick in ASTM.[7] The continuous flow bioassay tests are run in a similar manner to those of batch culture except for the following.

The culture medium is continually supplied by either gravity flow or by being pumped through the flasks. Instead of introducing the inoculum free into the culture medium the inoculum is introduced onto a Millipore™ filter. Vacuum is applied to the Millipore filter until the filter starts to turn white. This securely fastens the diatoms to the filter but does not intermesh them with the filter fibers. A very thin capillary tube is introduced through the stopper to the flask. The opening of this tube is immediately over the surface of the diatoms. It should be approximately 1 cm above the surface of the filter paper. The rate of flow of the new medium is between one and two liters per day if the volume of the medium is 100 cc or less, unless for some reason it is desirable to greatly increase this rate of flow. The medium should be made fresh each day. Watts and Harvey[8] have described this method.

Inoculum

The Millipore filters to be used in the experiments are numbered, weighed, and assigned to a flask of the same number. All filters must be moistened, dried and weighed in order to get an average weight and the amount of variation to be expected due to subsequent drying. The inoculum is homogenized as for batch cultures. It is also examined under the microscope to determine that the diatoms are distributed uniformly. Ten ml samples are carefully measured and placed in each of five vials (which have been previously weighed) and the mean weight and standard error of the mean determined. One milligram of inoculum is then added and the volume of inoculum to secure this weight is noted.

Sufficient inoculum to equal one mg is then placed in about 10 cc of water and vacuum filtered as described above. The inoculations are made one to each concentration until six duplicates for each concentration have been made. For example, if there are four dilutions, one inoculation for each dilution will be made, one for the control, and one for obtaining the mean weight of the inocula. Thus a total of 25 flasks will be inoculated, and five inoculated filters will be dried for determining the average size of the inoculum at the beginning of the experiment.

The experiment can run as long as desirable, but usually not to such a length of time that the diatoms grow out over the edge of the filter. At the end of the experiment or when it is desirable to determine the increase in growth, the filter is dried to constant weight and the amount of increase determined. This is related to the weight of the inoculum at the beginning of the experiment.

Analyses of Results

The results of the various test concentrations are compared to the control. The concentration which reduces the growth 50% as determined by weight is estimated from the graph. At least four concentrations should be tested in order to develop the graph. At the end of the experiment the cells should be examined before they are dried in order to determine the health or condition of the cells.

One should take care to remove all diatoms from the flasks. After removing the filter paper from the flask, the walls of the flask should be scraped down with a sterilized rubber spatula and all of the media filtered through a Millipore filter in order to concentrate the cells that might be on the surface of the flask onto the filter.

If desired, the diatoms can be washed off the filters by a jet of water; the water can then be evaporated and the diatoms weighed.

The effects of various dilutions of a given waste can also be assayed by determining relative rates of ^{14}C uptake, amounts of pigment/gm, respiration rates, and rate of production of ATP.

METHODS FOR USING DIATOMS AS BIOASSAY
ORGANISMS IN SEMI-NATURAL CONDITIONS

Laboratory bioassay procedures are usually
performed using unialgal cultures. However, in
nature algae usually compete with more than a hundred
species belonging to different algal groups. As a
result the success of a species is not determined
by survival but rather by its ability to out-compete
other species.

Methods for Performing Tests

In order to determine the effects of various
types of alterations in the chemical and physical
environment on species living together in a community, an experimental set-up was designed whereby
water from a stream with its seston could be diverted
through experimental boxes in a greenhouse which at
any moment in time supported similar diatom
communities.[9]
In these experiments stream water is pumped
into a circular plastic tank. The water is continually agitated by the direction of the water
pumped into the tank. In the center of the tank
inlets to each of eight experimental boxes are
supported so that the intakes are approximately six
inches above the floor of the tank. Water is pumped
through each intake into a separate mixing chamber.
The chemical being added is pumped into the mixing
jar, and the mixture then flows into the experimental
box. Four slides are held vertically in each box
to prevent the deposition of silt on the slide.
Temperature is constantly monitored by a probe in
each box. The pH is continually monitored in the
outflow from the box. Chemical tests are carried
out daily in order to maintain similar concentrations
of all chemicals known to be important for algal
growth. The light in each box is previously determined to be similar.
In these experiments usually all slides are
seeded with similar communities before the chemical
to be tested is added to the mixing jar. This
usually takes ten days to two weeks, but may take
longer depending on the season of the year.
Microscopical examination is made before the
experiments are started to make sure that the
communities are similar. Experimentation[9] has
shown that very similar communities develop in
the boxes, for we find that at any point in time

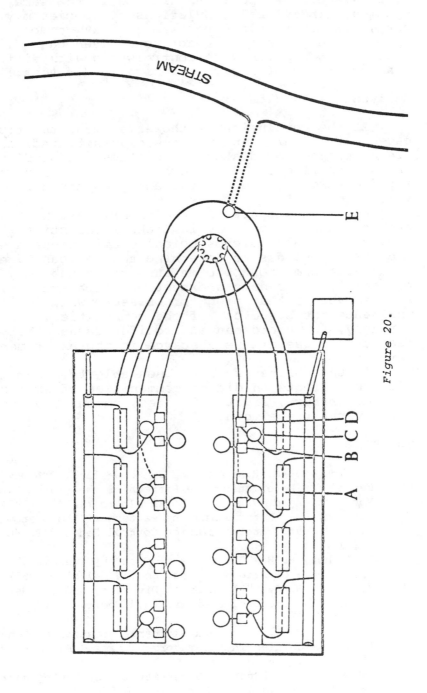

Figure 20.

95 to 97% of the specimens will be in the same species. Individual populations of species may vary, but nevertheless the same fairly common species are present to react to a given change. Experiments are run with controls and with different concentrations of a given chemical under test.

Analyses of Results

At the conclusion of the experiment the truncated normal curve and various diversity indices are developed for diatom communities. Depending on the experiment ^{14}C, O_2 evolution, volatile and residue weights, and pigment analyses are also carried out.

During the course of the experiments not only are the chemical and physical characteristics analyzed but also microscopical examinations are made of the slides to determine the various types of algae and microscopic invertebrates that may be present. All predators except protozoa and rotifers are removed so that predator pressure will not influence the structure of the communities. Fortunately, the protozoan and rotifer faunas that feed upon algae are very scarce in the experimental setups.

In this manner we have been able to experiment with the effects on diatom communities of shifts in pH, temperature, light, nitrates, ammonia, phosphates, and several trace metals.[9-12]

REFERENCES

1. Patrick, R., M. H. Hohn, and J. H. Wallace. "A New Method for Determining the Pattern of the Diatom Flora," *Not. Nat. Acad. Nat. Sci. Philadelphia No. 259* (1954).
2. Hohn, M. H. "Determining the Pattern of the Diatom Flora," *Jour. Water Pollution Control Fed. 33(1)*:48 (1961).
3. MacArthur, R. H. "Patterns of Species Diversity," *Biol. Rev., Cambridge Philos. Soc. 40(4)*:510 (1965).
4. McIntosh, R. P. "An Index of Diversity and the Relation of Certain Concepts to Diversity," *Ecology 48(3)*:392 (1967).
5. McIntire, C. D. "Structural Characteristics of Benthic Algal Communities in Laboratory Streams," *Ecology 49(3)*: 520 (1968).
6. Patten, B. C. "Species Diversity in Net Phytoplankton of Raritan Bay," *Jour. Mar. Res. 20*:57 (1962).

7. Patrick, R. "Evaluating Inhibitory Toxicity of Indus-
    trial Waste Waters," *ASTM Standards, Part 23*:517 (1964).

8. Watts, J. R. and R. S. Harvey. "Uptake and Retention
    of Cs^{137} by a Blue-green Alga in Continuous Flow and
    Batch Culture Systems," *Limnol. and Oceanog. 8(1)*:45
    (1963).

9. Patrick, R., N. A. Roberts, and B. Davis. 1968. "The
    Effect of Changes in pH on the Structure of Diatom
    Communities," *Not. Nat. Acad. Nat. Sci. Philadelphia,
    No. 416* (1968).

10. Patrick, R. "The Effect of Varying Amounts and Ratios
    of Nitrogen and Phosphate on Algae Blooms," Proc. 21st
    Ann. Industrial Waste Conf., Purdue University, Purdue,
    Indiana, pp. 41-51 (1967).

11. Patrick, R. "The Effects of Increasing Light and
    Temperature on the Structure of Diatom Communities,"
    Limnol. and Oceanog. 16(2):405 (1971).

12. Patrick, R., B. Crum, and J. Coles. "Temperature and
    Manganese as Determining Factors in the Presence of
    Diatom or Blue-green Algal Floras in Streams," *Proc.
    Nat. Acad. Sci. 64(2)*:472 (1969).

7.   ALGAL BIOASSAYS FOR ESTIMATING THE EFFECT OF
     ADDED MATERIALS UPON THE PLANKTONIC ALGAE IN
     SURFACE WATERS

Dee Mitchell.   Monsanto Company, St. Louis, Missouri.

ABSTRACT

    The Provisional Algal Assay Procedure (PAAP)
bottle test and the Microcosm Algal Assay Procedure
(MAAP) were compared. The PAAP was found to be an
excellent analytical tool for assessing the physiological effects of various materials upon algae, and
for determining the limiting nutrient in water
samples. However it lacks the capability, which
the MAAP does have, of delineating interations between the ecosystem and the material being assessed,
as well as between the various algal species making
up the natural community. Examination of sodium
nitrilotriacetate (NTA) showed no significant effects even at 200 mg/liter (at least 1000 times
the expected environmental levels) in the PAAP.
Likewise, NTA showed no effects in the MAAP at the
lower levels, but at 200 mg/liter a stimulation of
algal growth and loss of diversity was observed.

    There is an urgent need for laboratory pro-
cedures to evaluate the ecological significance of
various materials added to surface waters in terms
of their capabilities to stimulate algal growth and
to influence the predominance of undesirable species
in the planktonic population. Over the past 20
years this need has been satisfied provisionally
by various investigators, usually by measuring the
growth rates of pure algal cultures grown in defined medium or by observing the changes of

biological populations caused by pollution intrusions into the natural environment. The former methods show the effects on algal growth (cell numbers or cell mass) caused by the added material as compared to controls in which the material is lacking. The latter methods yield results both in terms of the cell growth induced and the changes in the natural community structure. Results are usually expressed in terms of changes in the species diversity. Ecosystems with diverse populations are usually more desirable and more stable than those with only a few species present.

The observation of population dynamics in nature has the advantage of being an exact measure of the response of the aquatic biota; it has a disadvantage in that results are obtained only after some change has occurred in the natural aquatic population. Thus, a receiving water may become degraded before design criteria to prevent that degradation can be developed from the data.

On the other hand, the response of unialgal cultures grown in defined medium may yield little indication of their response to the same stimulus in the natural environment. Hence, both methods have severe shortcomings and require a great deal of experience and judgment to extrapolate these results to the environmental acceptability of added materials.

The Provisional Algal Assay Procedure (PAAP) - Bottle Test[1] has been developed by the Joint Industry-Government Task Force on Eutrophication. In the bottle test, candidate algae--*Selenastrum capricornutum, Anabaena flos-aquae,* or *Microcystis aeruginosa*--are seeded into the water sample maintained under constant light and temperature conditions. Growth is determined by assaying the resulting algal crop on a routine schedule. Maximum yield and growth rates can be determined from the resultant growth curves, depending on the experimental design and the nutrient conditions of the test water.

The Microcosm Algal Assay Procedure (MAAP) which we have developed to meet this problem of estimating environmental responses of added materials involves the use of laboratory microcosms and identification and enumeration of the developing planktonic algal populations.[2] The resulting diversity is calculated as Shannon's diversity index:[3]

$$H_i = -\Sigma P_i \log P_i$$

where $P_i = N_i/N$; N_i = population of the ith genus; and N = population of the total planktonic algal community.

The index ranges from zero for unialgal populations to near unity for the most diverse planktonic algal communities encountered during several years of observation in nature and in the laboratory. Oligotrophic lakes usually have diversity indices of from 0.7 to 1.0 and as they become eutrophic, the diversity drops to 0.3 or less.

The following study, made in examining the environmental acceptability of trisodium nitrilotriacetate (NTA), compares the results obtainable by these two algal bioassays.

PAAP bottle tests were made on 0.45μ millipore filtered waters from three lakes of various trophic levels. These waters are: Table Rock Lake, an oligotrophic impoundment in Southwestern Missouri, North Tyson Twin Lake, a small mesotrophic lake located in the Washington University Research Center at Tyson Valley near St. Louis, and Lake 34, a eutrophic fishing impoundment at the Busch Wildlife Area near St. Louis. Triplicate water samples of 40 ml volume were placed in 125 ml Pyrex Erlemeyer flasks, seeded with 1000 cells/ml *Selenastrum capricornutum* grown in the PAAP medium, and spiked with 0, 0.2, 2.0, 20, or 200 mg/l NTA. The cultures were exposed to 400 ft-candles of light, continually agitated at 100 oscillations per minute, and kept at 24° ± 2°C throughout the growth period.

The growth of these cultures was monitored by observing the chlorophyll-a concentrations on a Turner fluorometer. The 21-day yields are shown in Figure 21. Each value shown represents the mean of the triplicate cultures. The oligotrophic lake water supported approximately 250 cells/ml. The NTA, even at 200 mg/l, appears to have no effect upon the growth of *Selenastrum* in this water. The mesotrophic and eutrophic waters gave yields of approximately 8000 and 2,000,000 cells respectively. In no case was the NTA, even at 200 mg/l observed to either stimulate or inhibit algal growth in these waters.

For the MAAP study laboratory microcosms were prepared from 0.5 liters of bottom mud and 6.0 liters of water taken from these same three lakes. To maintain the natural diversity a species invasion pressure was exerted on the microcosms. This is done by withdrawing 250 ml of water and replacing it with 250 ml of fresh lake water with its natural

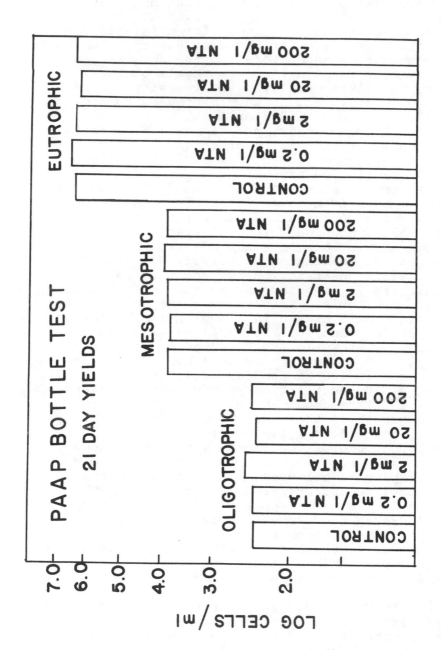

Figure 21. Yields of Algal Cultures Exposed to Various Levels of NTA

algal community (plus appropriate amounts of NTA) every two weeks.

Figure 22 shows the mean diversity indices of duplicate microcosms treated with 20, 60, or 200 mg/l NTA for the oligotrophic and mesotrophic microcosms and with 15, 80, or 500 mg/l for the

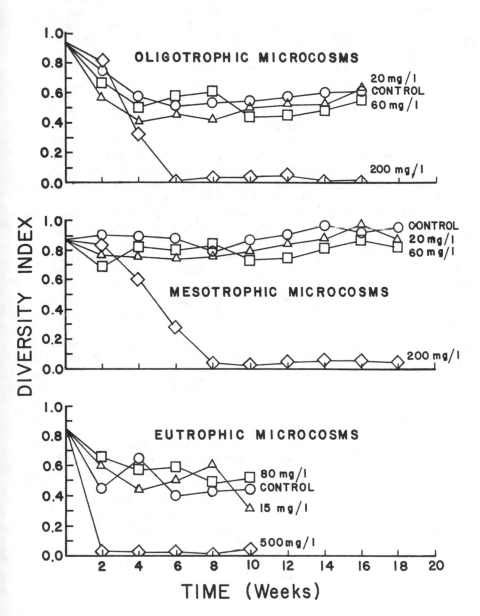

Figure 22. *Diversity Indices of Microcosms treated with NTA*

eutrophic microcosms. Lower concentrations show no significant difference from the controls. There was no apparent effect of NTA on the planktonic algal community, even at concentrations up to 80 mg/l, which is very much higher than expected environmental levels. Significant effects become evident only when the anticipated levels of NTA were exceeded by a factor of at least 1000. In all cases 200 mg/l NTA or more stimulated algal growth and caused the diversity to drop to near zero.

Theoretically, bioassays of various added materials by these two different methods should lead the researcher to the same conclusions. However, from these results it is obvious that the different types of algal bioassays may at times give contradictory results. In the MAAP there is presumably an interaction between the extremely high level of NTA and the sediments which makes nutrients available and induces growth of the algae population. The PAAP on the other hand is not designed to reveal this added material-sediment-algae interaction and hence did not show an algal stimulation at these high levels of NTA. The ultimate conclusion drawn from results of both NTA bioassays would be similar in that NTA at environmentally significant concentrations has no effect on the planktonic algal populations. Modifications could be attempted in the PAAP to allow for sediment interactions, possibly eliminating this factor as a cause for disagreement between the two methods. In any case, extrapolation of laboratory results to the environment requires a great deal of experience and judgment in the interpretation.

REFERENCES

1. Bartsch, A. F. *Provisional Algal Assay Procedures (PAAP)*, United States Government Memorandum, Pacific Northwest Laboratory, Federal Water Pollution Control Administration, Corvallis, Oregon (1968).
2. Mitchell, Dee, and James C. Buzzell, Jr. "Estimating Eutrophic Potential of Pollutants," *Journal of the Sanitary Engineering Division ASCE 97(SA4)*:453 (August, 1971).
3. Shannon, C. E., and W. Weaver. *The Mathematical Theory of Communication* (Urbana, Ill.: University of Illinois Press, 1963).

8.   PLANT CONTENT OF ELEMENTS AS A BIOASSAY OF
      NUTRIENT AVAILABILITY IN LAKES AND STREAMS

G. C. Gerloff and K. A. Fishbeck.   Institute of
Plant Development and Department of Botany, University of Wisconsin, Madison, Wisconsin.

ABSTRACT

    The technique of plant analysis was developed
as a relatively simple procedure for evaluating
nutrient supplies for nuisance aquatic plants.
The critical concentration (minimum plant content
for maximum yield) for each of ten essential elements was established in index segments of the
macrophyte *Elodea occidentalis*. Analyses of *Elodea*,
periodically collected from eight Wisconsin lakes,
were compared with the critical concentrations for
indications of nutrient deficiency. A growth-
limiting role of an element in a lake was indicated
by *Elodea* concentrations of that element below the
critical level. N, P, Ca, or Cu concentrations
were at or close to critical levels in plants from
one or more lakes. P or N supply was not in general
growth-limiting. Although confirming evidence is
required, the data indicated Cu deficiency in
several lakes sampled.

    One approach to the control of undesirable
aquatic plants in lakes and streams is to reduce
available supplies of an essential nutrient element
to growth-limiting levels. Evaluations of the
effectiveness and practicality of this approach
are facilitate if reliable procedures are available for determining nutrient supplies relative to

159

the requirements of the nuisance plants. Chemical analyses of water samples and enrichment bioassays involving growth or photosynthesis of aquatic organisms following additions of nutrients to water samples are two recognized procedures for nutrient assay of natural waters.[1] It is difficult to interpret water analyses in terms of potential for nuisance plant growth and in terms of concentrations at which specific elements become growth limiting.[1-3] Enrichment bioassays are subject to errors associated with extending data obtained from isolated samples to evaluations of nutrient supplies in entire lakes over a period of time. As alternatives, several bioassays based on plant rather than water samples have been proposed. For example, the amount of orthophosphate extracted from plant samples with boiling water and the uptake of ammonium nitrogen in the dark correlated with the adequacy of the phosphorus (P) and nitrogen (N) nutrition, respectively, of the plants sampled.[4] Also, Gerloff and Krombholz[5] used the concentrations of elements in macrophytes to indicate the availability of the elements in the environments in which the plants had grown. This was an application of the technique known in agriculture and horticulture as tissue or plant analysis. Plant analysis had found widespread practical use in evaluating nutrient supplies in soils for maximum crop production.[6,7]

The concept of plant analysis is based on the observation that the concentration of each nutrient element in a plant varies over a considerable range, and that a primary factor determining the concentration in a healthy plant is the availability of the element in the environment. To develop plant analysis for a particular species, the critical concentration for each potentially growth-limiting essential element must be determined.[8,9] The critical concentration is the minimum concentration of an element, or more often slightly less than the minimum concentration, in a plant or plant part which permits maximum growth. Critical concentrations usually are established in laboratory or greenhouse experiments in which plant yield and plant content of an element are related when plants are grown under external supplies of that element ranging from severely limiting to well above optimal concentrations. If a plant collected from the field contains less than the critical concentration of an element, the supply of that element probably was limiting growth at the time of sampling.

The purpose of this study was to develop and apply plant analysis in evaluating available nutrient supplies in lakes and streams for the growth of nuisance macrophytes. This involved establishing the critical concentrations for most of the essential elements in the macrophyte *Elodea occidentalis* and the collection and analysis of *Elodea* samples throughout the growing season from eight northern Wisconsin lakes. In earlier work,[5] only N and P were considered as possible growth-limiting factors and entire plants were analyzed for these elements. As in agricultural applications, analysis of entire plants was unsatisfactory. Therefore, in this study the critical concentration for each element was established in a plant part (index segment) which more accurately reflects environmental supply than does the entire plant.

EXPERIMENTAL PROCEDURES

Establishing Critical Concentrations

The nutrient medium and the methods for laboratory culture of *Elodea occidentalis* were in general as described in an earlier publication.[5] In the current study, iron (Fe) in a concentration of 0.56 ppm was provided as a chelate of EDDHA (ethylene di-(o-hydroxyphenylacetate) rather than as FeEDTA (iron salt of ethylene diamine tetraacetic acid). The cultures were bubbled continuously with air enriched to 1% CO_2 and filtered through cotton to remove microorganisms and through activated charcoal to remove air pollutants. Culture room temperatures were maintained at approximately 23°C. Treatments in all experiments involved at least duplicate cultures. In experiments to establish trace element critical concentrations, treatments were in triplicate and sometimes quadruplicate.

To develop *Elodea* deficient in manganese (Mn), zinc (Zn), boron (B), or molybdenum (Mo) and to establish critical concentrations of these elements, standard procedures for reducing environmental contamination were used, that is acid-washing culture and media containers, purifying water used to prepare culture solutions, and purifying culture media salts.[10,11] Even with these precautions, it has been impossible thus far to obtain Cu deficient *Elodea*.

Two procedures were employed in determining critical concentrations of the elements. In most cases, the *Elodea* was grown in nutrient cultures similar in all respects except for the concentration of the element under investigation which varied from below to well above optimal. Plants were harvested after a culture period of approximately four weeks. At harvest, and prior to drying to constant weight at 65-70°C, plants in each culture were divided into three segments: the terminal and second one-inch of growth on main branches and laterals, and the remaining leaves and stems. The critical concentration for an element was established from curves relating total oven-dry yields from the two-liter cultures representing treatments of an experiment and the element concentrations in the plant segments from those treatments. The critical concentration was the minimum content in the index segment associated with the maximum yield to slightly less (approximately five per cent) than the maximum yield.

The generally recognized mobility of an element in plants was a primary factor in selecting the first or second one-inch as the index segment for that element. Nitrogen, for example, is a relatively mobile element which is re-exported from older to younger tissues, particularly under conditions of N deficiency, but also as a result of senescence and non-nutritional abnormalities. In an unreported experiment, in plants grown for a considerable period under severe N deficiency, the N content of terminal one-inch segments was actually above the critical level, apparently due to re-export from extensive regions of dying tissues. It seemed inadvisable, therefore, to include the oldest or the youngest plant parts in the N index segment. The second one-inch seemed the most suitable for N and other mobile elements. In contrast, the terminal one-inch was considered the most appropriate to reflect nutrient availability for elements such as calcium (Ca) and B which are not re-exported from older, senescent tissues.

To verify critical levels established by the batch procedure, a technique was developed in which the rate of growth of *Elodea* was followed by daily measurements of increase in plant length. Plants were harvested and analyzed as soon as a decrease in growth rate indicated the critical concentration of the limiting element had been reached in the *Elodea*. For this procedure, the plants were cultured in aerated, sterilized media in pyrex trays of approximately two-liter capacity covered with glass plates. Some trays contained complete culture medium; other trays contained concentrations of an element which would support a maximum rate of growth only for a limited period. Daily measurements of length included the main shoots and all laterals.

Analyses of Lake Plants

Plant samples collected from lakes were lifted from the water with a garden rake, immersed in water, and taken to the laboratory. Within 3-4 hours, sufficient first and second one-inch segments to represent 3-4 g of oven-dry material were cut from the main branches and laterals of healthy, green plants. After repeated rinsing in lake water to eliminate debris, the samples were soaked 30-45 seconds in 0.2N hydrochloric acid, rinsed with lake water, rinsed in distilled water, placed in nylon bags, and dried to constant weight at 65-70°C.

The lakes sampled are located in or near Vilas County in northern Wisconsin in non-agricultural areas. They were relatively low in nutrients, the water hardness varied from 14 to 65, and the pH of the water was in the range of 7.1 to 8.0.[12]

Analyses of plant samples from experiments to establish critical concentrations were by quantitative procedures in general use for inorganic plant analysis.[13] Total nitrogen analyses were by a semi-micro Kjeldahl procedure.[14] Phosphorus determinations were made with a vanado-molybdate (yellow complex) procedure following dry-ashing of tissues at 550°C.[15] Potassium analysis was by emission flame photometery (Coleman Model 21) on 1N ammonium acetate extracts of oven-dry tissue samples.[16] Iron was determined by an o-phenanthroline colorimetric procedure.[17] Following dry-ashing at 550°C and acid solution of the residues,[18] calcium, magnesium (Mg), Zn, Mn, and Cu were determined by atomic absorption using a Jarrell-Ash instrument. Boron was determined as a curcumin complex[19] and Mo as a complex with thiocyanate following stannous reduction.[20]

Total N analyses on all field samples also were by the Kjeldahl procedure. In addition, all samples were analyzed for 10 elements by the Jarrell-Ash Multichannel Emission Spectrometer at the Wisconsin Alumni Research Foundation Laboratories.[21] In all lakes in which there were indications from the emission spectrometer analyses that a specific element had become limiting, or was close to limiting for plant growth, analyses were verified by the quantitative procedures mentioned. This included analyses for Ca, P, and Cu.

Samples to be analyzed by the Jarrell-Ash Spectrometer were ground in a Wiley Mill equipped with a stainless steel screen. Samples to be analyzed by conventional quantitative procedures were ground with an agate mortar and pestle.

RESULTS

The data in Table 14 and Figure 23 present the results from an experiment to establish the critical concentration of N in a suitable index segment of *Elodea*. The average N content in the second one-inch segment varied over a considerable range, from 1.14 to 4.32%, as the nitrate-nitrogen (NO_3-N) concentration in the culture medium varied from 10.5 to 42.0 ppm. As shown in Figure 23, plant N contents of less than approximately 1.60% were associated with external supplies of N inadequate to provide the amounts of this element necessary for maximum growth. The critical concentration of N for *Elodea* therefore was established as 1.60% in the second one-inch segment. Plant N contents in

Table 14
Amount of Plant Growth and Total Nitrogen Content of Sections of Stems and Branches of Elodea occidentalis When Cultured in Solutions Differing in Nitrogen Content to Establish the Critical Nitrogen Concentration

Initial N conc. of medium (mg/l)	Wt. of oven-dry plants (g/2l)			Total N content (%)			
	Sample 1	Sample 2	Ave.	Plant Segment	Sample 1	Sample 2	Ave.
10.5	1.583	1.575	1.579	1st 1"	1.45	1.49	1.47
				2nd 1"	1.14	1.15	1.14
14.0	1.694	1.928	1.833	1st 1"	2.04	1.75	1.89
				2nd 1"	1.52	1.35	1.43
21.0	2.428	2.252	2.340	1st 1"	2.13	2.05	2.09
				2nd 1"	1.63	1.55	1.59
31.5	1.933	2.371	2.152	1st 1"	4.10	3.21	3.67
				2nd 1"	3.19	2.45	2.82
42.0	1.861	1.931	1.896	1st 1"	5.56	5.35	5.45
				2nd 1"	4.38	4.26	4.32

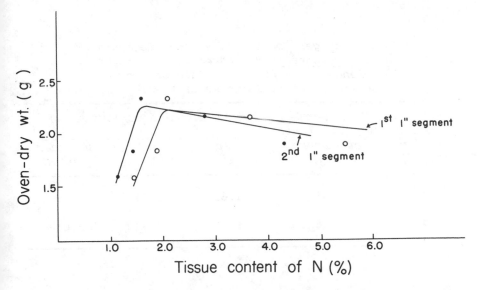

Figure 23. *The relationship between yield and total nitrogen content of the first and second one-inch terminal segments of main branches and laterals of Elodea occidentalis grown in solutions of varying nitrogen content to establish the critical concentration of nitrogen.*

excess of 1.60% represented luxury consumption of the element. The N contents of the first one-inch segment varied in the same pattern as in the second one-inch but were consistently higher. Because of the mobility of N in deficient plants, the second one-inch segment was considered a more reliable index segment.

The data in Table 15 and Figure 24 were obtained with tray cultures of *Elodea* in which daily increases in growth were determined. In one set of trays, the NO_3-N concentration in the nutrient medium was 42 ppm; in another set, the concentration was only 4.2 ppm. The total N content in the second one-inch segment of plants in the low N trays at harvest was 1.23%. This is slightly below the 1.60% critical concentration. However, as indicated in Figure 24, the *Elodea* plants probably were deficient in N several days before a decision to harvest the plants seemed justified. The dry weight of the low N culture plants was only 393 mg per tray in comparison with 504 mg under high N. The 14.0 ppm external N treatment in Table 14 represents a comparable yield decrease due to N deficiency. The N content of the

Table 15

The Total Nitrogen Content of Various Segments of
Elodea occidentalis Harvested When Nitrogen
Deficiency had Reduced the Rate of Growth
as Indicated by Increase of Length

Initial N conc. of medium (mg/l)	Oven-dry wt. of plants at harvest (mg)	Total N content (%)		
		1st 1"	2nd 1"	Remainder of plant
4.2	393	1.61	1.23	1.69
42.0	504	5.85	4.34	3.01

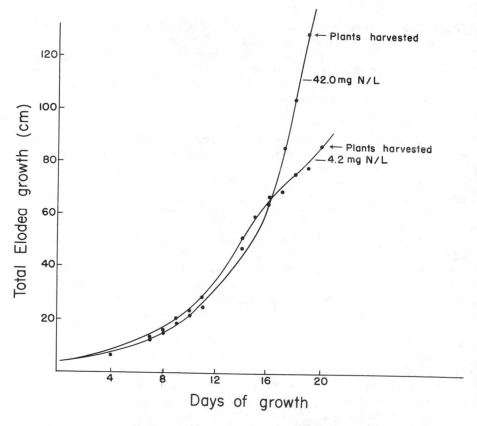

Figure 24.   Daily increase in total length of stems and
lateral branches of *Elodea occidentalis* when
grown in tray cultures under high and low levels
of *NO₃-N* to establish the critical nitrogen
concentration.

second one-inch segment from that treatment was
1.43% which is in reasonable agreement with the 1.23%
value from the tray experiment.

The results of additional experiments to establish the critical concentrations of most of the
other essential major and trace elements in index
segments of *Elodea* by the batch procedure are presented in Table 16. As anticipated, the critical
concentrations for the trace elements were much
lower than for the major elements, with a minimum
value of 0.15 ppm for Mo. The total range in concentration of each element also is presented. It
is apparent that the concentration of each essential
element in *Elodea* is not a fixed value but varies
over a considerable range.

Table 16

*Critical Concentrations and the Range of Concentrations
of Essential Mineral Elements in Index Segments of
Elodea occidentalis*

Element	Index segment	Critical concentration	Range of concentration
N	2nd 1"	1.60%	1.14-4.32%
P	2nd 1"	0.14%	0.06-0.35%
Ca	1st 1"	0.28%	0.17-0.62%
Mg	2nd 1"	0.10%	0.03-0.19%
K	2nd 1"	0.80%	0.25-2.51%
Fe	1st 1"	60 ppm	40-219 ppm
Mn	1st 1"	4.0 ppm	2.2-16.7 ppm
Zn	2nd 1"	8.0 ppm	3.6-34.4 ppm
Mo	1st 1"	0.15 ppm	0.04-6.4 ppm
B	1st 1"	1.3 ppm	0.3-11.2 ppm

Analyses of Lake Plants

Examples of the analyses of *Elodea* samples
routinely collected from eight Wisconsin lakes are
presented in Tables 17 and 18.

Table 17

Essential Element Content (Oven-dry Basis) of First and Second One-Inch Segments of
Elodea occidentalis Collected from Little John Lake During the Summer of 1970

Date Sampled	Segment	%					Ppm				
		N	P	K	Ca	Mg	Fe	B	Cu	Mn	Zn
June 16	1"	3.50	0.65	2.30	0.52	0.22	>1000	21.1	1.3	404	46
	2"	3.35	0.70	2.30	0.85	0.23	>1000	24.9	2.1	>410	35
July 6	1"	3.05	0.54	2.75	0.54	0.22	795	13.6	1.5	>410	71
	2"	2.66	0.54	2.55	0.91	0.19	>1000	14.9	2.6	>410	86
July 27	1"	2.81	0.47	2.45	0.68	0.21	665	13.5	2.2	386	56
	2"	2.22	0.40	2.45	1.07	0.19	855	16.6	<1.0	>410	74
Aug. 18	1"	3.58	0.65	2.80	1.09	0.25	905	14.9	3.4	254	95
	2"	2.84	0.54	3.25	1.55	0.20	820	13.3	1.8	>410	89
Sep. 8	1"	3.49	0.80	2.55	1.09	0.24	760	13.5	3.6	360	66
	2"	3.10	0.68	3.45	1.17	0.19	625	12.5	<1.0	>410	81

Except for N, all analyses were made with a Jarrell-Ash direct-reading, computer-programmed emission spectrometer.

Table 18

Essential Element Content (Oven-dry Basis) of First and Second One-Inch Segments of
Elodea occidentalis Collected from Salsich Lake during the Summer of 1970

Date Sampled	Segment	%					Ppm				
		N	P	K	Ca	Mg	Fe	B	Cu	Zn	Mn
July 7	1"	3.78	0.35	2.20	0.48	0.21	555	12.1	6.2	197	98
	2"	3.47	0.21	2.80	0.64	0.17	445	13.1	4.3	182	150
July 28	1"	3.55	0.25	2.20	0.42	0.19	450	12.4	5.0	160	116
	2"	3.14	0.18	2.65	0.66	0.16	480	13.5	3.8	224	190
Aug. 19	1"	3.50	0.33	2.65	0.50	0.20	545	11.5	4.7	162	124
	2"	3.02	0.15	2.80	0.62	0.16	430	8.8	1.8	156	185
Sep. 10	1"	3.47	0.40	2.35	0.64	0.18	485	8.4	3.1	177	137
	2"	2.83	0.40	2.10	0.91	0.19	580	12.5	4.0	177	225

Except for N, all analyses were made with a Jarrell-Ash direct-reading, computer-programmed emission spectrometer.

Concentrations of N and P in the lake plants are of particular interest because these elements most frequently have been considered limiting for aquatic plant growth. In all of the samples from Little John Lake, Table 17, the N and P concentrations are well above the critical concentrations of 1.60% and 0.14% in the second one-inch segment. The P concentration in every sample was at least double the critical concentration indicating very adequate supplies of that element during the entire growing season.

The lowest concentrations of elements in general were observed in samples from July 27. This sampling apparently was in the period of most rapid plant growth and greatest pressure on nutrient supplies. However, even in the July 27 samples, the concentrations of all elements, except N, for which critical values are reported in Table 16 were at least double those values.

The analyses in Table 18 are of *Elodea* samples from Salsich Lake. The N and P data contrast with the results from Little John. The N concentrations are in general slightly higher, but the P concentrations are considerably lower. For example, in the second one-inch segment from July 28 the P concentration was only 0.18% and in the August 19 sample 0.15%. Comparable P values in the Little John samples were 0.40 and 0.54%. The low values for Lake Salsich plants, which were verified with a standard colorimetric procedure, approach the critical concentration of 0.14%, and indicate that in this lake P supplies were close to limiting for *Elodea* growth. Further additions of P would be expected to increase the presently sparse weed beds in Salsich Lake.

Because of inadequate plant tissue, all samples from the eight lakes could not be analyzed for Mo. The range of Mo in seven samples analyzed was 0.50 to 0.73 ppm, in the terminal one-inch, with an average of 0.57 ppm. Even the lowest concentration was three to four times the critical concentration, thus eliminating the need to consider Mo as a possible growth-limiting factor.

The Cu analyses reported in Tables 17 and 18 are of interest because some are sufficiently low to suggest Cu deficiency. Several of the Cu concentrations in *Elodea* from Salsich and all of the values from Little John were less than 4 ppm. For a number of samples the copper data reported in Tables 17 and 18 were checked by atomic absorption analysis. The atomic absorption values were consistently higher. However, the Cu contents obtained by atomic absorption still were below the 4-7 ppm critical Cu concentrations established for corn, oats, soybeans and alfalfa[22] and for citrus.[23] These data confirm that either *Elodea* was Cu deficient in Little John Lake or that this species has a surprisingly low requirement for Cu. The low concentrations of Cu in the *Elodea* contrast sharply with the concentrations of the other trace elements considered in Tables 17 and 18 which were many times in excess of the critical values.

Analytical data on all the samples from the eight lakes will not be reported. However, in Table 19 data are presented for samples taken from each lake during the late July period of nutrient stress. Analyses are reported only for the four elements which critical concentration comparisons indicated might limit plant growth. When two different analytical procedures were used, data from both are presented.

Comparisons with analyses of samples from other lakes emphasize that the P concentrations in the plants from Salsich Lake were relatively low. Whitney Lake also seems close to P deficiency for *Elodea* growth, as indicated by the 0.17% value for the index segment. The data on Clear Lake are of interest because the P concentration was relatively high at 0.24% but the N concentration, 1.94%, in the second one-inch was the lowest of any sample, and only slightly above the 1.60% critical value. The 2.70% and 0.28% concentrations of N and P respectively in the Erickson Lake samples were high, but the 0.28% Ca concentration in the terminal one-inch index segment was at the critical level. It is not surprising that Ca might be a limiting factor in soft water lakes and that species with a higher requirement for Ca than *Elodea*, or less capacity to absorb Ca from the environment, might be eliminated from soft water lakes such as Erickson.

Probably the most consistently low values in Table 19 are to be observed in the Cu data. In *Elodea* from four of the eight lakes, Cu concentrations were low enough to suggest a growth-limiting role of that element.

In contrast to the other six lakes, Allequash and Little Spider Lakes seem relatively well supplied with all the essential nutrient elements.

Table 19

Concentrations of Potentially Growth-limiting Elements in *Elodea occidentalis* Collected from Northern Wisconsin Lakes During Late July, 1971

Lake Sampled	Date Sampled	Methyl Orange Alkalinity*	Plant Segment	Concentration of						
				N %	P (%) JAS**	P (%) Std.**	Ca (%) JAS	Ca (%) Std.	Cu (ppm) JAS	Cu (ppm) Std.
Allequash	July 27	39	1"	3.53	0.34	0.31	0.42		3.8	6.5
			2"	3.19	0.28	0.22	0.61		3.4	7.0
Clear	July 29	33	1"	2.41	0.42	0.34	0.69		2.2	
			2"	1.94	0.28	0.24	0.66		1.8	
Whitney	July 29	22	1"	2.34	0.23	.21	0.68		2.2	
			2"	2.02	0.21	0.17	0.89		1.8	
Erickson	July 28	24	1"	3.41	0.47	0.41	0.28		3.8	
			2"	2.70	0.29	0.28	0.44		2.2	
Big Kitten	July 29	65	1"	2.71	0.42	0.39	1.61		<1.0	3.9
			2"	2.16	0.40	0.34	1.51		2.2	2.7
Salsich	July 28	14	1"	3.55	0.25	0.26	0.42		5.0	
			2"	3.14	0.18	0.18	0.66		3.8	
Little John	July 27	45	1"	2.81	0.47	0.35	0.68		2.2	2.4
			2"	2.22	0.40	0.25	1.07		<1.0	3.3
Little Spider	July 28	--	1"	3.36	0.44	0.37	0.52		3.8	6.5
			2"	2.59	0.28	0.27	0.61		2.2	4.5

*Values for methyl orange alkalinity were reported by the Wisconsin Department of Natural Resources.

**JAS indicates analyses made with the Jarrell-Ash emission spectrometer; Std. refers to analyses by standard colorimetric, emission flame photometer, and atomic absorption procedures.

DISCUSSION

The primary importance of the work reported is considered to be in the development of a simple yet reliable procedure for assaying nutrient supplies in lakes and streams for nuisance plant growth. The results obtained suggest plant analysis can be a useful technique, and that it offers several advantages. Plant analysis minimizes the difficulties associated with obtaining representative samples of the aquatic environment. The plants become the sampling device, and the concentration of an element in the plants is an integrated expression of all the factors which affected the availability of that element in the microenvironments to which the plants were exposed during growth. Chemical analyses of water samples must be interpreted in terms of fractions available and unavailable to plants and concentrations and quantities which become limiting for growth. These problems are reduced in the plant analysis procedure, because plant analysis is based on values which have been correlated with plant growth and yield responses.

Two aspects of the field data seem worthy of comment. First, although N and P are the elements of most interest in practical pollution control, the data indicated that neither element was a general limiting nutrient for *Elodea* in the lakes sampled. In some lakes, N or P was near the critical concentration. In these ecosystems, reduction in available supplies of N or P, whichever was limiting, probably would reduce macrophyte populations; also, further eutrophication with these elements would accentuate nuisance growths. A second point of interest is the suggested Cu deficiency in several northern Wisconsin lakes. This must be verified by laboratory experiments establishing the critical concentration of Cu for *Elodea*. Nevertheless, the data obtained focus attention on the importance of trace elements in the field nutrition of aquatic plants.

From the standpoint of widespread occurrence in lakes of the area sampled, *Elodea occidentalis* was a suitable selection as a bioassay organism. However, *Elodea* is rooted in bottom muds, and nutrients are obtained both from that source and the water layer above. It undoubtedly would be desirable to have the assays primarily reflect nutrient availability in the water layer and to be relatively uninfluenced by the direct absorption from sediments that occurs when field collections involve natural populations as in this study. It is anticipated, therefore, that the most useful application of the plant analysis technique will be to locate the bioassay organism in porous, inert containers near the surface in lakes and streams in which nutrient supplies are to be evaluated. Samples of the index segments will be periodically removed and analyzed for elements suspected to limit growth. The analyses will be evaluated by comparisons with the critical concentrations.

Although the results presented suggest plant analysis is a useful bioassay, the need for additional studies to refine the technique is emphasized. Hundreds of investigations have provided the information necessary to firmly establish plant analysis as a procedure for nutrient assays in soils. Undoubtedly, there will be unique aspects in applications of plant analysis to aquatic plants and environments. Further investigations of the most suitable index segments, more precise establishment of critical concentrations, and the extension of the technique to other macrophytes and to algae are suggested for immediate study.

ACKNOWLEDGMENTS

This project has been supported and financed, in part, by a grant from the Water Quality Office of the Environmental Protection Agency.

The assistance of Marcia Wuenscher, Frann Hutchison, Ann Mickle, and Brian Marcks in various aspects of the work reported are gratefully acknowledged.

REFERENCES

1. Gerloff, G. C. "Evaluating Nutrient Supplies for the Growth of Aquatic Plants in Natural Waters," In *Eutrophication: Causes, Consequences, Correctives.* (Washington, D.C.: Nat. Acad. Sci., 1969).
2. Lee, G. F. "Analytical Chemistry of Plant Nutrients," In *Eutrophication: Causes, Consequences, Correctives.* (Washington, D.C.: Nat. Acad. Sci., 1969).
3. Lund, J. W. G. "Phytoplankton," In *Eutrophication: Causes, Consequences, Correctives.* (Washington, D.C.: Nat. Acad. Sci., 1969).
4. Fitzgerald, George P. "Field and Laboratory Evaluations of Bioassays for Nitrogen and Phosphorus with Algae and Aquatic Weeds," *Limnol. Oceanogr. 14:*206 (1969).

5. Gerloff, G. C., and P. H. Krombholz. "Tissue Analysis as a Measure of Nutrient Availability for the Growth of Angiosperm Aquatic Plants," *Limnol. Oceanogr.* 11:529 (1966).

6. Reuther, W. (ed.). *Plant Analysis and Fertilizer Problems.* Publ. No. 8 (Washington, D.C.: A.I.B.S., 1961).

7. Chapman, H. D. *Diagnostic Criteria for Crops and Soils.* (Univ. of Calif. Div. of Agricultural Sciences, 1966).

8. Ulrich, A. "Physiological Bases for Assessing the Nutritional Requirements of Plants," *Ann. Rev. Plant Phys.* 3:207 (1952).

9. Smith, P. F. "Mineral Analysis of Plant Tissues," *Ann. Rev. Plant Phys.* 13:81 (1962).

10. Hewitt, E. J. *Sand and Water Culture Methods Used in the Study of Plant Nutrition.* Tech. Commun. No. 22 of the Commonwealth Bureau of Horticulture and Plantation Crops, East Malling, Maidstone, Kent (1966).

11. Stout, P. R., and D. I. Arnon. "Experimental Methods for the Study of the Role of Copper, Manganese and Zinc in the Nutrition of Higher Plants," *Amer. Jour. Bot.* 26:144 (1939).

12. Black, J. J., L. M. Andrews, and C. W. Threinen. *Surface Water Resources of Vilas County.* Wisconsin Conservation Department (1963).

13. Johnson, C. M., and A. Ulrich. *Analytical Methods for Use in Plant Analysis.* Bulletin 766 (Berkeley, Calif.: California Agricultural Experiment Station, 1959).

14. Bremner, J. M. "Determination of Nitrogen in Soil by the Kjeldahl Method," *Jour. Agr. Sci.* 55:11 (1960).

15. Jackson, M. L. *Soil Chemical Analysis* (Englewood Cliffs, N.J.: Prentice-Hall, Inc., 1958).

16. Attoe, O. J. "Rapid Photometric Determination of Potassium and Sodium in Plant Tissue," *Soil Sci. Soc. Amer. Proc.* 12:131 (1947).

17. Sandell, E. B. *Colorimetric Determination of Traces of Metals.* (New York: Interscience Publishers, Inc., 1950).

18. Buchanan, J. R., and T. T. Muraoka. "Determination of Zinc and Manganese in Tree Leaves by Atomic Absorption Spectroscopy," In *Analysis Instrumentation.* (New York: Plenum Pub. Corp., 1964).

19. Dible, W. T., E. Truog, and K. C. Berger. "Boron Determination in Soils and Plants," *Anal. Chem.* 26:418 (1954).

20. Johnson, C. M., and T. H. Arkeley. "Determination of Molybdenum in Plant Tissue," *Anal. Chem.* 26:572 (1954).

21. Christensen, Robert E., F. B. Coon, and P. H. Derse. *Computer Application for Direct Concentration Print Out of Plant Tissue Analysis by Emission Spectroscopy.* Jarrell-Ash Technical Pamphlet (1968).

22. Melstead, S. W., H. L. Motto, and T. R. Peck. "Critical
     Plant Nutrient Composition Values Useful in Interpreting
     Plant Analysis Data," *Agron. Jour. 61:*17 (1969).
23. Chapman, H. D. "The Status of Present Criteria for the
     Diagnosis of Nutrient Conditions in Citrus," In *Colloquium
     on Plant Analysis and Fertilizer Problems.* Publ. No. 8
     (Washington, D.C.: A.I.B.S., 1961).

9. THE IMPORTANCE OF CHEMICAL SIGNALS IN STIMULATING BEHAVIOR OF MARINE ORGANISMS: EFFECTS OF ALTERED ENVIRONMENTAL CHEMISTRY ON ANIMAL COMMUNICATION

Jelle Atema, Stewart Jacobson, and John Todd. Woods Hole Oceanographic Institution, Falmouth, Massachusetts.

David Boylan. Hawaii Institute of Marine Biology, University of Hawaii, Honolulu, Hawaii.

ABSTRACT

The detection of certain chemicals in the environment by aquatic organisms occasionally stimulates specific behavioral acts that are essential functions in the life pattern of these organisms. Contaminants that affect the chemical sensing ability of fish could therefore have a strong impact on the normal activities of aquatic communities.

Kerosene and the branched chain-cyclic fraction of kerosene were found to stimulate feeding behavior in *Homarus americanus*. A dilute sea water extract of kerosene consisting primarily of benzenes and napthalenes (1 part/10^9 or ppb) was found to inhibit attraction of the mud snail *Nassarius obsoletus* to food extract. In yet another chemotactic response, attraction of the flatworm *Bdelloura candida* to its host *Limulus polyphemus* was inhibited by the presence of Hg Cl_2, Fe Cl_2, detergents and phosphate binders in detergents in concentrations ranging from 1 to 100 parts/10^6 or ppm.

The types of chemicals that may interfere with normal responses in fish can be predicted in part once the identity is known of the natural chemical substances being sensed. Using this approach, the characterization of natural stream substances that guide *Alosa pseudoharengus* (alewives) to their

spawning grounds has been attempted. The substances
are low molecular weight, polar, heat stable, non-
volatile and may contain acidic and basic centers
(not free amino acids).

INTRODUCTION

The importance of the chemical senses, smell
and taste, to the functioning of land based animals
has long been established.[1,2] It is thus not surprising that recent investigations disclose that
chemical sensing is just as important or more so
to the functioning of aquatic organisms.[3-5] Several
behavioral traits which could be provoked by chemical
stimulation include attraction and sex stimulation
of the male by a receptive female, recognition of
homestream waters by anadromous fish, recognition
of an individual of the same species or different
species by an organism, location of food and
schooling of fish. Decreases in the effectiveness
of chemical sensing could have drastic effects on
aquatic populations.

There are many ways in which pollution effects
on aquatic organisms can be measured. It is our
opinion that meaningful data can be collected only
through the development and use of sensitive bioassays that are paralleled as closely as possible
to conditions present in the field. We suspect that
the external senses, and in particular the chemical
senses, are most susceptible to pollution interference at very low concentrations.

The experiments described are divided into two
groups: (1) the effects of pollution on total animal
behavior and on chemotaxis; and (2) the identificatio
of behaviorally important chemicals as a prelude to
the prediction of some effects of water quality
deterioration on fish.

THE EFFECTS OF POLLUTION ON
TOTAL ANIMAL BEHAVIOR AND ON CHEMOTAXIS

Effects of Kerosene Fractions on
Homarus americanus and *Nassarius obsoletus*

The effects of kerosene fractions on the
lobster, *Homarus americanus*, and the mud snail,
Nassarius obsoletus, have been investigated. The
two studies differ in that the lobster study deals
with the effect of kerosene and kerosene fractions
on total behavior in lobster communities, whereas

the mud snail study is an experiment designed to
yield information concerning quantitative effects
of the aqueous soluble fraction of kerosene on one
behavioral act, attraction to food.

Kerosene was chosen as a test substance for
several reasons. Lobster fishermen have found that
kerosene-soaked bricks fished almost as well as red
fish, a common lobster bait. It was our hope to
isolate the attractive fraction. Another reason
for its choice was that kerosene is for the most
part a distillate fraction of crude oil and contains
many of the components of crude oil that would be
expected to penetrate the water column through dis-
solution.[6] Thus any action provoked by the presence
of kerosene should also be provoked by crude oil.
Because of the low boiling point range, fractionation
of kerosene into straight chain (normal), branched
chain, cyclic and aromatic hydrocarbons was easily
accomplished and identification of some of the
individual components was possible through gas
chromatography and mass spectrometry.

Kerosene fractionation

Kerosene (Gulf grade) was fractionated into
three parts: straight chain hydrocarbons, branched
chain and cyclic hydrocarbons, and aromatic hydrocarbons. Kerosene (2 ml) was applied to a column
containing 100 g of alumina (activated at 120° for
12 hr) and the column was washed with 500 ml of
pentane. The pentane was carefully concentrated
to 25 ml and the concentrate was placed on a column
containing 200 g of silica gel (activated at 120°
for 12 hr). Elution with pentane was continued
until monitoring of absorption at 254 nm indicated
the appearance of aromatic hydrocarbons. The pentane fraction was then carefully evaporated to 40
ml and subjected to urea clathration. The 40 ml
pentane solution containing only saturated hydrocarbons was diluted to 65 ml with acetone and added
slowly to a 30 ml methanol solution containing 2.2
g dissolved urea. Any precipitate formed was
dissolved by the addition of small amounts of
methanol. The solution was evaporated slowly in
vacuum, keeping the flask below room temperature.
The clathrate residue was washed four times with
25 ml pentane. The pentane washes were combined
and concentrated to 40 ml, followed by reclathration
with urea. This process was repeated four more
times. The final pentane wash contained predominantly

179

the branched chain and cyclic hydrocarbon fraction.
The urea clathrate was dissolved in water and extracted with pentane. This pentane extract contained
predominantly straight chain hydrocarbons. Further
fractionation of the branched chain and cyclic
hydrocarbons was achieved using thiourea clathration.
However, the separation was difficult and our
early studies indicated that the activities of the
branched chain hydrocarbons and the cyclic hydrocarbons were similar. The combined branched chain
and cyclic fraction was used in the reported bioassay. The aromatic fraction was prepared according
to the procedure described by Boylan and Tripp.[6]
Extraction of the kerosene with 2.5 l sea water
resulted in an aqueous extract containing predominantly aromatic material. After extraction with
pentane, the pentane fraction was concentrated under
vacuum (15 Torr, 10° C) to 0.5 ml. Since the yield
of aromatic hydrocarbons in the aqueous extract was
known [610 mg/l],[6] the concentration of oil in the
pentane solution was calculated to be 8%. This
solution was used as the aromatic fraction.

The kerosene fractions were analyzed by gas
chromatography (6' x 1/16" OD, SS column packed
with 100-120 mesh, textured glass beads coated with
0.2% Apiezon L; program rate, 50° C → 150° C @ 5°/
min) and the results are shown in Figure 25.

Homarus americanus bioassay

Three to five lobsters were housed together in
each of five 180 gallon observation tanks that were
designed to simulate some of the characteristics of
their natural environment. Whole kerosene and its
fractions (straight chain hydrocarbons, branched
chain/cyclic hydrocarbons and aromatic hydrocarbons)
were used in this test. Each community was exposed
to a particular kerosene fraction or whole kerosene
only once. The communities were fed normally during
the "rest" period of several days between trials.
Each test was performed during periods of the day
when the lobsters were normally least active.
Twenty eight social units, five body care units and
twenty-one feeding units were analyzed.

At the beginning of each trial three blank
cleaned asbestos strips (pre-extracted with
benzene/methanol) were dropped into the center of
the community tank and the responses of the lobsters
to the blanks were recorded for 15 min. The
dominant lobster would, on occasion, patrol its
home range after the blank strips were introduced;

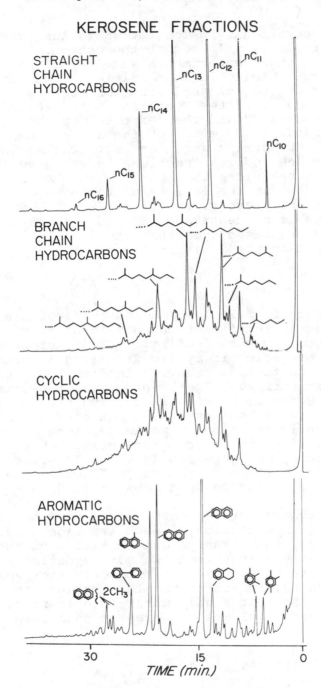

KEROSENE FRACTIONS

Figure 25.   Gas chromatograms of kerosene fractions
               column; 6' x 1/6" OD, SS packed with 100-
               120 mesh textured glass beads coated with
               0.2% Apiezon L, Temp. programmed 50° → 150°C
               at 5°/min.

and only occasionally a lobster would lunge at the
moving strip as it floated to the bottom. At no
time during these experiments were any of the blank
strips eaten. Following the initial 15 min observation period, two cleaned asbestos strips, each
containing 20 µl of test material, were introduced
into the test tank and the responses of the animals
were recorded for 30 min. At the end of the trial
period all the strips were removed. The effect of
whole kerosene and its fractions on the body care
and feeding behaviors is shown in Table 20A.
Table 20B depicts the effects of the same materials
on the community social behavior. The number of
communities is represented by n and the means ($\overline{X}$)
and ranges are based upon the actual number of times
a given unit appeared within a community.

Results

The data from Tables 20A and 20B indicate
that the most significant changes in behavior were
those associated with feeding. Whole kerosene
and the branched chain cyclic fractions induced
searching and feeding behavior leading to ingestion
of the test strips. The branched chain cyclic
fraction seemed to be more attractive than whole
kerosene perhaps because of the presence of the
polar aromatic fraction in whole kerosene. The
polar aromatics did induce a searching behavior
at a distance, but were repulsive to lobsters at
close range. The straight chain fraction did
not have any influence upon lobster feeding
behavior.
Although it was impossible to determine pre-
cisely the concentration of each fraction affecting
the lobsters, quantitative determination of the
amount of sample lost from the strips during tests,
combined with data collected from oil/water solubility studies, indicate that the maximum concentration of the most highly water-soluble fraction
(aromatic fraction) was not greater than 60 parts/
10^{-9} (ppb).

Nassarius obsoletus bioassay

The response of the mud snail *Nassarius obsoletus*
to a food extract, a food extract plus seawater-soluble
components from kerosene, and to sea water soluble

components from kerosene alone was measured. Since
this work will be published as a full paper,[7] only
minimum details will be presented here. The experiments were performed in a plexiglass chamber,
with equal angles among the arms. Each arm measured
42 cm long and 10 cm in height and width. One of
the upper arms was randomly selected as the stimulus
arm for each trial. Both upper arms received 1.5
l/min sea water (15°C); in the test experiments,
the chemical stimulus (2.5 ml/min) was completely
mixed with the background water before introduction
into the stimulus leg. The attractant described by
Jacobson and Boylan[7] is a purified fraction of an
oyster homogenate. The sea water extract of kerosene
was prepared according to the method described by
Boylan and Tripp.[6] The method yields approximately
610 mg of aromatic hydrocarbons per liter of seawater
extract of kerosene (25 ml). The final concentrations
of the kerosene (water extractables) and the oyster
extract in the stimulus arm amounted to 1 part/10^9
(ppb) and 0.33 parts/10^6 (ppm) respectively. The
results of this experiment are tabulated in Table 21.

Results

The results demonstrate that the water-soluble
part of kerosene in concentrations as low as 1 part/
10^9 can inhibit the attraction of the mud snail to
a food source.

*Discussion of Homarus americanus
and Nassarius obsoletus results*

Kerosene does affect normal lobster behavior,
the most unusual effect being the attraction to and
ingestion of kerosene, a most unlikely "food." The
stimulation of feeding behavior has been traced to
the branched chain/cyclic saturated hydrocarbons in
kerosene. The aromatic part "water-soluble fraction"
of kerosene seemed to produce stress-related behavior,
but this was difficult to quantify by the described
bioassay. For this reason the effects of the
seawater-soluble part of kerosene on the attraction
of *Nassarius obsoletus* to a food source was initiated.
This study conclusively showed that the seawater-
soluble fraction of kerosene inhibits attraction of
Nassarius obsoletus to food in concentrations of
1 part/10^9. Since the seawater extract of kerosene
is predominantly composed of benzene and napthalene

Table 20

Effect of Kerosene and Kerosene Fractions on the Behavior of the Lobster *Homarus americanus*

A

Detailed data (n, $\bar{x}$, s.e., Range) by fraction:

UNIT	Straight Chain n	$\bar{x}$	s.e.	Range	Branched-Cyclic n	$\bar{x}$	s.e.	Range	Polar Aromatic n	$\bar{x}$	s.e.	Range	Whole Kerosene n	$\bar{x}$	s.e.	Range
Body Care																
Agitate	3	.3	0.3	0 - 1	4				4	1.5	0.7	0 - 3	5	0.4	0.3	0 - 1
Antennule stroke	3	.3	0.3	0 - 1	4	3.3	1.1	1 - 6	4	0.8	0.3	0 - 1	5			
Antennae twitch	3	1.0	1.0	0 - 3	4				4				5	1.6	0.8	0 - 5
Groom	3	1.0		1	4	3.8	1.8	1 - 9	4	5.5	2.5	3 - 13	5	2.6	2.7	1 - 11
Swimmeret beat	3				4				4				5			
Feeding																
Anten. Freq.	3				4				4				5			
Ambul. sway	3	1.6	1.7	0 - 5	4	2.8	1.2	1 - 6	4	4.0	1.7	1 - 9	5	4.8	1.3	3 - 10
Approach	3	1	0.6	0 - 2	4	1.8	0.3	1 - 2	4	0.5	0.3	0 - 1	5	1.6	1.9	0 - 4
Carry	3				4				4				5	0.8	0.6	0 - 3
Chew	3				4	0.3	0.3	0 - 1	4				5	1.2	0.6	0 - 3
Drop	3				4	1.0	0	1	4	0.3	0.3	0 - 1	5	0.8	0.2	0 - 2
Drop down on	3				4	1.0	0.4	0 - 2	4				5	0.2	0.2	0 - 1
Exopodite wave	3	1.6	1.2	0 - 4	4	2.8	0.8	1 - 4	4	1.5	0.7	0 - 3	5	3.2	1.5	0 - 7
Ingest	3				4	1.5	0.5	1 - 3	4				5	1.0	0.5	0 - 3
Jump	3				4	0.3	0.3	0 - 1	4				5			
Lunge at	3				4				4				5			
3rd Maxilliped sway	3	0.3	0.3	0 - 1	4	1.3	0.8	0 - 3	4	0.5	0.5	0 - 2	5	0.4	0.3	0 - 1
3rd Max. grind	3				4				4				5	0.8	0.2	0 - 1
Pick up	3	0.3	0.3	0 - 1	4	1.5	0.3	1 - 2	4	0.8	0.8	0 - 3	5	1.6	0.4	1 - 3
Rake	3				4				4	0.3	0.3	0 - 1	5	0.2	0.2	0 - 1
Sit on	3	0.7	0.7	0 - 2	4				4				5			
Spit out	3	0.3	0.3	0 - 1	4				4				5	0.2	0.2	0 - 1
Touch antennae	3	0.7	0.7	0 - 2	4				4				5	0.2	0.2	0 - 1
Touch 1st amb. legs or chelopeds	3	2.0	2.0	0 - 6	4	1.0	0.6	0 - 2	4	0.5	0.3	0 - 1	5	3.2	2.7	0 - 13
Walk close to stimuli	3	0.7	0.7	0 - 2	4	0.3	0.3	0 - 1	4				5	1.0	0.6	0 - 3
Walk over stimuli	3				4	0.3	0.3	0 - 1	4	2.3	1.4	0 - 6	5	0.8	0.2	0 - 1

Ranges ($\bar{x} \pm 2 \times$ s.e.):

UNIT	Straight	Branched Cyclic	Polar Aromatic	Whole
Body Care				
Agitate	0 - 0.9		0.1 - 2.9	0 - 1.0
Antennule stroke	0 - 0.9	1.1 - 5.5	0.2 - 1.4	
Antennae twitch	0 - 3.0			0 - 3.2
Groom	1.0 - 1.0	0.2 - 7.4	0.5 - 10.5	0 - 8.0
Swimmeret beat				
Feeding				
Anten. Freq.				
Ambul. sway	0 - 5.0	0.4 - 5.2	0.6 - 7.4	2.2 - 7.4
Approach	0 - 2.2	1.2 - 2.4	0 - 1.1	0 - 5.4
Carry				0 - 2.0
Chew		0 - 0.9		0 - 2.4
Drop		1.0 - 1.0	0 - 0.9	0.4 - 1.2
Drop down on		0.2 - 1.8		0 - 0.6
Exopodite wave	0 - 4.0	1.2 - 4.4	0.1 - 2.9	0.2 - 6.2
Ingest		0.5 - 2.5		0 - 2.0
Jump		0 - 0.9		
Lunge at				
3rd Maxilliped sway		0 - 2.9	0 - 1.5	0 - 1.0
3rd Max. grind				0.4 - 1.2
Pick up	0 - 0.9	0.9 - 2.1	0 - 2.4	0.8 - 2.4
Rake			0 - 0.9	0 - 0.6
Sit on	0.3 - 1.1			
Spit out	0.1 - 0.5			0 - 0.6
Touch antennae	0.3 - 1.1			0 - 0.6
Touch 1st amb. legs or chelopeds	0.8 - 3.2	0.2 - 1.0	0 - 1.1	0 - 8.6
Walk close to stimuli	0.3 - 1.1	0 - 0.9		0 - 2.2
Walk over stimuli		0 - 0.9	0 - 5.1	0.4 - 1.2

184

B

Social

UNIT	Straight Chain				Branched Cyclic				Polar Aromatic				Whole Kerosene				(x̄ ± 2 x s.e.)			
	n	x̄	s.e.	Range	n	x̄	s.e.	Range	n	x̄	s.e.	Range	n	x̄	s.e.	Range	Straight	Branched Cyclic	Polar Aromatic	Whole
Antennae wave	3	0.3	0.3	0 - 1	4	0.3	0.3	0 - 1	4				5				0 - 0.9	0 - 0.9		
Approach lobster	3	2.6	1.5	0 - 5	4	1.3	0.8	0 - 3	4				5	3.6	1.8	1 - 10	0 - 5.6	0 - 2.9		0 - 7.2
Arched pose	3	0.3	0.3	0 - 1	4	0.3	0.3	0 - 1	4				5	1.8	0.5	1 - 3	0 - 0.9	0 - 0.9		0.8 - 2.8
Bulldoze	3				4				4				5							
Chase	3	0.3	0.3	0 - 1	4	0.3	0.3	0 - 1	4	0.5	0	0 - 2	5	1.6	1.1	0 - 6	0 - 0.9	0 - 0.9	0.5 - 0.5	0 - 3.8
Crouch	3				4				4				5	0.2	0.2	0 - 1				0 - 0.6
Face off	3	0.3	0.3	0 - 1	4				4				5	0.4	0.3	0 - 1	0 - 0.9			0 - 1.0
Flee	3	0.3	0.3	0 - 1	4	0.3	0.3	0 - 1	4	0.3	0.3	0 - 1	5	0.8	0.6	0 - 3	0 - 0.9	0 - 0.9	0 - 0.9	0 - 2.0
Flop down	3				4	0.8	0.3	0 - 1	4	0.3	0.3	0 - 1	5	0.4	0.2	0 - 2		0.2 - 1.4	0 - 0.9	0 - 0.8
Fold claws	3	0.7	0.3	0 - 1	4	0.5	0.3	0 - 1	4	0.3	0.3	0 - 1	5	0.4	0.3	0 - 1	0.1 - 1.3	0 - 1.1		0 - 1.0
Lunge at Lobster	3				4				4	0.3	0.3	0 - 1	5						0 - 0.9	
Open Claw	3				4	0.3	0.3	0 - 1	4	0.3	0.3	0 - 1	5	0.6	0.6	0 - 3		0 - 0.9	0 - 0.9	0 - 1.8
Open Claw stance	3				4				4				5	0.4	0.3	0 - 1				0 - 1.0
Raise up	3				4	0.5	0	0 - 2	4	0.3	0.3	0 - 1	5	0.2	0.2	0 - 1		0.5 - 0.5	0 - 0.9	0 - 0.6
Raise up on claws	3				4				4	0.3	0.3	0 - 1	5						0 - 0.9	
Rapid swim	3				4				4	0.3	0.3	0 - 1	5						0 - 0.9	
Rear up	3	2.3	1.2	0 - 4	4	2.3	0.6	1 - 4	4	0.5	0.3	0 - 1	5	2.0	0.6	1 - 4	0 - 4.7	1.1 - 3.5	0 - 1.1	0.8 - 3.2
Reverse	3				4				4				5	0.6	0.3	0 - 1				0 - 1.2
Roll over	3				4				4				5							
Run	3				4				4	0.3	0.3	0 - 1	5						0 - 0.9	
Side walk	3				4	0.3	0.3	0 - 1	4				5					0 - 0.9		
Still	3	4.3	3.4	0 - 11	4	6.3	2.1	3 - 12	4	9.0	2.7	4 - 15	5	7.8	3.5	2 - 21	0 - 11.1	3.9 - 10.5	3.6 - 14.4	0.8 - 14.8
Strut	3				4				4				5							
Tail curled under	3				4				4	0.3	0.3	0 - 1	5	0.2	0.2	0 - 1			0 - 0.9	0 - 0.6
Tail flip	3				4				4				5	0.2	0.2	0 - 1				0 - 0.6
Tail up	3				4	0.3	0.3	0 - 1	4				5					0 - 0.9		
Touch antennae	3	0.7	0.7	0 - 2	4				4				5	0.2	0.2	0 - 1	0 - 2.1			0 - 0.6
Upright stance	3				4				4				5							

Table 21

Mean ± Standard Error of Number of N. obsoletus in Stimulus
and Nonstimulus Arms in Each Treatment at 15 Minutes*

	Attractant (oyster extract)	Attractant + Kerosene extract (1 ppb)	Kerosene extract (1 ppb)	Control (seawater)
Stimulus arm x ± s.e.	4.4 ± 1.0	1.7 ± 0.3	1.0 ± 0.5	0.8 ± 0.8
Nonstimulus arm x ± s.e.	0.6 ± 0.4	1.3 ± 0.3	1.7 ± 0.7	0.3 ± 0.3
Number of trials (10 snails/trial)	5	3	3	4

*Means joined by underlining are not significantly different
at or below the 0.05 protection level using Kramer's modification of Duncan's multiple range test.

compounds[6] and kerosene usually contains 30-40%
aromatic material, it would be expected to produce
an effect in the same concentration region. From
previous studies[6] we can conclude that crude oils
and some fuel oils have water-soluble fractions
similar to that of kerosene. It is therefore expected that these oils will also disrupt the normal
feeding behavior of *Nassarius obsoletus*.
 Whether or not exposure of the lobster or the
mud snail to oil will cause long-term damage to
populations can only be answered by intensive
laboratory and field studies. However it is evident
that oil and oil products contain materials that can
exert subtle effects on marine organism behavior in
the part per billion range, well within the scope
of the solubility of some of the components of oil
in water.

Effects of Contaminants on the Attraction of *Bdelloura candida* to its host *Limulus polyphemus*

In order to test the effects of various contaminants on other chemically stimulated behavior patterns of marine organisms, the attraction of the symbiotic flatworm, *Bdelloura candida*, to its host the horseshoe crab, *Limulus polyphemus*, was investigated. Details concerning the bioassay used and the chemical identity of the host specific substances responsible for the attraction are to be published elsewhere.[8]

Limulus polyphemus bioassay

Water from glass aquaria containing horseshoe crabs were shown to be attractive to the flatworm *Bdelloura candida*. A Y-tube choice system containing two upstream legs made of 4 mm glass tubing (8 cm long) and a downstream leg made of 7 mm glass tubing (also 8 cm long) was used as the bioassay apparatus. Each of the upstream legs, one alternately selected as the *Limulus* water leg and the other selected as the seawater leg, had a flow maintained at 8 ml/min. The *Bdelloura candida* introduced in the downstream leg consistently selected the *Limulus* water over straight seawater. Contaminants of known concentration were mixed with the *Limulus polyphemus* water prior to testing. The contaminants included phosphate binders ($Na_3 PO_4$ - $Na_4 P_2 O_7$ - $Na_5 P_3 O_{10}$ in 1/1/1 mixtures), $HgCl_2$, $FeCl_2$, detergent, $Zn Cl_2$, NH_4OAc, dieldrin (in Et OH) and kerosene (both seawater soluble part and a water emulsion). Each *Bdelloura candida* was used only once. The effects of the contaminants on the attraction of the *Bdelloura candida* to *Limulus polyphemus* water is shown in Table 22.

Results and discussion

The seawater extract of kerosene had no effect on *Bdelloura candida* attraction at 0.6 parts/10^6 (ppm). However, a sample of *Limulus* water shaken with kerosene and allowed to settle for an hour proved to be inactive at 100% but regained activity at 12% dilution. Mercuric chloride and phosphate binders were active inhibitors at 100 parts/10^6 (ppm) and the activity did not recover at dilution to 12%. Ferrous chloride and detergents inhibited

187

Table 22

*Effects of Contaminants on the Attraction of Bdelloura candida to "Fingerprinted" Water from Its Host Limulus polyphemus**

Compound	1 ppm	100 ppm	1000 ppm	Comments
Kerosene (sea-water extract)	?	?	?	No interference at 0.6 ppm, interference when active water is shaken with kerosene (concentration unknown)
Phosphate Binders†	+	0		Activity not recovered at 12% dilution [12 ppm]
Hg Cl$_2$	+	0		Activity not recovered at 12% dilution [12 ppm]
Fe Cl$_2$	+	0		Activity recovered at 12% dilution [12 ppm]
Detergent	+	0		Activity recovered at 12% dilution [12 ppm]
Zn Cl$_2$	+	+	0	Activity partly recovered at 25% dilution [250 ppm]
NH$_4$ OAc	+	+	0	Activity not recovered at 25% dilution [250 ppm]
Dieldrin (in ETOH)	+	+	-	No interference

*Y Tube Choice Apparatus Used. Control is sea water.

†Na PO 1OH O, Na P O, Na P O - 1/1/1 mixture.

(+) *Limulus p.* water plus contaminants was attractive to *Bdelloura c.*

(0) *Limulus* water plus contaminant was not attractive to *Bdelloura c.*

chemotoxis at 100 parts/10^6 (ppm) but the activity
of the samples recovered at dilution to 12%. Zinc
chloride, dieldrin and ammonium acetate had no
effect at 100 parts/10^6 (ppm). Several of the metal
contaminants tested may have formed insoluble salts
with inactive components in *Limulus polyphemus* water
which could account for the relatively high concentrations of metals required for interference. The
mechanism(s) of interference has (have) not been
studied.

THE IDENTIFICATION OF BEHAVIORALLY IMPORTANT
CHEMICALS AS A PRELUDE TO THE PREDICTION OF
SOME EFFECTS OF WATER QUALITY DETERIORATION
ON FISH

Classification of Chemicals that
Attract *Alosa pseudoharengus* to
Homestream Waters

The return of adult alewives, *Alosa pseudoharengus*,
to their homestreams during the spawning season has
been investigated. Like salmon,[4] alewives seem to
be able to detect and do prefer the chemical odor of
their homestream.[9] Our goal in this study was to
determine the nature of the dissolved materials in
homestreams that were attractive to adult alewives.

*Concentration and fractionation
of homestream chemicals*

Fifteen gallons of stream water (Bourne River,
Bourne, Massachusetts) were collected in a non-toxic
plastic bucket and immediately stored in a freezer
(-20°C) for 16-20 hr. The liquid core was siphoned
and reconcentrated by freezing to approximately 200
ml. The water was removed using a rotary evaporator
(30° C water bath, 15 Torr). The residue was either
dissolved in 15 gal of well water and tested or was
fractionated, reconstituted to 15 gal with well
water, and tested. Since each test required fifteen
gallons of stream water or equivalent, it was necessary
to repeat this process many times.

1) Freeze concentrate. The residue resulting
from freeze concentration and evaporation was considered the freeze concentrate.

2) Boiled Bourne water. The freeze concentrate
was dissolved in a minimum amount of water, boiled
for 5 min, cooled, reconstituted to 15 gal and tested.

3) Distillate. The solution after freeze concentration was taken to dryness (15 Torr) and the distillate saved. One liter of triply distilled water was added to the residue and again the resulting solution was taken to dryness. This was repeated four times and all distillates were combined, reconstituted to 15 gal and tested.

4) Distillate residue. The residue from (3) was reconstituted to 15 gal and tested.

5) Polar fraction. The freeze concentrate residue was dissolved in a minimum amount of triply distilled water and passed through 200 g of Amberlite XAD-4 resin (prewashed with $CHCl_3$, acetone, MeOH and water). The same procedure was followed with another sample but this time the adsorbant used was 200 g of polyethylene powder. The filtrates from each column were tested for activity.

6) Non-polar fraction. The Amberlite XAD-4 used in (5) was washed with ethanol. The ethanol was removed (15 Torr) and the residue tested for activity.

7) Fractionation by molecular size. The freeze concentrate was dissolved in a minimum amount of water and boiled for 5 min in order to destroy active enzymes and bacteria. The solution was cooled and passed through a molecular filter (Milipore-Pelicon 1024A) at 50 psi. The solution above the filter was stirred continuously and the filter was never allowed to go dry. Only molecules of an approximate molecular weight of 1000 or lower penetrated the membrane. The filtrate contained the fraction with compounds of molecular weight less than 1000 and the solution above the filter retained the compounds of molecular weight greater than 1000.

8) Sephadex QAE[Cl⁻] filtrate. One hundred grams of Sephadex QAE was allowed to swell overnight in distilled water, followed by washings with 0.1N NaOH and then 0.1N HCl. The resin was then washed with triply distilled water until the eluate was neutral. The ion exchange, now in the chloride form, was packed into a glass chromatographic column.

The Bourne water freeze concentrate was dissolved in a minimum amount of water and boiled 5 min in order to destroy enzymes and bacteria. The solution was then passed through the ion exchange resin followed by a 100 ml wash with triply distilled water. The filtrate and the wash were combined, and adjusted to pH 6.8-7.8 using small amounts of HCl or $NaHCO_3$ if necessary. The filtrate

was then reconstituted to 15 gal and tested. The
Sephadex QAE[Cl⁻] treatment exchanged all the anions
in the freeze concentrate for chloride ion.

9) Sephadex CM[H+] filtrate. The procedure
for (9) was identical with that used in (8). The
Sephadex CM[H+] treatment removed bases from the
filtrate.

10) Sephadex CM[Na+] filtrate. The ion exchange
gel was allowed to swell overnight in distilled
water. It was then washed with 0.1N HCl and then
with 0.1N NaOH. The final wash with triply distilled water was continued until the eluate was
neutral. The rest of the process was the same as
in (9) and (8). Sephadex CM[Na+] treatment exchanged
cations in the freeze concentrate for Na+ ion in the
filtrate.

11) Sephadex DEAE[NH₂] filtrate. The process
described in (10) was used. Sephadex DEAE[NH₂]
treatment removed acids from the freeze concentrate.
Since the free amine groups of the DEAE Sephadex
are active towards aldehydes and other functional
groups, the lack of response of fish to the filtrate
may not reflect the importance of acids in homestream
response.

12) Importance of amino acids. The distillate
residue (4) was tested for the presence of free
amino acids with an amino acid analyzer.

Alosa pseudoharengus bioassay

A Y-maze system made of plexiglass (1' in height
and width and 2' long) with wooden leg extensions
(7' long) served as a bioassay apparatus. A continuous background flow of 5 gal/min was maintained
in each upstream leg. During the test period an
additional 1 gal/min was introduced into each upstream leg (stimulus and control water). The
temperature of the water in each of these legs was
monitored constantly and the test was rejected if
fluctuations greater than 0.5° C between legs
occurred. Most tests were run at night under low
light levels since, under these conditions, the
physical parameters of the test could be more
easily controlled. Five sexually mature alewives
(caught in the Bourne River fish ladder, Bourne,
Massachusetts) were allowed to swim freely in the
plexiglass portion of the Y-maze. Two of the three
legs were selected randomly as the stimulus and
control legs. After a 1 to 2 hr acclimation period,
observations were begun. Positions of the five fish

were recorded at 15 second intervals for 45 min or
180 observations. The 45 min consisted of a 15 min
pre-test, a 15 min test and a 15 min post-test
period.

The results of testing alewife preference for
stream water and stream water fractions are tabulated in Table 4. The number of times fish were
observed in the stimulus, control and downstream
legs during the pre-test (sums A, B, C respectively)
and the test (sums A^1, B^1 and C^1 respectively) are
summarized. In order to estimate the degree of
response, the equation $\dfrac{(A^1-A) + (B-B^1)}{A+A^1 + B+B^1}$ X 100 was
applied. Due to the experimental design, it was
difficult to statistically measure the significance
of the results.

Discussion

The results listed in Table 23 suggest that
alewives prefer homestream water (Bourne River,
Bourne, Massachusetts) over Oyster Pond and Siders
Pond water (Falmouth, Massachusetts) and Pocasset
water (Pocasset, Massachusetts). However, Bourne
fish did not show a preference for Bourne water
over Sandwich River water (Sandwich, Massachusetts).
The fact that Sandwich River is the only river that
empties into the Gulf of Maine may be significant.
Alewives responded to boiled Bourne freeze
concentrate, distillate residue, polyethylene filtrate and the fraction containing compounds of
molecular weights of less than 1000 approximately
as strongly as they did to Bourne River water itself.
Although the distillate was not attractive, it is
difficult to reason why the fish do not consistently
prefer the distillate residue when the distillate
is run against the distillate residue. When the
distillate and distillate residue are combined, the
response approaches that observed for Bourne water
and the distillate residue alone. We have concluded
from the above results that the active fractions of
homestream water chemicals are heat stable, mostly
non-volatile, polar (polyethylene and Amberlite
XAD-4 remove non-polar material) and smaller than
molecular weight 1000.

The results obtained with the ion exchange
treated water samples are doubtful and must be rerun.
Most of the tests were run at the end of the alewive
migration period (June, 1972) when the response of
fish to Bourne water was erratic. Many fish used

Table 23

Alewife (Alosa pseudoharengus) Response to Homestream Water and Homestream Fractions*

Stimulus Leg (A)	Control Leg (B)	Pretest A	B	C	Test A'	B'	C'	Change in Distribution $\dfrac{(A'-A) + (B-B')}{A + A' + B + B'} \times 100$
Bourne Water	Oyster Pond Water	43	119	138	77	120	103	9.8
Bourne Water	Siders Pond Water	109	38	153	141	68	91	17.4
Bourne Water	Pocasset Water	64	111	125	112	101	87	15.2
Bourne Water	Sandwich Water	124	110	66	129	132	39	3.4
		35	80	185	92	79	129	20.2
		122	67	111	137	89	74	-1.7
		42	113	145	6	218	76	-37.2 av. -5.5
Bourne Water	Well Water	71	122	107	131	105	64	17.9
		57	7	236	183	9	108	48.4
		105	58	137	153	25	122	23.8
		128	99	73	125	92	83	.9
		57	98	145	95	75	130	18.8
		108	74	118	144	30	126	22.4 av. +15.7
		111	99	90	126	87	87	6.4
		94	113	93	129	63	88	21.8
		99	135	66	145	125	30	11.1
		127	73	100	128	70	102	1.0
Well Water	Well Water	98	116	86	96	108	96	1.4
		96	108	96	104	101	95	3.8 av. +2.6
Boiled Bourne Water	Well Water	62	92	146	123	79	98	20.8

Table 23, Continued

Stimulus Leg (A)	Control Leg (B)	Pretest			Test			Change in Distribution $\frac{(A'-A) + (B-B')}{A + A' + B + B'} \times 100$
		A	B	C	A'	B'	C'	
Ice Less Freeze Concentrate	Well Water	107	37	156	214	37	49	27.1
Bourne Freeze Concentrate	Well Water	79	193	28	111	38	151	44.5
		90	99	111	133	55	112	25.0
		97	122	81	124	96	80	12.1 av. +21.7
		103	107	90	120	101	79	5.3
Distillate Fraction	Well Water	73	121	106	62	132	106	-5.7
		111	89	100	129	79	92	6.9 av. -6.6
		127	24	149	99	61	140	-20.9
Distillate Residue	Well Water	61	93	146	100	63	137	21.8
		60	98	142	110	90	100	16.2 av. +19.0
Distillate Residue	Distillate	124	91	85	137	78	84	5.4
		137	92	71	132	106	62	-4.7
		100	107	93	108	110	82	1.2 av. +3.9
		97	114	89	129	88	83	13.6
		147	67	86	150	69	81	.2
Distillate + Distillate Residue	Well Water	87	120	93	93	102	105	6.0
		113	80	107	174	42	84	24.2 av. +15.1
Mol. Wt. Compounds < 1000	Mol. Wt. Compounds > 1000	38	54	208	130	42	128	39.4
		97	124	79	133	88	79	16.3 av. +27.8
Amberlite (XAD-4) Filtrate	Well Water	113	114	73	133	97	70	8.1

Sample	Source	A	B	C	A'	B'	C'		Date
Polyethylene	Well Water	86	50	164	146	30	124	25.6	av. +24.8
Filtrate		61	145	94	102	105	94	23.9	
Ion Exchange (Sephadex CM[H+])	Well Water	123	96	81	123	89	89	-0.7	June 5
Filtrate		98	81	121	92	89	119	-3.9	May 28†
Ion Exchange (Sephadex CM[Na+])	Well Water	93	107	100	73	135	92	-11.8	June 8
		107	92	101	126	75	99	9.0	June 1†
		95	93	112	76	88	136	-3.9	June 4
Filtrate		116	93	91	93	86	121	-4.1	May 28†
Ion Exchange (Sephadex QAE[Cl⁻])	Well Water	84	123	93	113	99	88	12.7	May 28†
		116	90	94	112	95	93	-2.2	June 3
		94	91	115	95	79	126	3.6	June 5
Filtrate		92	113	95	81	118	99	-4.0	June 8
Ion Exchange (Sephadex DEAE[NH₂])	Well Water	116	77	107	115	77	108	-0.3	June 5
		112	58	130	93	78	129	-11.4	May 28†
Filtrate		107	91	103	99	80	122	+0.9	June 3

*The distribution of five free swimming fish in modified Y-test apparatus was recorded every 15 sec during pretest (15 min) and test (15 min) period. Totals A, B, C refer to number of times fish were observed in stimulus (A), control (B), and downstream (C) legs during pretest period. Totals A', B', C' were recorded during the test period.

†From June 2 on, the test fish responded only weakly or not at all to homestream water. Thus the starred test results are preferred.

in these tests were poorly developed sexually or had
already spawned (spent alewives have not shown a
preference--either negative or positive--to homestream
water). The situation is further complicated by the
presence of another species of alewife *(Alosa
estivalis*) in the June runs that was difficult to
differentiate from *Alosa pseudoharengus*. Despite
these problems, our results indicate that Sephadex
QAE (Cl⁻[and Sephadex CM [Na⁺] filtrates were
active in at least one case whereas the filtrates
of Sephadex CM [H+] and Sephadex DEAE [NH₂] were
never active. The large response to control water
that was observed in some tests suggested that ion
exchange impurities were present in the filtrates
and might have interfered in these tests. The
absence of detectable amounts of amino acids in
the active distillate residue indicated that free
amino acids were not involved in the alewife
responses to homestream water.

One can tentatively conclude that the materials
in Bourne water attractive to migrating *Alosa
pseudoharengus* were at least in part low molecular
weight acids and bases. If this is true (and additional studies are necessary in order to prove this
point conclusively), then it is possible that these
acids and bases are also important in homestream
recognition. It is tempting to propose that bacterial decomposition of organic detritus is the
source of these acids and bases. One attractive
feature supporting this view is that the relative
concentration of compounds arising in this way would
be much more constant during seasonal changes than
organics released by living biota other than soil
bacteria. Since the alewive migrations occur during
the spring when fauna and flora changes are at a
maximum, selective sensing of such a group of
chemicals must be considered necessary if the
chemosensory input is to serve an important function
during homestream recognition.

Once the chemicals guiding homestream selection
by alewives are identified, the prediction of the
types of contaminants that may interfere with this
process of selection can be made. The information
gained would allow the establishment of realistic
pollution guidelines that are needed in order to
enable us to preserve or restore spawning grounds of
anadromous fish. This work would also serve as a basis
for studies of other fish, especially estuarine
spawners, who may also locate spawning grounds through
the use of their chemical senses and whose environment is most susceptible to pollution effects.

ACKNOWLEDGMENTS

We should like to thank Brian Andressen for his valuable contributions in the early phases of the alewife work and Bruce Tripp, Laurie Stein and David Engstrom for their excellent technical assistance. This work was supported by E.P.A. Grant #28050, AEC Grant #AT(11-1)-3567, and the Charles E. Culpeper and Sarah Mellon Scaife Foundations. One of us, Stewart M. Jacobson, was funded through an N.S.F. traineeship G2-2019.

REFERENCES

1. Johnston, J. W. Jr., D. G. Moulton, and A. Turk (Eds.). *Communication by Chemical Signals*, Vol. I, *Advances in Chemoreception*. (New York: Appleton, Century and Crofts, 1970).
2. Sondheimer, E., and J. B. Simeone (Eds.). *Chemical Ecology* (New York: Academic Press, 1970).
3. Kleerekoper, H. *Olfaction in Fishes*. (Bloomington, Indiana: Indiana University Press, 1969).
4. Hasler, A. D. *Underwater Guideposts*. (Madison, Wisconsin: University of Wisconsin Press, 1966).
5. Todd, J. H. *Scientific American 224(5)*:99 (1971).
6. Boylan, D. B., and B. W. Tripp. *Nature 230(5288)*:44 (1971).
7. Jacobson, S. M., and D. B. Boylan. "Sea Water Soluble Fraction of Kerosene: Effect on Chemotaxis in a Marine Snail, *Nassarius obsoletus*," *Nature* (1972) in press.
8. Boylan, D. B. "Marine Chemotaxis: Recognition of the Host *Limulus polyphemus* by the Commensal *Bdelloura candida*," Submitted for publication, August, 1972.
9. Thunberg, B. E. *Anim-Behav. 19*:217 (1971).

10. ANALYSIS OF TOXIC RESPONSES IN MARINE POIKILOTHERMS

Gilles LaRoche, G. R. Gardner, Ronald Eisler, E. H. Jackim, P. P. Yevish, and G. E. Zaroogian.* National Marine Water Quality Laboratory, West Kingston, Rhode Island.

INTRODUCTION

Current evidence suggests that the environment has been considered as a sump of limitless capacity which could absorb, utilize or otherwise detoxify almost any material. The idea that only finite amounts of various substances may be handled by the environment is a concept which has only recently been recognized and accepted with some degree of certainty. There is no doubt that environmental balances are in danger and that, as a consequence, the age of a biogenic environment may well be numbered. A catastrophe would therefore appear imminent, but no one can knowledgeably establish how near it is. This doubt is fueling a fire of uncertainty which may in time give rise to panic or indifference unless adequate and realistic knowledge can be gathered.

At the moment two of the pressing questions are what and how much we can release to the environment with predictable harmful effects. Both questions can be answered in the laboratory and under simulated environmental conditions.

By assessing the damage resulting from exposure of species to likely pollutants, alone or in realistic combinations, it is possible to derive guidelines which would protect the environment. With this in mind, a program of systematic acute exposures was

*Present address: Consortium on Water Research, 6600, Cote-des-Neiges, Suite 211, Montreal 249, Quebec, Canada.

established to determine the lethal levels (96-hour
TL-50) of various pollutants to several marine or
estuarine species of poikilotherms. Following acute
studies, which are meant to furnish preliminary information, one initiates chronic exposures which
would reveal the effects of prolonged sublethal
concentrations. During and at the end of all exposures a number of adapted clinical tests have been
applied to evaluate the changes which might have
significance in the survival of the species studied.
In all instances, changes or differences are considered to be real when they occur or prevail in
association with pollutant exposures, through a
comparison of exposed organisms with appropriate
controls (which are maintained under identical conditions except for the treatment). Damage is
considered significant when the survival of the
species is improbable. For instance, determinations
that pollutant effects will distinctly modify the
behavior patterns of the species, its growth, its
development, or its reproduction might well furnish
the necessary evidence. For this purpose, several
morphological or physiological parameters are
measured to assess the changes which may reflect
significant functional anomalies.

In this presentation we will examine a few of
these preliminary observations which may ultimately
lead to the establishment of water quality requirements for marine or estuarine organisms.

A frustrating aspect of bioassays in such complex
mixtures as seawaters is that only limited comparisons
can be made with bioassays completed on fresh water
organisms. For obvious reasons, such as the different physico-chemical performance of pollutants
in these two media plus the differing physiology of
organisms in fresh or seawaters, only limited comparisons of the results are possible.

METHODS AND RESULTS

Except for a few field studies to be mentioned
later, the experimental exposures were carried out
at 20 o/oo salinity in a chemically defined seawater
maintained at 20°C. In this instance we refer to
the 96-hour TL-50 which has already been defined in
prior publications.[1,2]

For *Fundulus heteroclitus*, an estuarine teleost,
it has been observed that the 96-hour TL-50 for
Cu^{2+} is approximately 20 mg/liter (ppm) initial
concentration. This is high for freshwater organisms.

However, if this species is submitted to systematic histological inspection, it becomes apparent that significant damage may be observed very soon after exposure to concentrations as low as 0.5 mg Cu^{2+} per liter.

Figure 26 shows a portion of a sensory center within the cephalic extension of the lateral line in a control mummichog. This morphological appearance is indistinguishable from that of a similar

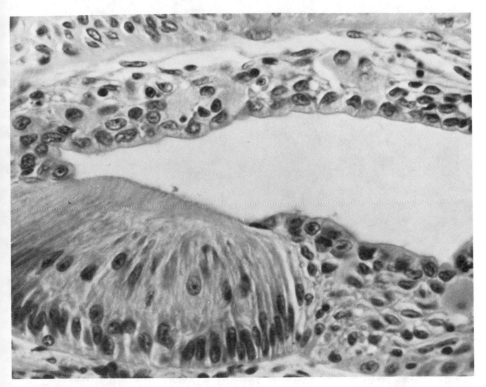

Figure 26.

specimen taken from nature. Note that the squamous epithelium lining the lumen of the lateral line canal appears relatively healthy. In addition the sensory unit, which includes sensory cells (neuromasts) and supporting cells, appears unaffected.

Figure 27 represents the effects of 0.5 mg/l of Cu^{2+} as $CuCl_2$ on the above structure after 24 hours of exposure. It is clear that the squamous epithelium lining the lumen is distinctly necrotic and contains pyknotic nuclei. Further, the germinating layers of this epithelium are also quite

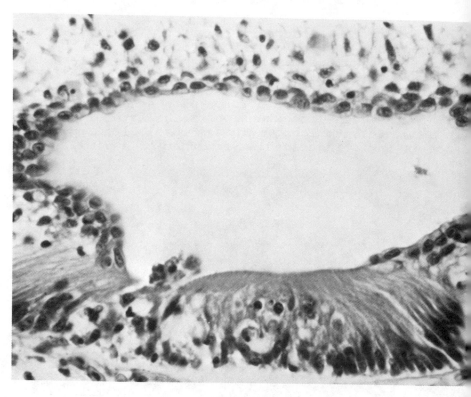

Figure 27.

vacuolated, suggesting that even in the absence of
a killing dose, deep cellular changes can be noted.
In addition to the damage on the squamous epithelium,
the sensory unit of this area is also severely
afflicted. Note that there are few, if any,
normal-appearing neuromast nuclei (sensory cells).
It is evident that the deeper layer of supporting
cells has also been affected by 0.5 mg/l of Cu^{2+}
within 24 hours. Mostly pyknotic nuclei can be
noted in the area. Under these conditions, it was
also observed that the epithelium of the olfactory
sacs showed visual degeneration which would result
in impaired olfactory responses. We have not yet
extended our studies on the probable behavioral
anomalies related to these findings, but it would
appear reasonable to suggest that this damage would
significantly impair the response of these fish to
olfactory stimuli.

In order to ascertain the probable route of
the corrosive action of Cu, experiments were set to
test the effects of $CuCl_2$ injection. Volumes of

0.1 ml at levels as high as 50 mg/liter Cu^{2+}, as $CuCl_2$ injected intraperitoneally daily for five days, failed to show any sign of sensory cell damage within the cephalic extensions of the lateral line or the olfactory sacs. It seems plausible therefore to suggest that effects of $CuCl_2$ are from external contacts and that Cu^{++} incorporation does not have to be invoked to explain these pathological conditions.

Figure 28 shows the gill areas of a control quahaug clam, *Mercenaria mercenaria*, illustrating

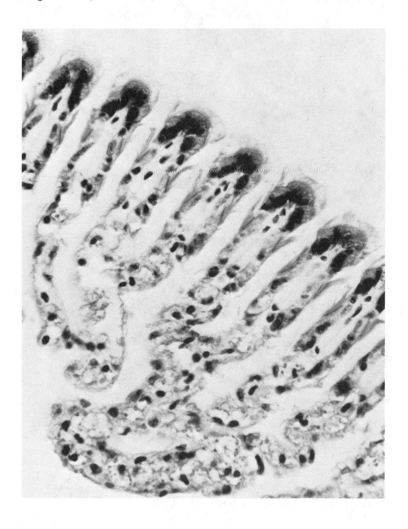

Figure 28.

the ciliated gill lamellae. Figure 29 represents
the changes produced by as little as 0.25 mg/liter
or 250 mg/liter (ppb) of $CuCl_2$ over a period of 96
hours. Under these conditions the ciliated columnar

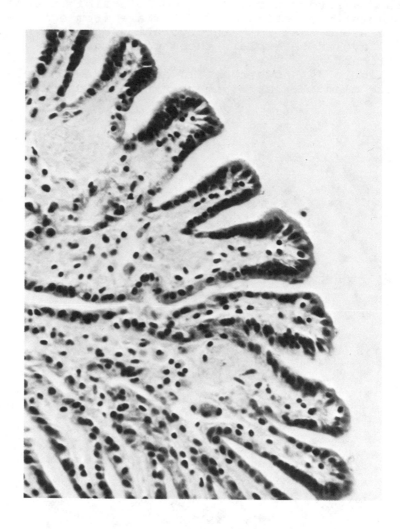

Figure 29.

cells at the tip of the gill lamellae have disappeared. The columnar epithelium of the lamellae
appears to have been replaced by cuboidal cells
which may be interpreted as an aspect of the regen-
erative process resulting from the destructive
action of Cu^{++}. In any event, with the absence of

cilia, it is realistic to conclude that this organism
will undoubtedly starve. Again, anatomical changes
suggest that loss of cilia would limit the survival
of a species exposed to sublethal concentrations of
$CuCl_2$.

Figures 30, 31, and 32 represent changes observed in the crested goby, *Lophogobius cyprinoides*,
a coastal fish collected from Florida. Gobies were
collected in an area of heated effluent from a
power plant where the total concentration of copper
had been measured at 100 mg/l. Histological studies
were then made to see if damage to the sensory system could be observed in the goby as noted for much
lower concentrations in *Fundulus*. No change could
be observed within the sensory unit of the cephalic
extension of the lateral line. However, very distinct damage could be identified in the liver of
these fish when compared to controls obtained from
an area less than 3 km distant.

Figure 30 shows the general histological aspect
of the liver in a control goby. Vacuoles containing

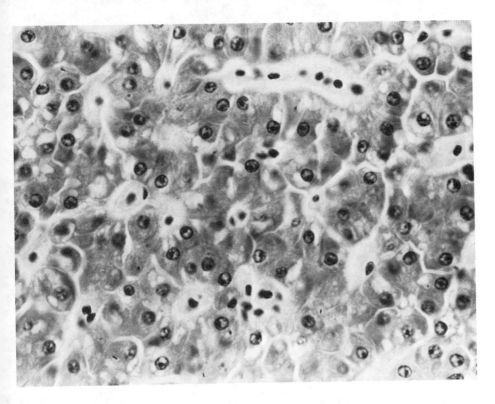

Figure 30.

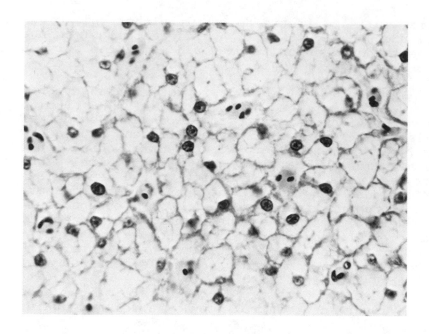

Figure 31.

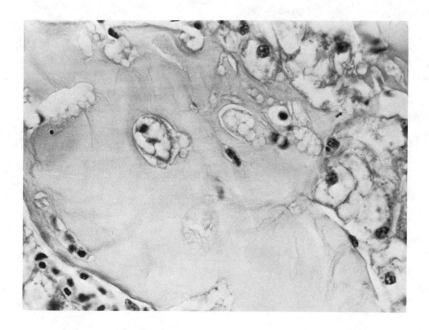

Figure 32.

lipids or glycogen can be seen along with nuclei which have typical appearances for this species.

Figure 31 shows the atypical condition of the liver from a crested goby collected in the effluent area. A condition of this type would not accompany prolonged survival in mammals. The cells are highly vacuolated and the nuclei appear at various stages of degeneration. In many cases (Figure 32), a type of coagulation necrosis can also be identified.

This type of observation is certainly informative in that it may guide us into an area of research where factors can be maintained in such a way as to identify the specific condition or conditions producing the damage.

At present a brief analysis of the conditions involved might be proper. The effluent described above has a slightly elevated temperature and an identified elevated total copper concentration. Are these the only variables or added variables to the otherwise normal environment?

An increase of total copper has been measured, but before this can be related to liver damage, controlled experimental demonstrations have to be made to answer a number of questions, including: (1) is it the copper alone that does the damage? (2) is it increased temperature and copper that produce the change? (3) what is the chemical state of the copper from the effluent? (4) how long are the gobies exposed? (5) how long does it take before a goby develops the pathology? (6) how well can the goby from the effluent area survive and possibly reproduce? and (7) is this liver condition produced by products other than copper or in combination with copper? In other words, copper has been identified and measured. However, presence and measurement of a product do not exclude the presence of unidentified damaging products or conditions. As can be noted, this very interesting partially-monitored ecological and histopathological observation furnishes an insight into what needs to be initated experimentally.

In another laboratory study, informative evidence was gathered through morphological observations of the gastrointestinal tract of an estuarine fish. In this instance we know that NTA (nitrilotriacetic acid trisodium salt) for *Fundulus* has a 168-hour TL-50 of 5500 mg/liter.

Figure 33 represents a section of the ileum in a control fish. Note the intestinal epithelium interspersed with a few mucous or goblet cells. It

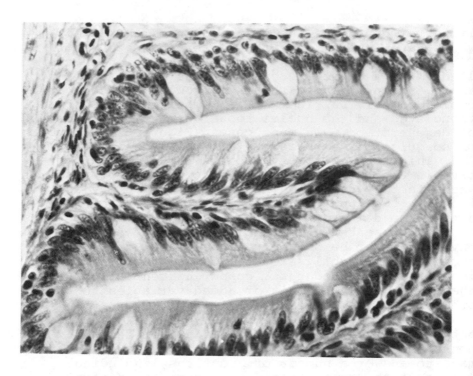

Figure 33.

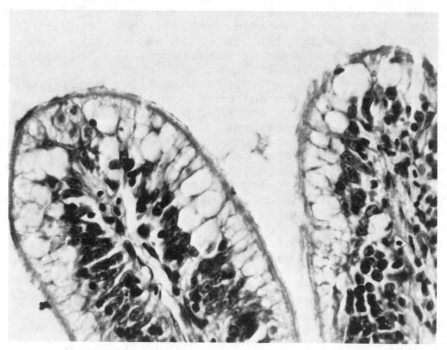

Figure 34.

is evident that these goblet cells are in a secretory mode and that the surrounding epithelium is not vacuolated.

Figure 34 represents a section through the ileum of a *Fundulus* exposed to 5500 mg/liter of NTA for 168 hours. At this concentration, a distinct vacuolization of the epithelium occurs, and only a few goblet cells are still discernible. The condition undoubtedly represents an extensive impairment of the digestive processes of the animal at an unrealistic NTA concentration. In this instance, the changes observed furnished the guidance which allowed the identification of subtle changes occurring at a much lower concentration. Figure 35 shows that detectable epithelial damage is evident at the crests of the ilial folds and again characterized by vacuolization. The fish demonstrating this anomaly had been exposed to only 1 mg/l of NTA for 168 hours. Intestinal changes suggest that continued exposure to 1 mg/l NTA may impair the survival of *Fundulus* by decreasing intestinal activity. Since the mummichog is one of the more resistant estuarine forms studied in our area, it is expected that

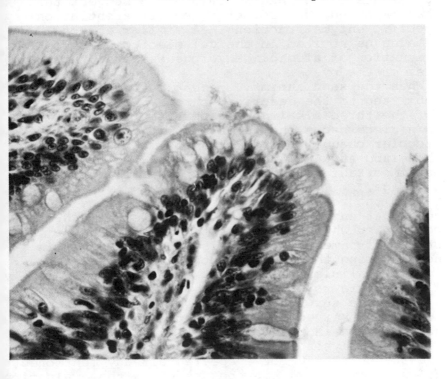

Figure 35.

significant functional damage may result from
exposure of more sensitive species.

With the use of a flow-through seawater system,
a 9-month study has shown that 10 mg/liter of NTA
will significantly modify the gel electrophoretic
behavior of a few enzymes. For instance, in gill
tissue of the lobster, *Homarus americanus*, one of
the two fractionable lactic dehydrogenase (LDH)
bands became extremely faint after exposure of
lobsters to NTA. It is difficult at this point to
postulate that irreparable damage to the species
would result from this notable change. Nonetheless,
it would be appropriate to suggest that a decrease
of one LDH isoenzyme has accompanied NTA exposure.
It is probable that such a difference may only be
temporary or otherwise unimportant in the survival
of the lobster, but this alteration suggests that
changes in carbohydrate metabolism have been associated with a chronic NTA exposure.

In the oyster, *Crassostrea virginica*, the
above mentioned NTA chronic exposure was associated
with the disappearance of two of four identifiable
bands exhibiting malate dehydrogenase activity.
Disappearance of the bands would again suggest that
NTA has impaired the production or differentiation
of specific enzyme fractions. A similar effect
could also be noticed in the gel electrophoretic
fractionation of alcohol dehydrogenase from various
tissues of the oyster.

When the sand shrimp, *Crangon septemspinosa*,
is submitted to the described chronic NTA treatment,
one of the three alkaline phosphatase fractions
shows a discernable change in mobility, thus implying
a molecular change. Since this enzyme plays a role
in the elaboration of the exoskeleton, it would
appear important to evaluate the significance of
this change.

From the response of these three species it is
evident that enzyme systems are affected by a NTA
concentration which is likely to be found in the
coastal environment. It is also clear that NTA is
either entering the cells or otherwise alters specific intracellular metabolites (isozymes). It is
still premature to conclude that the changes recorded
will limit the survival of these species. However,
the changes observed do indicate the need for additional research in this area.

In an analysis of toxic responses, an aspect
of the 96-hour TL-50 for a species already mentioned,
Fundulus heteroclitus, may be useful in the evaluation

of cadmium chloride alone or in combination with
other metals. It is already known that at a salinity
of 20 o/oo and a temperature of 20°C in a chemically
defined medium, the 96-hour TL-50 for *Fundulus* is
approximately 80 mg/liter of Cd^{++}. The survival of
this species for extended periods of time at cadmium
concentrations (as $CdCl_2 \cdot 2\frac{1}{2}H_2O$) ranging from 0 to
75 mg/liter was then tested.

The upper portion of Figure 36 is a composite
graph showing a shaded area which covers the 96-hour
period of treatment and the lighter area for the
period of time in days following exposure, where
fish were maintained in cadmium-free seawater. The
second series of graphs from the top show the survival of fish exposed to the same doses for periods
of 24 and 6 hours respectively. Besides indicating
the difficulties in obtaining chemically reproducible
measurements with bioassays, this study demonstrates
that *Fundulus* exposed to 75 mg/liter of Cd^{2+} as
$CdCl_2 \cdot 2\frac{1}{2}H_2O$ for 24 hours were just as impaired in
their 50-day survival as animals which had been
exposed to the same concentration for 96 hours.
In spite of the profound quantitative limitations
attached to the 96-hour TL-50, it is still possible
to derive valuable information which may guide
efforts in the elaboration of biochemical, physiological or toxicological mechanisms associated with
certain types of exposures.

Figure 37 is meant to reinstate the value of
the 96-hour TL-50 assay on *Fundulus* by demonstrating
the relative toxicity of varying combinations of
Cd, Zn, and Cu. Examination of the sixth column
shows that at 0 mg/liter of Zn^{++} and 0 mg/liter of
Cu^{++}, the addition of 0 to 10 mg/liter of Cd^{++} does
not cause death during 96 hours. It must be recalled
that the 96-hour TL-50 for Cu^{++}, as $CuCl_2$, alone is
20 mg/liter for *Fundulus* and that the 96-hour TL-50
for Cd^{++} alone is approximately 80 mg/liter. In
addition, the 96-hour TL-50 for Zn has been established at 60 mg/liter for this species. Examination
of the seventh column where fish are exposed to 0
mg of Zn and 1 mg/liter of Cu in combination with
0, 1, or 10 mg/liter of Cd, shows that there is a
detectable degree of synergism in the toxicity of
these elements within 96 hours. In the eighth
column the compounding effects of 10 mg/liter of
Cd^{++} plus 8 mg/liter of Cu^{++} are evident. Synergism
was observed at 3 mg/liter of Zn plus 3 mg/liter of
Cu, and the addition of 1 or 10 mg/liter of Cd^{2+} to
these Zn-Cu mixtures produced more-than-additive

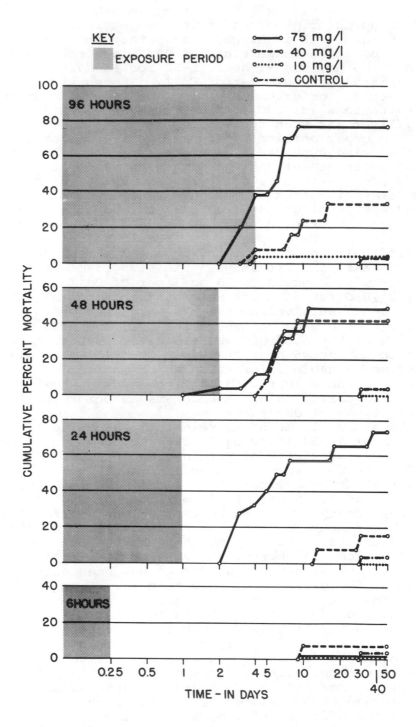

Figure 36.

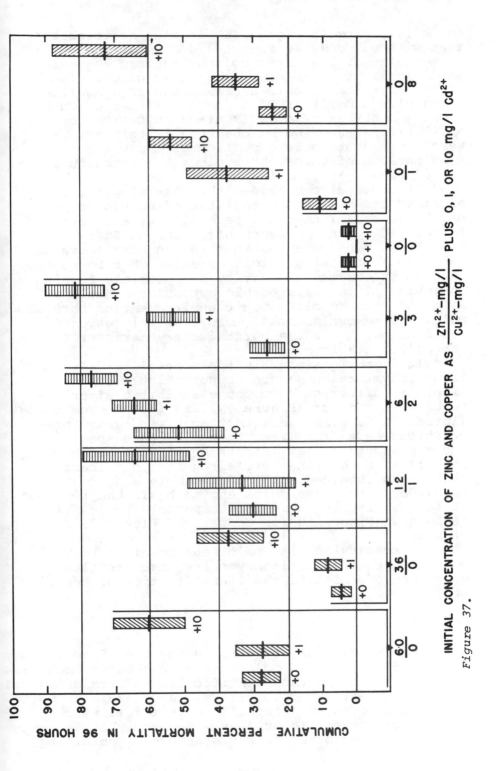

INITIAL CONCENTRATION OF ZINC AND COPPER AS $\dfrac{Zn^{2+}-mg/l}{Cu^{2+}-mg/l}$ PLUS 0, 1, OR 10 mg/l Cd^{2+}

Figure 37.

toxicity. From this graph it is also noticeable that without Cd (*i.e.* Figure 37, 0 under the bars) 60 mg/liter of Zn (first column) appears as toxic as 12 mg/liter of Zn plus 1 mg/liter of Cu. In addition, these amounts are as toxic as 3 mg/liter of Zn plus 3 mg/liter of Cu or 8 mg/liter of Cu alone. A number of other semi-quantitative toxic similarities can also be observed in this graph which represents only a portion of the various combinations of these three elements tested on *Fundulus*.

It could also be added that histological examinations have shown that the effects of copper on the lateral line canal persist, but are not aggravated by the presence of Cd and/or Zn.

Biochemical changes in fish can also serve as indicators of toxicological effects. For instance, changes in enzymatic activity can possibly furnish an insight into the probable metabolic functions affected by a specific toxic environment or through multi-enzyme evaluations. Differential responses would allow at least a partial characterization of the pollutants involved.

It is obvious that in the control *Fundulus* the liver enzyme responses to copper (Figure 38, *in vivo*) are different from those exhibited after exposures to cadmium (Figure 39, *in vivo*). From these figures it is also clear that specific enzyme preparations made from liver tissue of this species and exposed to copper (Figure 38, *in vitro*) show overall responses that are altogether different than those obtained with cadmium. In addition, it is evident that on a multi-enzyme basis the response of *in vivo* exposures is not comparable to *in vitro* responses (Figure 38 for copper and Figure 39 for cadmium).

In conclusion, it would seem possible to develop specific multi-enzyme response profiles which could assist in the identification of certain pollutants.

CONCLUSIONS

I. Short-term bioassays (including the 96-hour TL-50) and field observations can offer valuable assistance in the identification of toxicological responses. The short-term assays are also most valuable in the establishment of realistic chronic exposure studies. However, bottle tests should no longer be satisfactory in obtaining

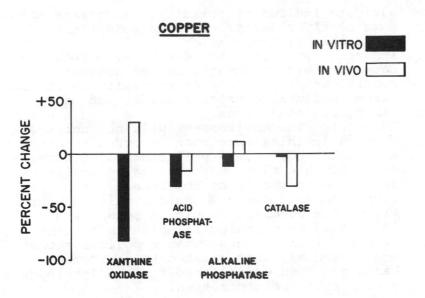

Figure 38.

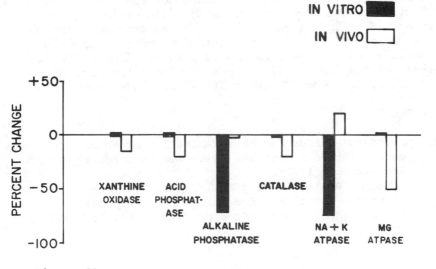

Figure 39.

Figures 38 and 39 represent effects of either copper or cadmium on the enzyme activities of isolated tissues exposed to either elements (*in* *vitro*) or of the intact organisms (*in* *vivo*) exposed to the same elements. Results are expressed as changes in activity from control values.

ultimate indices of toxicity. Bioassays should more nearly reflect realistic conditions of environmental exposure.

II. Inherent synergisms between heavy metals and between other pollutants and environmental conditions *prescribes* the necessity to initially screen pollutants under specific and well controlled conditions.

III. Monitoring the environment will not necessarily furnish anything more than recorded evidence of a catastrophe, unless well controlled experimentation has shown what should be monitored and how much should be tolerated.

IV. Subtle damage, such as could impair the survival of a species, should be sought and identified by pursuing all plausible methods of bioanalysis. Thus, when a pollutant concentration has been related to the damage, a basis has been established for the development of tolerable concentration.

The objective of this presentation was to demonstrate the importance of multidisciplinary approaches in the analysis of toxicological responses. It was also meant to give examples of limitations in the evaluations of field or laboratory studies. Finally, it should be indicated that this work is not a definitive technical presentation and that additional information on such matters can be obtained from individual authors or the Librarian, National Marine Water Quality Laboratory, P.O. Box 277, West Kingston, Rhode Island 02892.

REFERENCES

1. *Standard Methods for the Examination of Water and Wastewater*, 13th edition (New York: American Public Health Association, 1971).

2. LaRoche, Gilles, Ronald Eisler, and Clarence M. Tarzwell. "Bioassay Procedures for Oil and Oil Dispersant Toxicity Evaluation," *J. Wat. Poll. Cont. Fed. 42(11):* 1982 (1970).

11. CHEMICAL TOXICITY STUDIES WITH CULTURED MAMMALIAN CELLS

Alexander R. Malcolm and Benjamin H. Pringle.
Northeast Water Supply Research Laboratory,
Environmental Protection Agency, Narragansett,
Rhode Island, and
Harold W. Fisher. Biophysics Laboratory,
University of Rhode Island, Kingston, Rhode
Island.

ABSTRACT

Quantitative studies were conducted with an established bovine ovarian cell line to explore the usefulness of single-cell plating techniques for the assessment of acute chemical toxicity. The molar toxicities of several chemicals were determined by titrating relative plating efficiency as a function of toxicant concentration under conditions of continuous exposure. In kinetic tests, rates of cell killing were investigated by measuring relative plating efficiency as a function of exposure time to selected concentrations of ionic cadmium. Survival curves were constructed from kinetic data and the rates of cell killing defined by the slopes of the curves. A plot of cell inactivation rate against molar concentration was nonlinear and failed to indicate a discernible concentration threshold for cell killing by cadmium.

INTRODUCTION

Advancements in the development of *in vitro* techniques with mammalian cells over the past twenty years have made possible the routine application of these methods to a variety of problems in molecular biology and medicine. In particular, application of the single-cell plating techniques originally developed by Puck and co-workers[1] makes possible three important types of operations: first, it permits an accurate determination of the number of cells in a given population capable of sustained growth under defined conditions; second, it readily allows the isolation of rare mutants useful in the study of genetic processes; and third, it provides a highly quantitative approach to the measurement of cell growth and the effects thereon of various agents.[2] Accordingly, such techniques have proven extremely useful in studies of mammalian cell nutrition,[3-5] virology,[6-8] radiobiology,[9-10] and genetics.[11-14] The successful application of these techniques to a wide range of problems suggests the potential value of the same methods for the routine screening of acute and chronic chemical toxicity. Highly toxic substances could then be assigned priority for more adequate testing.

Although *in vitro* cell techniques have been utilized previously to investigate the toxic effects of selected environmental chemicals,[15-17] the present study was undertaken specifically to evaluate the sensitivity of single-cell plating techniques for the assessment of both acute chemical toxicity and the kinetics of toxic action. In relation to kinetic studies, we were particularly interested in determining if concentration thresholds could be detected for the killing of cells in culture in the same way that such thresholds can be detected for the killing of whole organisms in various assay systems.

MATERIALS AND METHODS

Cell Line

A continuously-cultivated bovine cell line designated "T4"[18] and originally isolated from normal ovarian tissue[19] was used in all experiments. When grown *in vitro* as a monolayer, the cell is epithelial in appearance and has a generation time of 12 hours,[20] where generation time is defined as the average time required for the cell population to double in the exponential growth phase. The cell line has an absolute plating efficiency (defined as the ratio of macroscopic colonies produced to initial cells inoculated) approaching 100% under ideal growing conditions. The properties of fast growth and high plating efficiency make this par-ticular cell line useful in tests utilizing colony formation.

Growth Media, Solutions, and
Test Chemicals.

Ham's nutrient mixture F-12[21] was the growth
medium for all experiments and for maintaining stock
cells. This medium was obtained in powdered form
(Grand Island Biological Co.) and supplemented with
1.2 g/l sodium bicarbonate (Baker & Adamson) for pH
control. Other supplements were 100 U/ml penicillin
(Calbiochem), 100 µg/ml streptomycin sulfate
(Calbiochem), and 5% (by volume) prefiltered fetal
calf serum (Grand Island Biological Co.). Sterilization of the medium was by filtration (Millipore
filter holder YY3014200; membrane PHPW 14250,
Millipore Corp.) and storage was in 250-ml media
bottles (Wheaton) at 4°C.

Stock 20X saline Dl solution[1] was prepared by
dissolving 160.0 g NaCl (Baker & Adamson, biological
grade) and 8.0 g KCl (Allied Chemical, biological
grade) in 500 ml triple-distilled water. A second
solution, prepared by dissolving 0.9 g $Na_2HPO_4 \cdot 7H_2O$
and 0.6 g KH_2PO_4 (Fisher, primary standard grade)
in 300 ml of triple-distilled water, was added to
the first solution and the total volume brought to
one liter. For cell washing, the 20X saline Dl
solution was diluted 1:20 with triple-distilled
water, 1.0 g/l d(+) dextrose added, and the pH
adjusted to 6.8 with 0.1N NaOH. This solution was
sterilized by autoclaving in 250-ml media bottles
and stored at 4°C.

Trypsin Dl solution[1] was prepared by dissolving
1.0 g trypsin (1:300, Nutritional Biochemical Corp.)
and 1.0 g d(+) dextrose in one liter of 1X saline
Dl solution at 37°C for one hour. Sterilization was
by the Millipore filtration procedure described
above and storage was in 125-ml media bottles at
-20°C.

Stock solutions of 0.1 molar cadmium(II),
copper(II) and zinc(II) were prepared by dissolving
the purified nitrate salts (Fisher, Certified grade)
in triple-distilled water. A 0.1 molar chromium(VI)
solution was similarly prepared from crystalline
CrO_3 (Mallinckrodt, 98.0%). Selenium stock solutions
(0.1 molar) were prepared by dissolving selenium
powder (Fisher, Certified grade) in concentrated
nitric acid, then diluting with triple-distilled
water. Stock 0.1 molar nitrilotriacetate (NTA)
solutions were made from the recrystallized sodium
salt (Monsanto, 99.8%) by dissolving in triple-
distilled water. Rotenone stock solutions (0.1

molar) were made by first dissolving the rotenone
(City Chemical Co., 95.0%) in acetone, then diluting
with absolute ethanol containing 0.01% (by volume)
Tween-80 (Atlas Chemical Industries). Working
solutions were prepared from the stock solutions by
diluting with 1X saline or complete growth medium.
Test solutions were made from working solutions by
further dilution with growth medium.

Experimental Procedures

Stock cultures of cells were maintained as
monolayers in 75 cm^2 plastic tissue culture flasks
(Falcon, #3024) containing 15 ml of growth medium.
Experimental cells were prepared from a stock
culture by dispersing the monolayer into a population of single cells with 5 ml of 0.1% trypsin Dl
solution. Cell concentration was determined with
a hemocytometer and the cell population diluted
serially with growth medium to a density convenient
for quantitative cell plating.[1]
Two types of single-cell plating experiments
were performed. Titration tests were first carried
out in which the acute toxicity of each test chemical
was evaluated by plating experimental cells (200 per
dish) into a set of 100 x 20 mm plastic petri dishes
(Falcon, #3003) containing 10 ml of growth medium
to which the test chemical had been added at different concentrations. Control cells were plated
into dishes containing nontoxic growth medium. The
dishes were placed into perforated aluminum trays
and incubated six to seven days at 37°C in a water-
jacketed CO_2 incubator (Forma, model 3282) under
atmospheric conditions of saturated humidity and
5% CO_2. Cell colonies were fixed in 10% formaldehyde,
stained with crystal violet and the macroscopic
colonies[9,22] counted.
In the titration experiments, relative plating
efficiency (defined for any experimental dish as the
ratio of colonies produced to colonies obtained in
control dishes) was measured as a function of
toxicant concentration. The absolute plating
efficiency was used to evaluate the quality of the
growth medium and the overall incubating conditions.
Experiments in which control plates yielded an
average absolute plating efficiency of less than
70% were discarded.
Kinetic studies were performed with ionic
cadmium in which the rates of cell killing were
investigated by measuring relative plating efficiency

as a function of exposure time to single concentrations of the chemical. Cells were initially dispersed, counted, diluted, and quantitatively plated into equal numbers of experimental and control dishes containing fresh nontoxic growth medium. All dishes were incubated 12 hours to permit cell attachment to the plastic surfaces. The medium was then removed and toxic medium added to all experimental plates (zero time). At the same time, nontoxic medium was added back to all control plates. Control dishes were treated in this manner to correct for any loss of cells arising from the experimental operations (*e.g.*, medium removal or cell washing). All dishes were returned to the incubator. At selected time intervals, two experimental and two control plates were removed from the incubator and the medium removed from each dish. Fresh toxic medium identical to that previously added to the experimental plates was added back to each experimental and each control dish, then immediately removed so that the total exposure received by the control cells was only a few seconds. This step was necessary to control the effect of any residual test chemical not washed out of the experimental dishes. Each dish was rinsed once with 1X saline solution, fresh growth medium added back, and incubation resumed for six to seven days. The decrease in relative plating efficiency with exposure time was used to determine the rate of cell killing by single concentrations of ionic cadmium.

RESULTS AND DISCUSSION

Titration Tests

Figure 40 shows some of the dishes from a titration experiment with ionic cadmium. The number of colonies per dish decreased rapidly at cadmium concentrations greater than 4×10^{-6} molar. In all plating efficiency tests of the titration or kinetic type, the points shown on the curves are the means of values determined in a series of 3 to 5 separate experiments carried out over a period of several months. At each dose or concentration, the standard error of the mean was within $\pm 25\%$, except in those experimental situations yielding very low survival values, where the reproducibility was not as good. Variation of this magnitude may be expected in experiments of this type.[9]

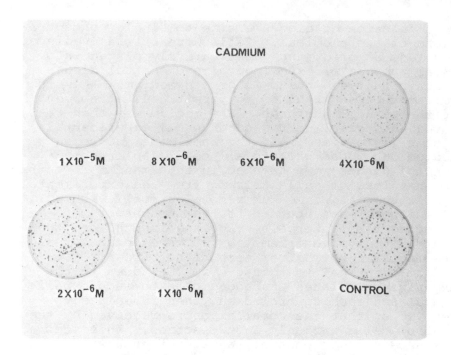

Figure 40. Colony development from single T4 cells exposed
continuously to different concentrations of ionic
cadmium.

The titration plot defines two important parameters: first, it defines the range of concentrations
toxic to the cells in an acute type of test with
respect to cell killing; second, it serves to establish the titration end point (defined as that
concentration permitting 50% relative plating
efficiency under conditions of continuous exposure).
For cadmium, this occurred at about 7×10^{-6} molar
(Figure 41). If the titration end point differed
significantly from one toxicant to another, then
this parameter might be useful for comparing acute
relative toxicities (defined by the ratio of
titration end points) under standardized conditions.
To determine what degree of difference might
be expected between the titration end points of
various chemicals, experiments were conducted with
six additional chemicals. The resulting titration
curves are shown in Figure 42 and the titration end
points are given in Table 24. The titration data
clearly indicate that substantial variation in

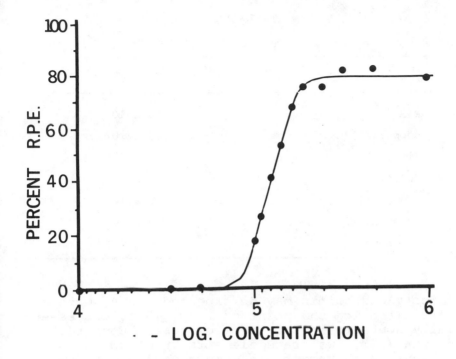

Figure 41. *Titration curve for ionic cadmium (relative plating efficiency vs negative log of the molar concentration).*

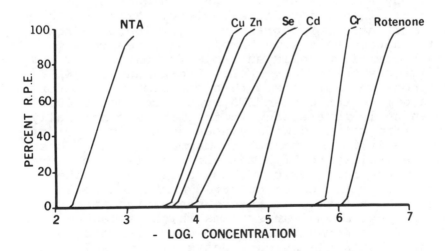

Figure 42. *Comparative titration curves for seven different chemicals.*

Table 24

Comparative Titration End Points

Chemical	Molar Titration End Point
NTA	2.3×10^{-3}
Copper(II)	8.0×10^{-5}
Zinc(II)	6.0×10^{-5}
Selenium	2.5×10^{-5}
Cadmium(II)	7.0×10^{-6}
Chromium(VI)	1.0×10^{-6}
Rotenone	3.5×10^{-7}

toxicity end points may be expected. NTA, with a molar titration end point of 2.3×10^{-3}, was essentially nontoxic in this type of test. Copper(II) and zinc(II), two trace elements known to be nutritional requirements in some mammalian systems, showed similar toxicities on the basis of their titration end points. Selenium was more toxic than copper and zinc by factors of 3.0 and 2.5, respectively. Although cadmium, as previously noted, was substantially toxic to the cells, chromium was more toxic by a factor of 7.0. Why ionic chromium was so toxic is not immediately clear. Experiments are currently in progress to determine if chromium (VI) induces this same highly toxic response in other mammalian cells suitable for use in plating efficiency tests (*e.g.* HeLa, Hep-2, and CHO-Kl). Rotenone was more toxic than cadmium by a factor of 20 and more toxic than NTA by a factor of 6,600. Rotenone is interesting in that it is a naturally occurring rotenoid widely used as an insecticide because it is "nontoxic" to mammals.[23] Yet, based on its ability to kill mammalian cells in culture, it was the most toxic substance tested.[20] The effect of rotenone on cultured mammalian cells suggests that this substance has a high intrinsic toxicity to mammalian systems which is masked by the defense mechanisms of the animal. This may be compared with NTA which appears to have a low intrinsic toxicity on the basis of the titration test. Of course, metabolic processes occurring *in vivo* which alter the intrinsic toxicity of given chemicals must be recognized. The salient point, however, is that cell assay systems may reveal a need for additional investigations not suggested by animal testing alone.

Kinetic Tests

At the present time, extensive kinetic studies have been carried out only with ionic cadmium. In Figure 43, kinetic data for five different concentrations of cadmium are presented. These curves

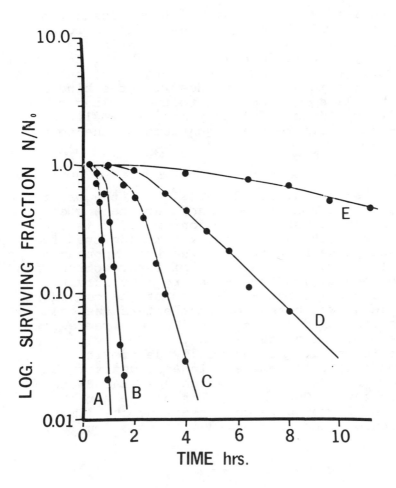

Figure 43.   *Survival curves for cell killing by ionic cadmium. The molar concentrations are: A, 2.00×10^{-4}; B, 1.00×10^{-4}; C, 5.00×10^{-5}; D, 2.50×10^{-5}; E, 1.25×10^{-5}*

were constructed from plating efficiency data by
plotting the log of the fraction of surviving cells
against exposure time. The family of curves in
Figure 43 demonstrates the following:
1. Each survival curve is characterized by an
   initial shoulder (induction time) representing
   an exposure time over which no significant cell
   inactivation occurs.
2. A straight line relationship is obtained for
   exposures longer than the induction period.
3. The slopes of the survival curves are proportional to the molar concentration of cadmium.

The form of the survival curves describing the
kinetics of cell killing by chemicals is identical
to that obtained by Chen and Selleck in a study of
chemically-induced lethality in fish.[24] Considering
each cell to be an independent organism, the same
quantitative model used to describe the kinetics
of fish killing was applied to the killing of cells
in culture. This model consists of two linear,
first-order differential equations of the form

$$dN/dt = 0 \qquad ; \quad 0 < t < t_i \qquad (1)$$

$$dN/dt = -AC^n N + BN ; \qquad t > t_i \qquad (2)$$

where N is the number of cells, C is the molar concentration of toxicant, A and B are rate coefficients,
n is the inactivating reaction order, t is exposure
time, and t_i is the induction time (defined for each
survival curve by extrapolating the exponential portion of the curve to a point of intersection with
the horizontal line representing 100% survival).
For a cell, the $-AC^n N$ term may be considered
the rate at which the toxicant enters the cell and
exerts its toxic effect. The BN term represents
the rate at which the cell detoxifies by pumping
the toxicant out, metabolically destroying it, or
by repairing cell damage. Indeed, detoxification
may not occur at all, in which case the BN term
becomes zero. Upon separation of the variables,
equation (2) may be integrated over the limits in
equation (3) and rearranged into equation (4).

$$\int_{N_o}^{N} dN/N = -AC^n \int_{t_i}^{t} dt + B \int_{t_i}^{t} dt \qquad (3)$$

Expansion of equation (4) yields equation (5). Since
the induction time (t_i) is a constant for each

$$\ln N/N_O = (-AC^n + B)(t - t_i) \tag{4}$$

$$\ln N/N_O = (-AC^n + B)t - (-AC^n + B)t_i \tag{5}$$

concentration, the last term in equation (5) may be set equal to a single constant (D). Equation (6)

$$\ln N/N_O = (-AC^n + B) + D \tag{6}$$

represents a straight line with a slope of $(-AC^n + B)$ and describes the exponential portions of the survival curves.

As stated above, the principal idea underlying the kinetic experiments was to determine if kinetic data could be used to detect concentration thresholds for the killing of cells in culture. Threshold plots were constructed by plotting the rate of cell killing against toxicant concentration. The two types of threshold plots postulated are illustrated in Figure 44. Curve A intersects the concentration axis at some concentration greater than zero, thus defining

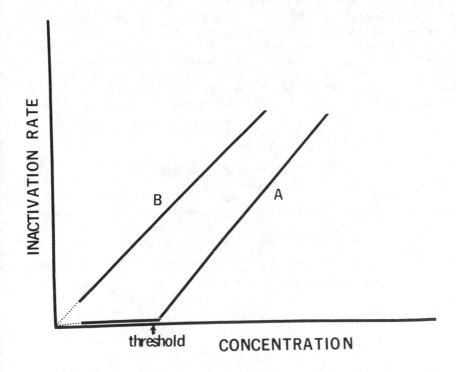

Figure 44.   *Conceptual threshold plots for cell killing by chemicals.*

227

the threshold as the largest detectable concentration yielding no detectable rate of cell inactivation. This is the type of threshold plot obtained by Chen and Selleck using the fish bioassay system referred to above. Curve B describes the situation where measurable rates of cell killing occur for all detectable concentrations of toxicant. This curve extrapolates through the origin suggesting the absence of any threshold for cell killing.

If the relationship between the rate of cell killing and toxicant concentration were linear, as in the curves of Figure 44, then one could assign a value of unity to the order (n) of the inactivating reaction. Using known values of concentration and the corresponding rates of cell killing, values for the rate coefficients A and B could be obtained from a system of simultaneous equations. The threshold concentration could then be obtained analytically by setting equation (2) equal to zero and solving for C as indicated in equation (7). The kinetic

$$C = (B / A)^{1/n} = (B / A) \qquad (7)$$

data obtained with ionic cadmium will now be considered in relation to this model.

Values for the slopes of the survival curves shown in Figure 44 were calculated from equation (4) using exposure times (t) yielding 10% survival. The pertinent kinetic data are summarized in Table 25 and the threshold plot is given in Figure 45.

Table 25

Summary of Kinetic Parameters for Survival Curves

Curve	Concentration	T_i (hrs.)	$T_{10\%}$ (hrs.)	$(-Ac^n + B)$
A	2.00×10^{-4}	0.53	0.82	-8.52
B	1.00×10^{-4}	0.83	1.28	-5.12
C	5.00×10^{-5}	1.80	3.20	-1.70
D	2.50×10^{-5}	2.10	7.40	-0.43
E	1.25×10^{-5}	3.40	28.00	-0.09

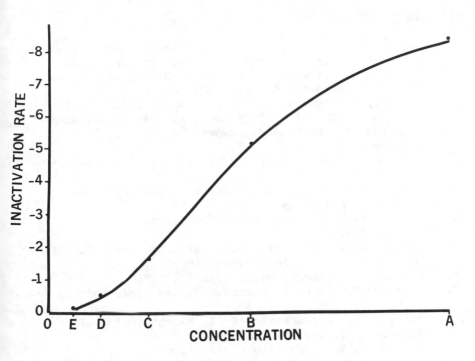

Figure 45.   Threshold plot for cell killing by ionic cadmium.

The nonlinear nature of the curve represents a marked deviation from the conceptual plots shown in Figure and prohibits any assumptions about the order of the inactivating reaction; consequently, the rate coefficients cannot be determined and the threshold concentration cannot be obtained from equation (7).

In summary, the titration and kinetic data suggest that single-cell plating techniques of the type discussed above are sufficiently sensitive to yield measurable differences in acute chemical toxicity for small variations in molar concentration of the same toxicant and to resolve toxicity differences between various chemical species.   In addition, sufficient precision and sensitivity were obtained to permit a quantitative assessment of the kinetics of toxic action.   In the case of ionic cadmium, no distinct concentration threshold for the killing of cells in culture was found.   Kinetic experiments with other substances are currently in progress to determine if any chemicals show distinct concentration thresholds for the killing of cells *in vitro*.

REFERENCES

1. Ham, R. G. and T. T. Puck. "Quantitative Colonial Growth of Isolated Mammalian Cells," In *Methods in Enzymology*, Colowick, S. P. and N. O. Kaplin, eds. (New York: Academic Press, 1962), p. 90.

2. Puck, T. T. "The Mammalian Cell as an Independent Organism," In *Cellular Biology, Nucleic Acids, and Virology* (New York: N.Y. Academy of Science, 1957), part 3, p. 291.

3. Sato, G., H. W. Fisher, and T. T. Puck. *Science 126:* 961 (1957).

4. Fisher, H. W., T. T. Puck, and G. Sato. *Proc. Nat. Acad. Sci. U.S. 44:*4 (1958).

5. Fisher, H. W., T. T. Puck, and G. Sato. *J. Exp. Med. 109:*649 (1959).

6. Marcus, P. I. and T. T. Puck. *Virology 6:*405 (1958).

7. Marcus, P. I. *Virology 9:*546 (1959).

8. Marcus, P. I. *Bact. Revs. 23:*232 (1959).

9. Puck, T. T. and P. I. Marcus. *J. Exp. Med. 103:*653 (1956).

10. Puck, T. T., D. Morkovin, P. I. Marcus, and S. J. Cieciura. *J. Exp. Med. 106:*485 (1957).

11. Chu, E. H. Y. *Genetics 62:*359 (1969).

12. Chu, E. H. Y. and H. V. Malling. *Proc. Nat. Acad. Sci. U. S. 61:*1306 (1968).

13. Puck, T. T. and F. T. Kao. *Proc. Nat. Acad. Sci. U.S. 58:*1227 (1967).

14. Kao, F. T., and T. T. Puck. *J. Cellular Physiol. 74:* 245 (1969).

15. Gabliks, J. and L. Friedman. *Proc. Soc. Exp. Biol. Med. 120:*163 (1965).

16. Litterst, C. L., E. P. Lichtenstein, and K. Kajiwara. *J. Agr. Food Chem. 17:*1199 (1969).

17. Litterst, C. L. and E. P. Lichtenstein. *Arch. Environ. Health. 22:*454 (1971).

18. Cooper, T. W. and H. W. Fisher. *J. Nat. Cancer Inst. 41:* 789 (1968).

19. Puck, T. T., C. A. Waldren, and C. Jones. *Proc. Nat. Acad. Sci. U.S. 59:*192 (1968).

20. McManus, A. T. M.S. Thesis, University of Rhode Island, Kingston, Rhode Island (1969).

21. Ham, R. G. *Proc. Nat. Acad. Sci. U.S. 53:*288 (1965).

22. Puck, T. T., P. I. Marcus, and S. J. Cieciura. *J. Exp. Med. 103:*273 (1956).

23. Hendrickson, J. B. *The Molecules of Nature.* (New York: W. A. Benjamin, Inc., 1965), p. 73.

24. Chen, C. W. and R. E. Selleck. *J.W.P.C.F. 41(R):*294 (1969).

12. ELECTROPHYSIOLOGICAL RESPONSES OF TROUT TO DISSOLVED OXYGEN AND CYANIDE

Thomas G. Bahr. Institute of Water Research,
Michigan State University, East Lansing, Michigan

ABSTRACT

Studies of lateral-line nerve and heart responses
in rainbow trout to selected pollutional stresses
are presented. Spontaneous and evoked neural discharges and electrocardiograms were measured *in situ*
from curarized fish exposed to hypoxic and asphyxic
conditions, and cyanide poisoning.
Asphyxiation and cyanide poisoning reduce the
heart rate, change wave forms of the electrocardiogram, and depress both spontaneous and evoked
activity from the lateral-line nerve. Evoked
responses persist longer under stress than spontaneous activity. Normal function of the lateral
line appears to be very sensitive to blood circulation and it is believed that ischemic conditions
in the lateral line were responsible for the
depressing effects of the stresses.
The methodology developed for this study proved
to be a useful tool for analyzing neurotoxic effects
of pollution-related stresses.

INTRODUCTION

Laboratory bioassays have been designed to
measure the effects of toxic agents on survival,
growth and reproduction of many aquatic organisms.
Although many of these tests are quite sensitive in
detecting damage to sublethal stressing agents a
serious drawback to using such end points as growth
and reproduction is the large amount of time required
to obtain useful information. Many months and even

years may be required to assess the impact of only one pollutant on one species of aquatic organism. This fact assumes increasing relevance when one considers the fact that management decisions for water quality will probably be made even though adequate scientific information is not available. Thus, it is increasingly urgent to supplement long-term bioassay experiments with "short cut" methods in order to obtain interim information as quickly as possible.

This paper will focus on a technique that provides rapid information on organism responses to impaired water quality by the use of physiological measurements on fish.

If an animal is experiencing stressful environmental conditions that may ultimately reduce his chances for survival, it is likely that many physiological mechanisms will adjust to combat the insult. Recognition of these adjustments gives one an immediate indication that long-term damage may be done. One of the first physiological systems to experience a functional change is the nervous system. Abiotic environmental changes are detected by transducer elements of the sensory nervous system and information is subsequently carried along nerve fibers to various integrating centers of the body.

If the sensory endings are sensitive to the stressing agent and if one is able to monitor neural information in fibers arising from these endings, there is then a convenient biological transducer to quickly measure the presence of the agent. Moreover, since the function of any segment of an animal's nervous system is highly dependent on normal function of ther physiological systems, a neural monitor should be sensitive to a wide variety of other stressors not directly affecting the nerve being monitored.

In addition to using nerve preparations to determine some of the sublethal effects of pollution, it is believed that a more detailed understanding of neural mechanisms in fish, coupled with knowledge of effects of various toxic pollutants, are necessary preliminaries for understanding the complexity of biological changes occurring in inhabitants of water receiving toxic wastes.

Most of the information regarding effects of toxicants on the nervous system has been obtained from experiments on animals other than fish, in particular, mammals. There is a growing awareness that applying this information to fish can be wholly

inadequate because of many biochemical and physiological differences existing between the two classes
of animals.

The objective of this study was to develop
electrophysiological methodology for investigating
one aspect of the nervous system in trout--the
lateral-line system--and to demonstrate effects of
selected stresses encountered in a pollutional
situation. Reasons for selecting the lateral-line
system and rainbow trout are discussed by Bahr.[1]
This paper will present data on spontaneous and
evoked neural discharge from the lateral-line nerve
and electrical activity from the heart (ECG's) of
rainbow trout. These measurements were observed
before and after subjecting fish to (1) exposure
to reduced partial pressures of dissolved oxygen
in the water, (2) exposure to conditions causing
asphyxiation, and (3) exposure to the metabolic
inhibitor, potassium cyanide.

Experimental Animals

Fish used in this study were two-year old
rainbow trout, *Salmo gairdnerii*, weighing from 170
to 340 grams each. About 50 fish were kept in each
of two 150 gallon circular holding tanks supplied
with a continuous flow of dechlorinated tap water
at a temperature of 13°C. Fish were fed daily on
a diet of dry pellet trout food (Formula 65-W,
Glencoe Mills, Glencoe, Minn. and Strike Fish Feed,
Country Best, Agway, Inc., Syracuse, N.Y.) furnished
by the Michigan Conservation Department. From time
to time this diet was supplemented with chopped beef
liver. Fish were apparently maintained in very good
health on this diet and noticeable growth was observed during the study period.

Holding tanks were illuminated with incandescent
lamps electrically switched on and off to coincide
with the natural photoperiod.

Fish Chamber and Water Supply

For purposes of obtaining electrophysiological
recordings under controlled conditions, fish were
individually transferred from the holding tanks to
a small rectangular Plexiglas® chamber. The chamber
was slightly larger than the fish itself and was
provided with a continuous supply of fresh oxygenated
water. The chamber provided a convenient place to
conduct surgery, to confine fish after attachment

of recording electrodes, and to expose fish to toxic chemicals. The design of the chamber and associated equipment allowed the following conditions to be satisfied: (1) fish could be completely covered with water, (2) the water temperature could be held constant throughout an experiment, (3) a constant flow of water could be passed over the gills of the fish, and (4) oxygen tension in the water could be carefully controlled and monitored during an experiment.

Immobilized fish were placed on their side in the chamber and the water inlet tube was placed in the mouth and secured by a steel pin which passed through the tube and through the lower jaw of the fish. Fish were further secured by fixing the dorsal fin to the side of the chamber with the aid of a springloaded clamp. Water left the chamber through an adjustable siphon at the rear. A constant flow of water into the chamber inlet (500 ml/min) was accomplished by first passing it into a 2-liter head tank located about 2 feet above the level of the chamber. The constant head was maintained by passing an excess flow of water into the tank and allowing it to overflow through a drain port located near the top. Pure oxygen was bubbled into the head tank through an air stone. Spherical float flow-meters were placed in line with the water and oxygen inputs to the head tank and oxygen tension in the water could be quickly changed to any desired level by adjusting the oxygen and water flow rates to predetermined values. As the dissolved oxygen level of the dechlorinated water entering the head tank was less than 1 ppm, it was not necessary to use oxygen stripping techniques on the water supply to the chamber to achieve low oxygen levels.

A polarorgraphic oxygen sensor (Beckman Model 777) was placed between the head tank and the fish chamber to monitor the dissolved oxygen during the course of an experiment. Introducing toxicants to the fish chamber was accomplished by metering the chemical from a reservoir into the inflowing water supply. A Beckman Model 746 Solution Metering Pump was used for this purpose. To prevent pulses of concentrated toxicants from entering the chamber as a result of the pulsating nature of the pump, a mixing chamber was placed between the inlet of the fish chamber and the juncture of the pump inlet to the water line. The experimental set-up is shown in Figure 46. Water temperatures in all experiments were maintained at 13°C.

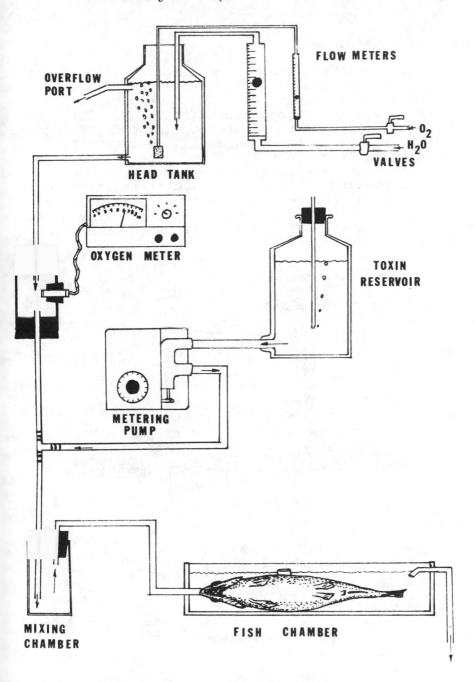

Figure 46. Diagram of the experimental fish chamber, head tank, oxygen metering equipment, and toxicant metering system. Water flow through the system is indicated by arrows.

Electrophysiology

The surgical methods and electrophysiological recording techniques used in this study are described in detail by Bahr.[1,2] Briefly, fish were anesthetized with tricaine methanesulfonate (MS-222) and then immobilized with d-tubocurarine chloride before being placed into the chamber. A piece of skin over the lateral line was removed to expose the lateral-line nerve and a flanged nerve chamber was positioned under the skin. Silver wire electrodes attached to the nerve chamber were connected to the nerve as shown in Figure 47. The electrical activity of the nerve was then amplified and displayed on a cathode ray oscilloscope.

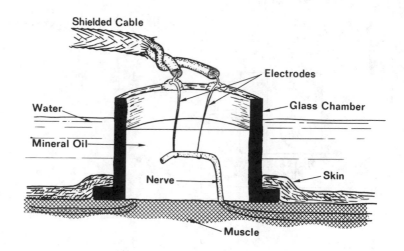

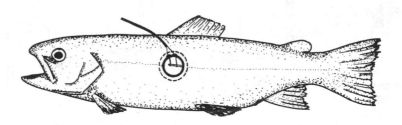

Figure 47.   *Electrode apparatus used for recording neural activity from lateral-line nerve in rainbow trout. Flanged glass cup inserted under skin excludes water from electrode area. Top diagram is an enlarged cross section of area indicated in lower figure.*

Electrocardiogram (ECG) measurements were routinely taken during most of the experiments as a check on heart function. ECG electrodes were constructed by soldering steel pins to insulated wire, then insulating the pin and wire with several coats of insulating varnish. The insulation at the sharp point of the pin was then scraped off, leaving a few millimeters of electrically conductive metal at the end of the pin. The pin was inserted at the base of the pectoral fin and pushed just under the skin until the end of the pin reached the midline one centimeter posterior to the heart. ECG activity was also displayed on a cathode ray oscilloscope.

Evoked Responses

The lateral-line system of trout functions as a mechanoreceptor.[3-6] Thus, to evaluate effects of water quality on the integrity of this system, it was necessary to mechanically stimulate the receptors to evoke a burst of neural activity. A standard stimulus used for this purpose was to direct drops of water onto the surface of the water overlying the lateral-line receptors. The water drop apparatus was located in a fixed position over the fish chamber so that drops would always hit the surface of the water at the same place and fall from a given height. So that the evoked neural response resulting from this mechanical stimulation could be easily seen on the oscilloscope it was necessary to synchronize the sweep of the oscilloscope beam with the falling drop of water. This was done by placing two silver wires immediately below the opening of the glass tube. As a water drop fell from the end of the tube it would simultaneously make contact with the end of both wires, completing a circuit consisting of a 22.5 v battery, placed in series with the external D.C. synchronizing input of the oscilloscope. When the falling water drop closed the circuit, the beam of the oscilloscope would sweep across the screen and the nerve impulse burst would always appear in the same position on the screen.

RESULTS

Control Experiments

Spontaneous Activity from Lateral Line

A continuous discharge of high-frequency nerve impulse activity was typically observed in unstimulated lateral-line nerves, neural activity apparently being the result of independent spontaneous firing of individual neurons within the main nerve bundle. In recordings from over 180 different fish less than 10 were found lacking spontaneous activity. Counting the number of nerve impulse peaks (spikes) over a given time interval from oscilloscope photographs revealed impulse frequencies exceeding 200 spikes per second. This figure represents a conservative estimate because many of the larger spikes counted as single impulses may actually have been the summated effect of two or more smaller spikes.

In fish used for lateral-line frequency measurements, the nerve was pared down by cutting through it about 3 cm distal to the recording electrodes, reducing the frequency to about 500 spikes per second. By paring the nerve in this manner, the problem of summation was greatly reduced, enabling the observer to accurately distinguish individual impulses and detect subtle frequency changes which might otherwise have been hidden. The effects of progressive paring can be seen in Figure 48.

Evoked Activity from the Lateral Line

A number of mechanical disturbances were found to elicit bursts of neural activity which could be clearly seen above the background of spontaneous activity. This evoked activity was present in intact as well as in pared nerve preparations. Bursts of evoked neural activity could be induced by gently tapping the side of the fish holding chamber, by dropping water on the surface of the water overlying the lateral line, by rippling the water within the chamber, by tapping the fish with a probe, and by numerous other methods. Examples of some of these responses can be seen in Figure 49.

The fish preparation was found to remain viable for as long as three days with no visible deterioration in either spontaneous activity or evoked neural responses.[1,2]

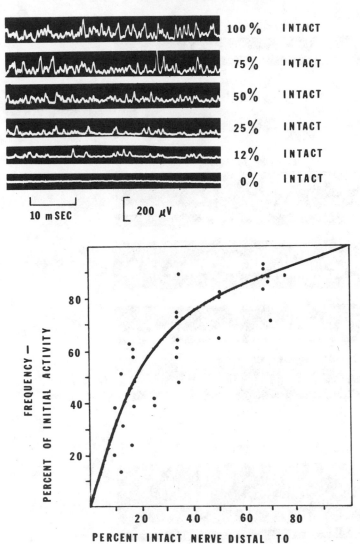

Figure 48.   *Spontaneous lateral-line activity in control fish*
demonstrating reduced frequency upon progressively
cutting the nerve. Top--Oscilloscope traces of
neural activity corresponding to percent of re-
maining original nerve length distal to recording
electrodes. Recordings from one fish. Bottom--
Plot of spontaneous activity (percent of initial
frequency) vs. percent of intact nerve length
distal to recording electrodes. Results taken
from ten different fish.

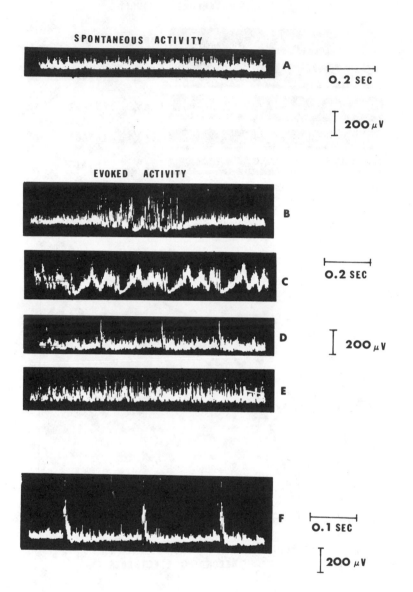

Figure 49. *Examples of various evoked responses produced by stimulating the lateral line. (A) Normal background of spontaneous activity; (B) Lateral line stimulated by scraping probe on rough edge of fish chamber; (C) Stroking finger across skin over lateral line; (D) Gently tapping head of fish with probe; (E) Directing a jet of water directly on the lateral line; (F) Tapping side of fish chamber with fingernail.*

Electrical Activity from the Heart

The top oscilloscope trace in Figure 50 represents a normal ECG from a curarized rainbow trout showing five consecutive heart beats. An enlarged

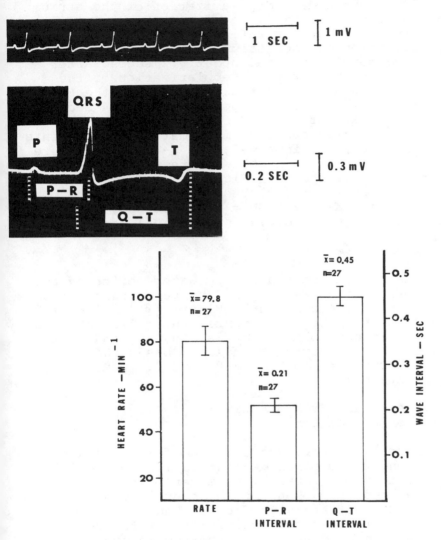

Figure 50. *Typical electrocardiogram (ECG) from rainbow trout using a unipolar electrode. Top-- Oscilloscope traces of ECG indicating positions of the P, QRS, and T waves, and the P-R and Q-T intervals. Bottom--Histogram of average rate, P-R interval, and Q-T interval. Data obtained from 27 separate fish. Vertical bar indicates one standard deviation.*

241

trace of the electrical activity from one beat is shown in the second trace and on it can be seen the three main waves of the ECG; the P-wave, QRS complex, and the T-wave. The P-wave is the first wave of the ECG and corresponds to electrical activity created by myocardial depolarization of the atrium. The QRS complex is the wave of electrical activity resulting from depolarization of the ventricle and repolarization of the atrium. The beginning of this wave complex started about 0.2 second after the start of the P-wave. Repolarization of the ventricle gives rise to the last wave of the ECG, the T-wave, which begins approximately 0.4 second after the start of the QRS complex.

To establish baseline information measurements of heart rate, P-R interval, and Q-T interval were made on 27 curerized fish. The results of the 27 experiments are shown in Figure 50. The mean heart rate was 79.8 (SD = 6.2) beats per minute, the mean P-R interval was 0.21 (SD = 0.03) second, and the mean Q-T interval was 0.45 (SD = 0.04) second.

Hypoxia Experiments

The oxygen tension in the water perfusing the gills of the fish was lowered by reducing the flow of oxygen to the head tank (Figure 46). Nine fish were used in these experiments, three fish at each of three reduced oxygen levels, 50, 40, and 30 mm Hg (pO_2). On a parts per million basis these values correspond to 3.5, 2.8, and 2.1 ppm, respectively.

Spontaneous and Evoked Activity
from the Lateral Line

Neural activity was recorded for two hours under conditions of oxygen saturation (pO_2 = 160 mm Hg) and then for three hours after oxygen tension had been reduced. None of these experiments were able to demonstrate significant changes in the frequency or other characteristics of the spon- taneous or evoked activity from the lateral-line nerve.

Electrical Activity from the Heart

A pO_2 of 30 mm Hg had the greatest effect on the heart rate. At this level the rate decreased over 20% during the first hour. Following this initial decrease, the rate stabilized at approximately the same level for the second hour and then began to gradually decrease again, approaching a rate of 60% of its pretreatment value after three hours. Results of these experiments are summarized in the plot shown in Figure 51.

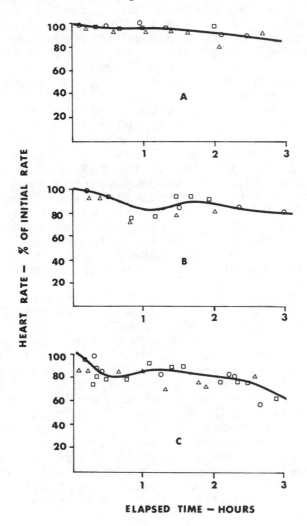

Figure 51.   *Effect of reduced oxygen tension on heart rate. Top--Oxygen partial pressure of 50 mm Hg; Center--Oxygen partial pressure of 40 mm Hg; Bottom--Oxygen partial pressure of 30 mm Hg. Three fish used in each experiment. Heart rate is expressed as a percent of the initial rate in the unstressed condition.*

ECG characteristics during the course of these experiments demonstrated some interesting changes. Within the first 20 minutes of exposure to a pO_2 of 30 mm Hg the P-R interval of the ECG increased from a normal value of about 0.2 second to nearly 0.4 second. During the same time period the wave form of the QRS complex changed from a typically large monophasic wave to a much smaller biphasic wave. The T-wave also changed during this period, becoming progressively larger both in amplitude and duration. The Q-T interval was very difficult to measure because the end of the T-wave was not well defined.

The delay of ventricular depolarization and repolarization created by the prolonged P-R interval caused the T-wave to nearly coincide with the P-wave of the next heart beat, although this event did not appear to change the basic heart rate. For the duration of the first hour, the P-R interval remained unchanged, the QRS complex continued to demonstrate a lower-than-normal voltage, and the T-wave continued to slightly increase in size. Occasionally the QRS complex would regain its monophasic form, but only for a few minutes at a time and at a subnormal voltage. Some of these changes can be seen in the recordings shown by Figure 52.

The decrease in heart rate during the first hour did not appear to be caused by decreased pacemaker rate because consecutive P-waves maintained normal intervals on the ECG. However, an entire beat would occasionally be missed, the ECG showing no activity from the end of a T-wave to the beginning of a later P-wave. This silent period would continue for the duration of one cardiac cycle and then the P-wave would reappear exactly one beat later. Less frequently, the silent period would continue for two beats. The apparent decrease in heart rate was primarily due to missed beats, not to a reduced pacemaker rhythm. This was largely true for the first hour. During the second hour, however, the pacemaker rhythm did decrease. This was evidenced by an increase in the interval between consecutive P-waves. The heart also became more regular during the second hour, missing very few beats. The regularity of the heart during the second hour accounted for the apparent plateau in Figure 51C. The decrease in rate during the third hour resulted from a combination of reduced pacemaker rhythm and reappearance of missed beats.

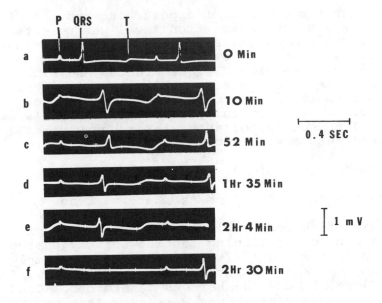

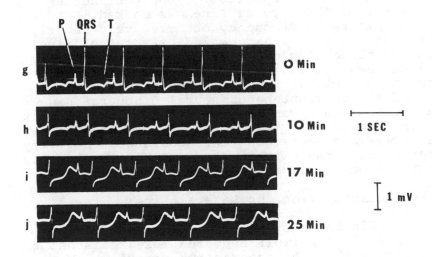

Figure 52.   *Examples of ECG changes during hypoxia*
                (pO₂ = 30 mm Hg). Top--(a), ECG at 160 mm
                Hg pO₂; (b-f), ECG at times after the pO₂
                in the water had been reduced. Bottom--
                Same experimental conditions upon a different
                fish. (A), ECG at 160 mm Hg pO₂; (h-j), ECG
                at times after pO₂ was reduced.

After three hours of exposure to reduced dissolved oxygen, the saturation level was re-established and the ECG showed signs of recovery within minutes. The biphasic QRS complex returned to its normal monophasic shape within eight minutes and within 20 minutes the ECG appeared normal except for a slower rate. In most cases, it required approximately one hour for the rate to fully recover. The ECG's were carefully observed for about three hours after complete recovery and no ill effects were noted.

Asphyxiation Experiments

Spontaneous Activity from the Lateral Line

Nine fish were subjected to conditions of asphyxia by discontinuing the flow of oxygenated water over their gills. This was done by simply clamping the water input tube to the chamber. The water level in the chamber was maintained. Conditions of these experiments were such that oxygen tensions in the vicinity of the gill epithelium quickly approached zero.

The effect of asphyxiation on spontaneous activity from the lateral line was marked. Reduction of discharge frequency occurred very shortly after the water flow had ceased; six out of nine fish exhibited complete loss of spontaneous activity in less than ten minutes. The other two fish lost spontaneous activity in about 20 minutes.

After the spontaneous neural discharge had ceased, the clamp was removed from the input tubes and oxygenated water was again allowed to perfuse the gills. In every case the spontaneous activity returned to normal within 40 minutes, the majority attaining normal levels in 20 minutes.

Evoked Activity from the Lateral Line

The discharge of neural activity resulting from water drop stimuli appeared more resistant to this stress than the spontaneous activity. When spontaneous activity had disappeared, about 50% of the evoked response still remained. With the spontaneous activity gone, the evoked response was clearly seen rising above the silent base line. Figure 53 shows six oscilloscope traces of the typical response of spontaneous and evoked activity to conditions of asphyxia. Neural activity to the

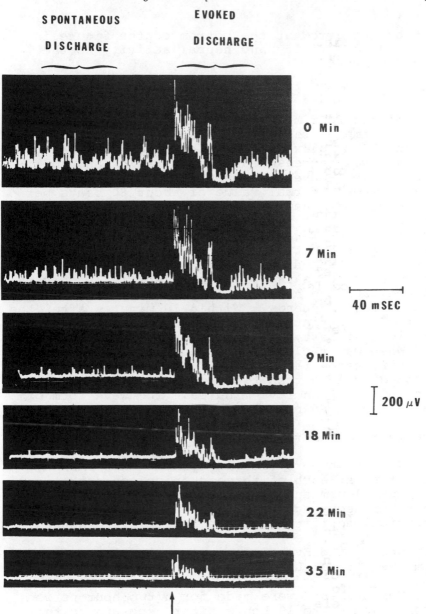

SPONTANEOUS

DISCHARGE

EVOKED

DISCHARGE

O Min

7 Min

├─────┤
40 mSEC

9 Min

18 Min

⊺
⊥ **200 μV**

22 Min

35 Min

Figure 53.   *Effect of asphyxiation on spontaneous and evoked neural discharge from lateral-line nerve. Numbers to the right represent the elapsed time in minutes after water flow over the gills was discontinued. Arrow at the bottom marks the time when the water drop stimulus hit the water over the lateral line. Spontaneous activity to the left of the arrow and evoked activity to the right.*

left of the arrow at the bottom of the figure is
spontaneous activity and neural activity to the
right is evoked.

Electrical Activity from the Heart

Within four minutes after the onset of asphyxiation (clamping the water inlet tube) the amplitude
of the QRS complex decreased by more than 70%. The
P-R interval increased and the duration and amplitude of the T-wave increased. The heart rate decreased from a normal value of about 80 beats per
minute to less than 40 per minute. After seven
minutes the heart rate fell to about 20 beats per
minute and the T-wave of the ECG was almost gone.
Electrical activity from the heart at this time was
very erratic and difficult to interpret. In the
following minutes the waves of electrical activity
weakened and the heart would beat only once every
five to seven seconds. Within 10 minutes after the
onset of asphyxiation it was judged that heart
function, at best, was very poor.

When water flow across the gills was resumed
the heart recovered rapidly. Regular rhythm was
restored in less than two minutes and the normal
rate in less than five minutes. The wave form of
the ECG appeared normal within two minutes.

Cyanide Experiments

Nine fish were used in cyanide experiments,
three fish at each of three cyanide concentrations
(0.3 ppm, 1 ppm and 4 ppm KCN). In each experiment
the spontaneous activity, evoked activity, and
electrical activity from the heart were measured.

Spontaneous and Evoked Activity
from the Lateral Line

Measurements were made for a two-hour period
before, and up to 10 hours after exposure to the
cyanide. Results of these experiments are shown
in Figure 54. Spontaneous activity was plotted
against the amount of cyanide having passed the
gills per gram of fish rather than time because of
size differences encountered between fish. Small
fish (less than 200 g) reacted much quicker to the
poison than the larger fish (over 300 g) because a
constant water flow (500 ml/min) was being forced
over the gills irrespective of the size of the

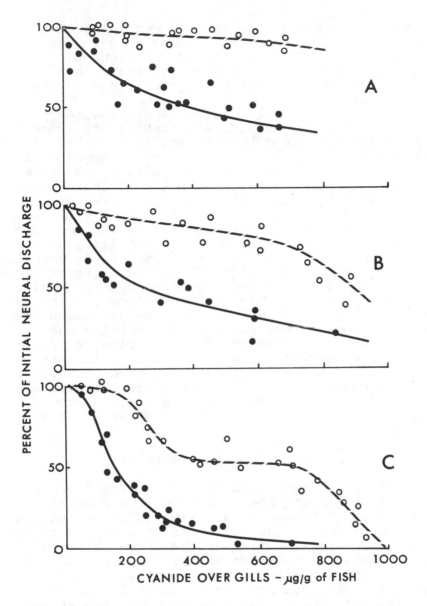

Figure 54. *Effects of 0.3 (A), 1 (B), and 4 ppm KCN (C) on spontaneous and evoked neural discharge from the lateral-line nerve. Abscissa represents the amount of KCN passed over the gills per gram of fish. Ordinate represents the lateral-line neural discharge expressed as a percent of the initial (pretreatment) activity. Top curves (dashed) are the evoked discharges and the bottom curves (solid) are the spontaneous discharges.*

fish. A small fish was thus forced to receive a larger amount of cyanide per unit weight than a larger fish. From a toxicological standpoint it was more meaningful to take fish weight into consideration after studying the conditions of these experiments.

As can be seen in Figure 54 cyanide had profound effects on both the spontaneous and evoked neural activity from the lateral line nerve. In every case severe reductions in nerve impulse frequency occurred. Comparing the effects of different cyanide concentrations reveals that higher concentrations of the toxicant have more marked effects on the functions of the lateral-line system than do the lower concentrations, even though the same quantity of cyanide had passed the gills. One might expect that in these experiments the magnitude of the physiological effect would be a direct function of the accumulative amount of cyanide having passed the gills rather than the concentration in the water. This phenomenon will be discussed later.

Recovery of the spontaneous activity and evoked response occurred after the flow of cyanide to the preparation was discontinued. All three fish exposed to the 4 ppm level recovered, both the spontaneous and evoked neural activity returning to normal levels after about one hour. Recovery experiments were not done at the other two cyanide levels.

Electrical Activity from the Heart

Figure 55 shows the effect of cyanide on the heart rate. Again the percent of the initial rate was plotted against the accumulated amount of KCN which passed over the gills per unit weight of the fish. Nearly every fish in the three series of experiments demonstrated an initial rate increase of 5 to 10%. Following this initial increase the heart rate dropped sharply to about 20% of the initial rate in the experiments conducted at 4 ppm KCN, and to approximately 40% in experiments at 1 ppm and 0.3 ppm. The low point of this rapid rate decrease occurred at nearly the same point on the abscissa in each series of experiments; after about 100 μg of cyanide per gram of fish had passed the gills. After this period of reduced activity the heart rate increased achieving a new level before again decreasing. As was observed with the nerve recordings the magnitude of the

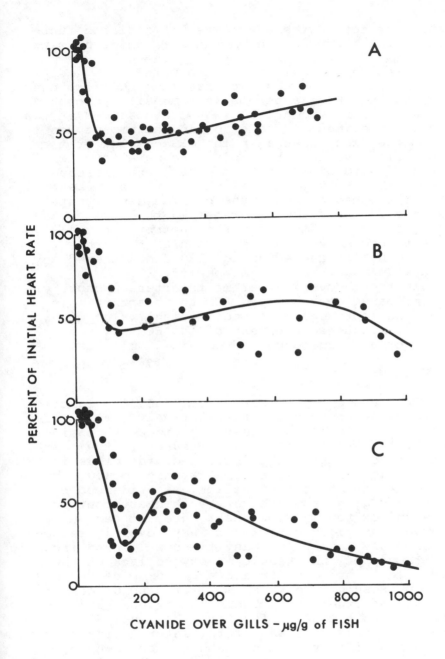

Figure 55.   *Effects of 0.3 (A), 1 (B), and 4 ppm (C) of cyanide on heart rate. Abscissa represents the amount of KCN passed over the gills per gram of fish. The ordinate represents the heart rate expressed as a percent of its initial value.*

cyanide effect on the heart was not a direct function of the accumulative amount of cyanide passing
the gills, but was more closely related to the
concentration of toxicant in the water.

The wave forms observed from the ECG recordings
taken during these experiments demonstrated changes
that were nearly identical to those observed in the
hypoxia experiments; reduction in the amplitude of
the QRS complex; change in the QRS complex from a
monophasic wave form to a biphasic wave form; increased length of the P-R interval; and increased
size of the T-wave. The slight increase in heart
rate at the beginning of the cyanide experiments
resulted from a slight increase in the rate of the
pacemaker, as evidenced by the shorter intervals
between successive P-waves. None of the other
measurements conducted on the ECG demonstrated any
changes during this initial period.

In the recovery experiments after exposure of
fish preparations to 4 ppm KCN the heart rate in
each fish approached its pretreatment value within
about 40 minutes. Recovery of the ECG waves to
their normal shape occurred within 20 minutes.

DISCUSSION

The severe impairment of lateral line function
on asphyxiation compared to relatively little effect
with hypoxia is perhaps best explained by considering the effects of these two stresses on the heart.
Anoxic conditions at the surface of the gills in
asphyxiation experiments were rapidly followed by
deterioration of heart function, as evidenced by
ECG's. Judging from the very weak heart contrac-
tions and extremely slow heart rate, it can be
assumed that blood circulation was severely reduced
and probably lacking in many tissues. These ischemic
conditions not only result in oxygen lack at the
tissue level but also create a build-up of metabolic
biproducts such as potassium, lactic acid, carbon
dioxide and others.[7,8] These conditions have been
shown to inhibit nerve function.[9-11] Thus, the high
metabolic oxygen demand of the lateral-line system
in the face of oxygen lack, and the build-up of
metabolites with no means for their removal could
easily explain the depressing effects of asphyxia
on lateral-line function. This is further supported by the observation that cardiac recovery
after cessation of asphyxia was quickly followed
by recovery of lateral-line function.

During hypoxia the heart continued to function throughout the test period, hence a build-up of metabolic products in the vicinity of the lateral-line nerve was probably prevented. This would explain the inability of the hypoxic stress to depress neural activity in these experiments.

The effects of cyanide on the lateral line are probably several. There are the well recognized inhibitory effects on the electron transport system such that oxygen cannot be biochemically utilized by the animal. Cyanide is also suspected of directly affecting nerve cell membranes[12-14] altering their ability to conduct impulses. These direct effects of cyanide on nerve tissue are also accompanied by secondary effects created by similar effects on the heart. As just discussed, cardiovascular integrity can have a profound influence on nerve function. Heart function was indeed impaired during cyanide poisoning and the rapid reduction in heart rate during early exposure periods was probably due to anoxia of myocardial tissue. The heart has a very high metabolic oxygen demand and cyanide could easily block its ability to use it. Because the myocardium cannot tolerate an oxygen debt it is not surprising that it reacted so quickly to cyanide. Thus, a combination of direct and indirect factors contributed to the overall effects of cyanide on the lateral line.

Reductions in both spontaneous and evoked activity from the lateral line were found to be greater at higher cyanide concentrations, even though the same amount of cyanide had passed over the gills per unit weight of fish. This might be explained by an excretion or detoxifying process if the following assumptions are made: (1) the detoxification or excretion process proceeds at a relatively fixed rate below that of the cyanide uptake rate, and (2) the rate of uptake of cyanide across the gills is a relatively linear function of the cyanide concentration in the water. If these assumptions are true, a greater fraction of the cyanide crossing the gills would be detoxified or excreted at lower than higher cyanide concentrations. The net effect of such a process is that higher concentrations of cyanide would "load" the fish with the poison faster than at lower concentrations. This could explain why higher concentrations of the chemical have a greater physiological effect than lower concentrations even though the accumulated dose per gram of fish was the same.

The fact that lateral line and heart function recovered after cyanide exposure supports the premise that there is some type of effective detoxification or excretion process occurring at a significant rate, and that physiological damage was reversible. Quantification of such a rate and relating it to the magnitude of physiological effect at different flux rates of cyanide across the gills offers an interesting and perhaps significant follow-up to this study.

CONCLUSIONS

The technique described here and results of preliminary experiments were very useful in assessing some biological effects of cyanide poisoning and hypoxic conditions. The lateral-line nerve provided a stable baseline of spontaneous and evoked neural activity which was markedly altered by changing the chemical environment of the fish. Of major significance is the fact that experimental animals can be fully prepared for this type of experimentation within minutes and a measure of physiological impact of a stressful condition can be determined in a matter of hours.

Analysis of lateral-line responses combined with electrocardiogram data support the earlier assumption that the functioning of the nervous system is sensitive to the integrity of other physiological systems in the animal and that stressing agents need not be specific neurotoxins for their impact to be recognized in a nerve monitoring setup. Although the results presented here are by no means definitive in describing all major physiological responses to the test conditions, they serve well to illustrate the type of data that can be obtained using this approach. With very little modification this technique can be applied to other fish and other nerves and a wide variety of chemical compounds can be tested.

The applicability of this technique and its results to the development of water quality criteria is more qualitative than quantitative. It promises insight into questions concerning the action of toxic materials and may offer a valuable technique to rapidly assess the bioactive components in an array of chemical species. It might be argued that the utility of this technique is restricted to the identification of short-term effects that have no relevance to the long-term well being of the animal. Only further work with this preparation can provide a satisfactory answer to this.

ACKNOWLEDGMENTS

    I wish to thank Drs. R. C. Ball, R. A. Pax, and P. O. Fromm of Michigan State University for their excellent help on the work described here, and to the Federal Water Pollution Control Administration for support of the research project.

REFERENCES

1. Bahr, T. G. Ph. D. Thesis, Michigan State University (1968).
2. Bahr, T. G. *Prog. Fish-Cult. 34(1)*:59 (1972).
3. Dijkgraaf, S. *Biol. Rev. 38*:51 (1962).
4. Hoagland, H. *J. Gen. Physiol. 16*:695 (1933).
5. Hoagland, H. *J. Gen. Physiol. 17*:77 (1934).
6. Hoagland, H. *J. Gen. Physiol. 17*:195 (1934).
7. Hashimura, S., and E. B. Wright. *J. Neurophysiol. 21*: 24 (1958).
8. Holeton, G. F., and D. J. Randall. *J. Exptl. Biol. 46*: 317 (1967).
9. Adrian, R. H. *J. Physiol. London 133*:361 (1956).
10. Fox, J. L., and P. I. Kenmore. *Exptl. Neurol. 17*:403 (1967).
11. Heinbecker, P. *Am. J. Physiol. 89*:58 (1929).
12. Schoepfle, G. M. *Am. J. Physiol. 204*: 77 (1963).
13. Schoepfle, G. M., and F. E. Bloom. *Am. J. Physiol. 197*: 113 (1959).
14. Schoepfle, G. M., and E. A. Eikman. *Am. J. Physiol. 212*: 1273 (1967).

SECTION IV

CONDITIONS AND SUPPORTING TECHNIQUES

FOR BIOASSAY

CONTRIBUTORS TO SECTION IV

Herbert E. Allen. Department of Environmental and Industrial Health, School of Public Health, The University of Michigan, Ann Arbor, Michigan.

John P. Barlow. Ecology and Systematics, Division of Biological Science, Cornell University, Ithaca, New York.

Jeffrey A. Black. Department of Environmental and Industrial Health, School of Public Health, The University of Michigan, Ann Arbor, Michigan.

Dennis T. Burton. Department of Limnology, Academy of Natural Sciences of Philadelphia, Philadelphia, Pennsylvania.

Frank deNoyelles, Jr. Ecology and Systematics, Division of Biological Science, Cornell University, Ithaca, New York.

Carlos Fetterolf. Bureau of Water Management, Michigan Department of Natural Resources, Lansing, Michigan.

Walter A. Glooschenko. Fisheries Research Board Detachment, Canada Centre for Inland Waters, Burlington, Ontario, Canada.

David M. Johnson. Department of Environmental and Industrial Health, School of Public Health, The University of Michigan, Ann Arbor, Michigan.

J. O. Johnson. United States Geological Survey, Denver Federal Center, Denver, Colorado.

Khalil H. Mancy. Department of Environmental and Industrial Health, School of Public Health, The University of Michigan, Ann Arbor, Michigan.

David D. Minicucci. Department of Environmental and Industrial Health, School of Public Health, The University of Michigan, Ann Arbor, Michigan.

James E. Moore. Fisheries Research Board Detachment, Canada Centre for Inland Waters, Burlington, Ontario, Canada.

Mike Newton. Bureau of Water Management, Michigan Department of Natural Resources, Lansing, Michigan.

Bruce J. Peterson. Ecology and Systematics, Division of Biological Sciences, Cornell University, Ithaca, New York.

Raymon F. Roberts. Department of Environmental and Industrial Health, School of Public Health, The University of Michigan, Ann Arbor, Michigan.

William R. Schaffner. Ecology and Systematics, Division of Biological Sciences, Cornell University, Ithaca, New York.

Arthur Scheier. Department of Limnology, Academy of Natural Sciences of Philadelphia, Philadelphia, Pennsylvania.

L. L. Thatcher. United States Geological Survey, Denver Federal Center, Denver, Colorado.

Mark Wuerthele. Bureau of Water Management, Michigan Department of Natural Resources, Lansing, Michigan.

John Zillich. Bureau of Water Management, Michigan Department of Natural Resources, Lansing, Michigan

13. THE SIGNIFICANCE OF PHYSICOCHEMICAL VARIABLES IN AQUATIC BIOASSAYS OF HEAVY METALS

Jeffrey A. Black, Raymon F. Roberts, David M. Johnson, David D. Minicucci, Khalil H. Mancy, and Herbert E. Allen. Department of Environmental and Industrial Health, School of Public Health, The University of Michigan, Ann Arbor, Michigan

The establishment of water quality criteria and standards for heavy metals is almost universally based on the total amounts of metals to be permitted in the system. Perhaps the only exception to this at present is the standard for chromium, where the hexavalent oxidation state is specified. Recently we have seen, in the case of mercury, that the form of the metal is extremely important; the toxicity of methyl mercury is much greater than that for the free metal ion. With other metals it is probably equally true that we should be concerned with the concentrations of the specific forms of the metals rather than solely the total concentration. It should be kept in mind that not only may metals be introduced in different forms to the aquatic system, but that by chemical, physical, and/or biological processes they may also transform from one species to another after their introduction.

In laboratory bioassays for metal toxicity, investigators have rarely been concerned with controlling or identifying the species of toxicant present. Rather, the metal is added to water from a stream or lake or it is added to a municipal water which has been treated to remove chlorine or other objectionable substances. The investigator may analyze the water to determine the values of common water quality parameters, but usually does not relate these to the observed toxicity of the metal being studied. These facts may explain the wide range of toxicities which have been reported.

A simplified picture of the relation between water quality and toxicity is shown in Figure 56. The biological response, or toxicity, is related to the form of the aqueous metal. It has been suggested

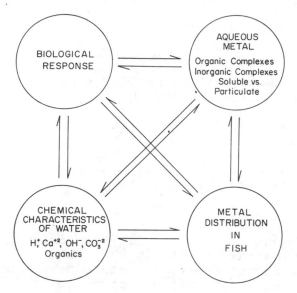

Figure 56.

Relationship of aquatic chemistry and biology for heavy metals.

that NTA or EDTA be added to water to reduce the toxicity of metals following accidental industrial discharges.[1] This suggestion is based on the fact that the NTA or EDTA complexes have greatly reduced toxicities compared to those of the inorganic metal species. Not only do the metal species present depend on the chemical characteristics of the water, but also the biological response is directly related to the aqueous chemistry. It is generally accepted that metals are less toxic in hard than in soft waters. This probably is a direct effect of the chemical characteristics of the water on the physiological status of the organism. Similarly, distribution of metal in the organism is related to the form of the metal presented to the organism--and the response of the organism is related to the distribution within the organism.

Many chemical characteristics of the water are important in defining both the chemical species of the metal present and the physiological status of the organism being tested. The concentrations of both dissolved oxygen and calcium ion affect the response of the organism. Some chemical constituents which react with and affect the metal species include hydrogen and hydroxyl ions (pH), carbonate (alkalinity and pH), ammonia ($NH_3 + NH_4^+$ and pH), and such naturally occurring organics as amino acids, proteins and humic materials.

In bioassay analyses the investigator must take into account and try to control not only variation due to biological factors, but also physical and chemical variables, some of which are listed in Table 26. While some of these variables, such as temperature and light, are taken into account (or their effects at least considered by most investigators), others are rarely considered. This paper does not review all the physical and chemical variables affecting toxicity, but rather provides some information on selected variables.

Table 26

Physicochemical Variables in Metal Toxicity Bioassays

Metal Concentration (Toxicant)	Alkalinity
Temperature	Ligands (Inorganic and Organic)
Dissolved Oxygen	Hardness (Calcium)
Flow	Particle Size
Light	Synergistic and Antagonistic Chemicals
Ionic Strength	
pH	

Most bioassays have been conducted in one of two ways: (1) in static chambers using various dilutions of industrial wastes to establish the minimum dilution of the waste needed to insure survival of fish or other aquatic organisms or (2) by adding a toxicant to a natural water, either in the laboratory or the field to determine the toxicity of the material tested. In neither approach, especially where metals are concerned, has the investigator sufficiently controlled the chemical and physical variables to permit his results to be applied to other situations. It would be virtually impossible for an investigator to control all the variables of significance in a laboratory bioassay. On the other hand, field experiments are not the complete answer. While the test indicates the toxicity of the metal under consideration in the environment, it is impossible for the investigator to use this method to understand the role of each of the important physicochemical variables in producing the observed effect.

Temperature

In almost any test the temperature must be controlled to permit meaningful interpretation of the results. Not only does the temperature control the metabolic rate of the test organism, but it also governs such physical and chemical properties of the system as equilibrium, adsorption and the kinetics of chemical reactions. Many laboratory bioassays are run without temperature control and field tests, of course, cannot be controlled.

Numerous studies have been made to determine the effects of temperature on fish, and many of them are interrelated with the effects of dissolved oxygen content on organisms, since an increase in water temperature results in a decrease of oxygen. Fry and Hart[2] showed that there is a relationship between water temperature and swimming speed of goldfish. The optimum temperature for a steady swimming speed occurred between 20° and 30°C. They postulated that a decrease in sustained swimming rate at higher temperatures was caused by a decrease in energy to perform external work. There have been many other experiments which showed that temperature can stress fish. In addition to stress, temperature influences the toxicity of certain poisons to fish.

Although it has been generalized that at a given poison concentration an increase in water temperature of 10°C halves an organism's survival time,[3] this is not always the case. Lloyd[4] used zinc as a toxicant to rainbow trout to calculate that by increasing temperature from 12° to 22°C, the median survival time was decreased by a factor of 2.35.

Rising temperatures increase an organism's CO_2 production which itself raises the lowest oxygen content which fishes can tolerate. This factor influences the response of an organism, rendering it more susceptible to a poison. In contrast to the results of the above experiments, phenols were found to be more toxic to fish at lower temperatures than at higher.[5] This may be due to a decrease in detoxification and excretion rates of the organism.

No assumptions should be made about temperature effects on toxicity. Tests should be continued whenever possible until a complete toxicity curve of survival time versus concentration can be achieved.[6]

Dissolved Oxygen

As dissolved oxygen decreases, alkyl aryl sul-
fonate detergents, cyanides, and cresols become more
toxic to fish.[7] This effect might result from impaired fish resistance to lower oxygen tensions.[8]
The apparent increase of toxicity of a poison with
decreasing dissolved oxygen was related to the increased exposure of an organism to the toxicant.
This increased exposure is due to an increase in
the water pumping rate through the gills.[9]

Many fresh water fish require a minimum of 5
mg/l oxygen to function at optimal capacity. In
bioassay tests, it is possible to reduce fish sensitivity to low dissolved oxygen if they are acclimatized first.[4] A fish shows increased oxygen demands
during certain periods in bioassay experiments.
These changes occur after feeding[10] and during
transferance of fish from one container to another.[11]

For fish eggs lying at the bottom of a stream
or lake, it is not only the concentration of dissolved
oxygen, but also the flow of water past the egg which
is of importance. The supply of oxygen to an egg is
governed by the same factors which control the response of an oxygen electrode since, in the abstract,
both deal with similar membrane transport phenomena.
The response of an oxygen electrode (which is directly
related to the rate of oxygen permeation through the
membrane) is related to the membrane permeability,
the diffusion of oxygen from the bulk of the solution
to the membrane and the activity of oxygen in the
solution. For the egg a similar situation exists;
the supply of oxygen is due to both the oxygen concentration and the velocity of the water past the
egg. It is, therefore, insufficient to relate
hatchability solely to oxygen concentration. Below
a certain flow the hatching success, at a fixed
"threshold" oxygen concentration will be related to
the flow, while above this flow, where the diffusion
path length is fixed, the hatching success will not
be improved by increased flow. Thus, an oxygen
electrode which is a physical-chemical analog of the
complex system could be used to provide information
regarding the hatchability of the eggs in the
laboratory or natural environment.

Flow

Flow or constant renewal bioassays are suggested
in *Standard Methods*[12] to maintain adequate dissolved

oxygen and to prevent changes in the media's toxicity. Flow will have effects on the respiratory rate of the organism and therefore the stress to which it is exposed, to the concentration of toxic metabolic wastes present in the system, to the species of metal present which will be affected by the organism's release of complexing agents and by the alteration of the system's pH due to respiration and metabolism, and to decreases in toxicant concentration due to adsorption onto the walls of the bioassay chamber or uptake of the toxicant by the organism.

In designing a bioassay chamber the investigator must provide a flow great enough to prevent chemical change yet not so great as to stress the organisms. Chemical change can take place, not only through reactions with the container and with metabolic products but also through changes in equilibrium due to exposure of the water with the atmosphere. The control of total carbonate and pH by the atmosphere is described later in this paper.

Like lakes, bioassay chambers can be described in terms of their flushing or residence times. For an inert chemical species this time is the ratio of the volume to the flow, though this predicts only the average residence time and dye studies should be conducted to insure that a uniform flow pattern exists throughout the chamber. The flushing of fecal wastes is extremely important as these may bind trace metals. Likewise, it is important that there be no areas of high turbulence which would stress the fish.

Ionic Strength and Hardness

Inorganic ions which appear to be necessary for a normal functioning of fresh water fish include Na^+, K^+, Mg^{2+}, Ca^{2+}, NH_4^+, Cl^-, HCO_3^-, PO_4^{3-}, and SO_4^{2-}. The ions Na^+, Cl^-, HCO_3^-, Ca^{2+}, K^+, PO_4^{3-}, Mg^{2+} are involved in osmotic balance of the cell. Sodium and potassium regulate the active transport mechanism.

The ions K^+, Mg^{2+}, Ca^{2+}, and Cl^- are necessary for activation of various enzymes of the body. Chloride ion is one of the more significant because even the heartiest of the carp cannot exist in water deplete of chloride for more than several days.[13] After a fish is "washed out" of Cl^- by placing it in distilled water, it may recover by absorbing Cl^- from water that contains chloride.

Many of the ions are taken up by feeding, although others are absorbed by the fish directly. Ca^{2+}, PO_4^{3-}, Cl^-, and SO_4^{2-} and the trace metals

Sr^{2+} and Li^+ are taken up by the fish directly through the gills.[14] Following abrupt decreases in solution ionic strength, calcium decreases the permeability of the fish's membrane to actively oppose loss of ions to the surrounding water. For this reason, calcium is one of the most important ions in aquatic systems.

There have been numerous studies investigating the toxicity of metals in waters of different hardness. It has been generally accepted that increasing hardness decreases the toxicity of metals to fish. Zinc was found to be 10 times more toxic to rainbow trout in soft water (12 mg/l $CaCO_3$) than in hard water (320 mg/l $CaCO_3$).[4]

Mount and Stephan[15] found similar results in their tests of copper toxicity to fathead minnows. They concluded that a concentration of copper between 10.6 µg/l and 18.4 µg/l in soft water (30 mg/l $CaCO_3$) would produce results equivalent to those found in hard water (300 mg/l $CaCO_3$) with 33 µg/l copper.

Pickering and Henderson[16] noted that copper had a much higher TLm in hard water than in soft water for both fathead minnows and bluegills. This indicates copper has a lower toxicity in hard water. The decreased toxicity of solutions of high hardness was attributed to complexation and precipitation of metal ions from solution as carbonates or hydroxides.

Another theory for decreased toxicity is that calcium reduces the toxicity of metals by affecting gill permeability. In support of this theory, Lloyd[4] has shown that preconditioning of fish in hard water may significantly increase an animal's resistance to toxic metals.

In contrast to these studies, Mount[17] found that complexation may increase the toxicity of metal solutions. This study showed that at any given hardness level, the solutions became more toxic as pH increased. He attributed this effect to the increased toxicity of zinc complexes formed at higher pH levels.

Tests have been run to determine what effects salinity has upon toxicity of certain compounds to fish. The results of these studies are variable. Experiments on fish show that as salinity increases to the isotonic point, zinc becomes less toxic;[18] ammonia, somewhat less toxic;[19] and phenol, more toxic.[5] The ammonium ion is three times less toxic to yearling rainbow trout in water with 30‰ salinity than in fresh water. Above 30‰ salinity, trout become less resistant to NH_4^+. Yearling rainbow trout and Atlantic salmon can withstand 15 and 13 times, respectively, more $ZnSO_4$ in 30-40°/$_{oo}$ sea water than they can in fresh water. It is significant that these species are more resistant to zinc poisoning in solutions that result in a minimal inward diffusion of water since it is postulated that zinc causes deterioration of gill epithelium.[4] In the phenol test, no explanation was given for the increased toxic effects in saline waters.

Investigations involving bioassay experiments have shown that marine animals can be acclimatized to lower salinity than the level in sea water. McLeese[20] allowed lobsters to acclimatize to reduced salinity for 4-7 days and found that the lower lethal salinity is raised by an increase in thermal acclimatization and a decrease in the level of dissolved oxygen acclimatization. The lower lethal salinity is lowered by acclimatization to reduced salinity.

Acclimatization modifies the physiological responses of fish to various environmental poisons. Results of a bioassay will not have precise meaning unless acclimatization is considered in relation to factors being studied.

Chemical Species of Metal Toxicant

Few bioassays have been conducted in a fashion which enables interpretation of the results in terms of the toxicity of specific forms of the metal. No studies have been conducted which controlled pH, alkalinity and hardness independently.

Mount[17] investigated the effect of pH and total hardness on the acute toxicity of zinc to fish. The hardness was controlled at values of 50, 100, and 200 mg/l $CaCO_3$ by mixing limestone spring water and deionized water. The pH was controlled by the addition of strong acid or base. Addition of the strong acid or base will, besides controlling the pH, change the alkalinity of the system.[21] For example, at a hardness of 100 mg/l the alkalinities were 16, 56, and 102 mg/l $CaCO_3$ at pH 6, 7, and 8, respectively.

Most investigators consider alkalinity and hardness to be related since, in nature, waters of high alkalinity tend also to be high in hardness. These two parameters are, however, basically not related since alkalinity refers to the system's acid neutralizing capacity and hardness to its capability to complex with EDTA. A solution of calcium chloride will have a high level of hardness and a low level of alkalinity while a sodium carbonate solution will have a low hardness and a high alkalinity.

Metal ions can react with both hydroxide and carbonate to form hydroxy, carbonato, or mixed complexes. Investigators must, therefore, distinguish between changes due to alterations in pH and carbonate levels to identify the toxicity of specific forms of metals.

The pH and alkalinity of a system can be independently controlled.[22] This involves control of the partial pressure of carbon dioxide in the atmosphere in equilibrium with the test water. For the control of pH and total alkalinity in synthetic aqueous systems, the concentration of sodium ion added in the form of sodium bicarbonate, carbonate or hydroxide fixes the total alkalinity while the partial pressure of carbon dioxide in the gas phase controls the pH.

The alkalinity, or acid neutralizing capacity of the system is given by equation 1:

$$\text{Total alkalinity} = 2[CO_3^{2-}] + [HCO_3^-] + [OH^-] - [H^+] \qquad (1)$$

The total carbonate concentration is given by equation 2:

$$\text{Total carbonate} = [CO_3^{2-}] + [HCO_3^-] + [H_2CO_3] \qquad (2)$$

Figures 57 and 58 show the partial pressures of carbon dioxide and the amounts of base necessary to maintain total alkalinities (eq/l) or total carbonate concentrations (moles/l) of 10^{-2} and 10^{-3}. There is a minimum pH which can be maintained for any designated total carbonate concentration and a maximum pH which can be obtained for any desired alkalinity. The maximum pH decreases by one unit as the alkalinity decreases ten-fold. In the Figures 57 and 58 the maximum pH decreases from 12 to 11 as the total alkalinity decreases from 10^{-2} to 10^{-3} eq/l. The minimum pH which can be maintained in the case of total carbonate control is not directly proportional to the total carbonate concentration. As the total carbonate decreases from 10^{-2} to 10^{-3} M the minimum pH increases from 4.0 to 4.7.

Equations for the control of pH and total alkalinity or total carbonate in synthetic or natural water systems have been developed by Roberts and Allen.[22] The control of a system through adjustment of the partial pressure of carbon dioxide in the atmosphere requires less equipment and is more reliable than the control of pH by addition of acid or base by a pH controller.

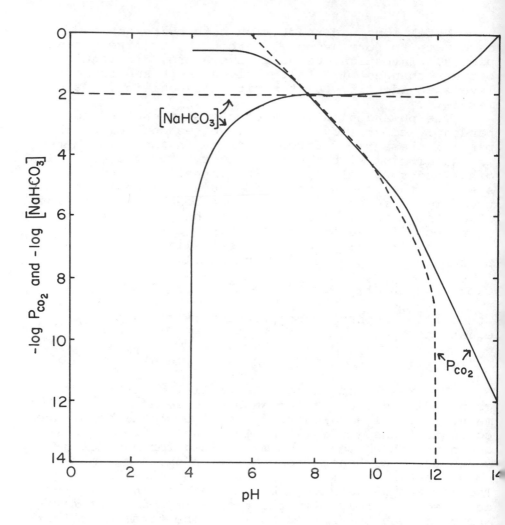

Figure 57.   *Partial pressure of carbon dioxide and moles/liter sodium bicarbonate required to maintain total carbonate = 10^{-2} molar (solid line) or total alkalinity = 10^{-2} equivalents/liter (dashed line). Total carbonate cannot be controlled below pH 4.0 and total alkalinity cannot be controlled above pH 12.0.*

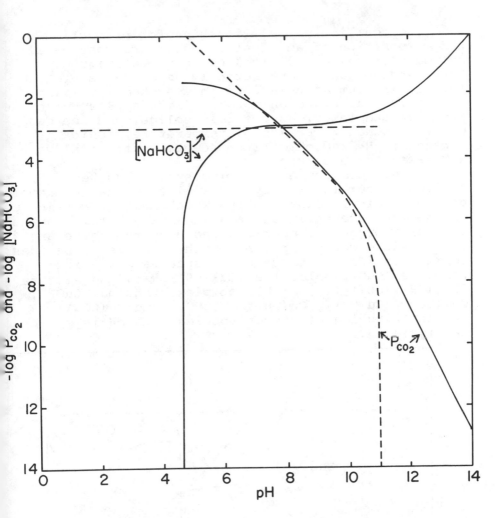

Figure 58.   Partial pressure of carbon dioxide and moles/liter
sodium bicarbonate required to maintain total
carbonate = 10^{-3} molar (solid line) or total
alkalinity = 10^{-3} equivalents/liter (dashed line).
Total carbonate cannot be controlled below pH 4.7
and total alkalinity cannot be controlled above
pH 11.0.

The equilibrium relationships for prediction of the composition of an aquatic system are most easily solved by numerical digital computer techniques. Morel and Morgan[23] have recently developed a program for equilibrium calculations in a system involving 20 metals plus hydrogen ion, 31 ligands plus hydroxide ion, 738 complexes, 83 possible solids, and one gas-phase component. Pressure, temperature, ionic strength, pH, and oxidation-reduction potential were all taken to be constant.

A less complex equilibrium system has been studied in our laboratory and will serve as an example of the chemical relationships which must be considered in the interpretation of metal toxicity data. In the aqueous inorganic copper-carbonate system there are ten possible copper species which may exist. Besides cupric ion (Cu^{2+}), three solid species may exist: tenorite (CuO), malachite [$Cu_2(OH)_2CO_3$], and azurite [$Cu_3(OH)_2(CO_3)_2$]. Soluble complexes include $CuOH^+$, $Cu_2(OH)_2^{2+}$, $Cu(OH)_3^-$, $Cu(OH)_4^{2-}$, $CuCO_3$, and $Cu(CO_3)_2^{2-}$. The distribution of copper species with pH is shown in Figure 59.

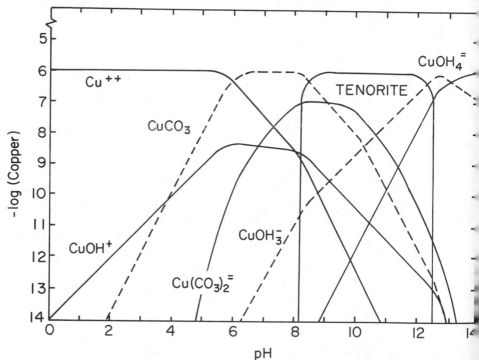

Figure 59. *Equilibrium diagram for copper. Total copper concentration equals 10^{-6} M and total carbonate concentration equals 10^{-2} M.*

In this example the dimeric copper species, $Cu_2(OH)_2^{2+}$, was neglected since under the designated conditions of 10^{-6} moles/liter total copper (63.54 µg/liter) and a total carbonate concentration of 10^{-2} moles/liter, its concentration would be negligibly small at all values of pH. Within the range of greatest interest in natural waters, pH 5 to 10, the dominant forms at equilibrium are Cu^{2+}, $CuCO_3$, and tenorite. Varying the total carbonate or alkalinity of the system will shift the predicted equilibrium distribution. Equilibrium calculations must be used with caution, however. The equilibrium constants used in the calculations are generally obtained at higher metal concentration and ionic strength than present in the natural aquatic environment. Often predicted stable solid species are not found under experimental conditions. For example, while tenorite is the predicted stable solid, we have not found it to form within one week in laboratory systems. Rather, solid copper hydroxide $Cu(OH)_2$ is present.

In the system described, the fraction of the metal present as a given species will be independent of the total metal concentration at any pH and carbonate value so long as the total metal concentration does not become so great as to permit dimeric metallic complexes and providing that no solid species are present. If metal precipitates form, increasing the total metal concentration will not change the concentration of metal in solution or the species of the metals. Rather, only the quantity of solid will increase. The pH region over which the solid is stable is dependent on the total amount of metal in the system as shown in Figure 60. The range of stability increases as the concentration increases. Since the dominant form of the metal at pH values less than that at which tenorite forms is soluble copper carbonate, it should be expected that the carbonate concentration added to the pH would influence the lower limit of tenorite stability. This is shown in Figure 60 by the curved portion of the lower stability boundary. The effect of carbonate is more pronounced as the total copper decreases. At high copper and carbonate concentrations malachite becomes a stable species, but will transform to tenorite by decreasing either the total copper or carbonate concentration.

Great care should be exercised in extending this equilibrium approach to the prediction of metallic species present in natural waters. While it is

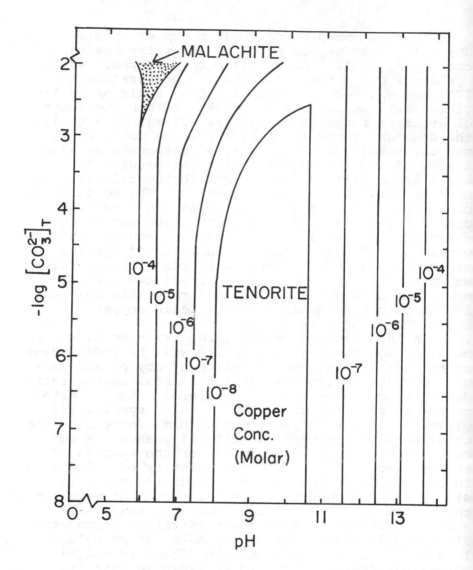

Figure 60. *Stability field diagram for tenorite and malachite as a function of pH and total carbonate concentration at various total copper concentrations.*

possible to measure the concentrations of inorganic complexing agents such as hydroxide, carbonate, bicarbonate, sulfate, chloride and ammonia and to incorporate these into equilibrium calculations, prediction of the effect of the organic constituents is not possible at the present time. A number of investigators have demonstrated that high molecular weight complexes of metals exist in natural aquatic systems.[24-27] Up to 50% of the copper in sea water is extractable into chloroform indicating that this copper is complexed or is in colloidal association with organic material.[28] Williams[29] found that 5 to 28% of the copper in sea water was present as complexes with stability constants greater than 10^{18}.

SUMMARY

   The role of physicochemical variables in aquatic bioassays has often been neglected. While most investigators have considered such variables as dissolved oxygen and temperature, other extremely important parameters such as the chemical composition of the test water and the factors involved in the design of flow bioassays have received scant attention. This paper has attempted to describe some of the physicochemical variables which can influence the observed toxicity of metals to aquatic organisms.

ACKNOWLEDGMENT

   This work was supported in part by Contract No. 14-12-591 with the Federal Water Pollution Control Administration, Department of the Interior. Jeffrey A. Black and Raymon F. Roberts were supported by Department of Health, Education and Welfare predoctoral fellowships No. 1 T01 ES00138-01.

REFERENCES

1. Sprague, J. B. "Promising Anti-Pollutant: Chelating Agent NTA Protects Fish From Copper and Zinc," *Nature* *220*:1345 (1968).
2. Fry, F. E. G., and J. S. Hart. "The Relation of Temperature to Oxygen Consumption in the Goldfish," *Biol. Bull. Wood's Hole* *94*:66 (1948).
3. Herbert, D. W. N., K. M. Downing, and J. C. Merkens. "Studies on the Survival of Fish in Poisonous Solutions," *Verh. Int. Ver. Limnol.* *12*:789 (1955).
4. Lloyd, R. "The Toxicity of Zinc Sulfate to Rainbow Trout," *Ann. Appl. Biol.* *48*:84 (1960).
5. Brown, V. M., D. H. M. Jordan, and B. A. Tiller. "The Effect of Temperature on the Acute Toxicity of Phenol to Rainbow Trout in Hard Water," *Water Res.* *1*:587 (1967).
6. Sprague, J. B. "Measurement of Pollutant Toxicity to Fish. II. Utilizing and Applying Bioassay Results," *Water Res.* *4*:3 (1970).
7. Herbert, D. W. N., G. H. J. Elkins, H. T. Mann, and J. Hemens. "Toxicity of Synthetic Detergents to Rainbow Trout," *Wat. Waste Treatm. J.* *6*:394 (1957).

8. Merkens, J. C., and K. M. Downing. "The Effect of Tension of Dissolved Oxygen on the Toxicity of Unionized Ammonia to Several Species of Fish," *Ann. Appl. Biol. 45*:521 (1957).

9. Chen, C. W. "A Kinetic Model of Fish Toxicity Threshold." Ph.D. Thesis. University of California, Berkeley (1968).

10. Davis, H. S. *Culture and Disease of Game Fishes.* (Berkeley, Calif.: University of California Press, 1953).

11. Schaeperclaus, W. *Textbook of Pond Culture.* U. S. Fish and Wildlife Serv., Fishery Leaflet *331:* 240 pp. (1933).

12. *Standard Methods for the Examination of Water and Wastewater.* APHA, AWWA, WPCF. 11th Ed. (1960).

13. Lagler, K. F., J. E. Bardach, and R. R. Miller. *Ichthyology* (New York: Wiley Press, 1962).

14. Hoar, W. S. (Ed.). *Fish Physiology.* Vol. 1: *Excretion, Ionic Regulation, and Metabolism* (New York and London: Academic Press, 1969).

15. Mount, D. I., and C. F. Stephan. "Chronic Toxicity of Copper to the Fathead Minnow in Soft Water," *J. Fish. Res. Bd. Can. 26*:2449 (1969).

16. Pickering, Q. H., and C. Henderson. "Acute Toxicity of Some Heavy Metals to Different Species of Warm Water Fishes," *Air and Wat. Poll. Int. J. 10*:453 (1966).

17. Mount, D. I. "The Effect of Total Hardness and pH on Acute Toxicity of Zinc to Fish," *Air and Wat. Poll. Int. J. 10*:49 (1966).

18. Herbert, D. W. M., and A. C. Wakeford. "The Susceptibility of Salmonid Fishes to Poisons Under Estuarine Conditions: I. Zinc Sulfate," *Air and Wat. Poll. Int. J. 8*:251 (1964).

19. Herbert, D. W. M., and D. S. Shurben. "The Susceptibility of Salmonid Fishes to Poisons Under Estuarine Conditions: II. Ammonium Chloride," *Air and Wat. Poll. Int. J. 9*:89 (1965).

20. McLeese, D. W. "Effects of Temperature, Salinity, and Oxygen on Survival of the American Lobster," *J. Fish. Res. Bd. Can. 13*:247 (1956).

21. Stumm, W., and J. J. Morgan. *Aquatic Chemistry.* (New York: Wiley-Interscience, 1970).

22. Roberts, R. F., and H. E. Allen. "The Control of pH and Total Alkalinity or Total Carbonate in Aquatic Bioassays," *Trans. Am. Fish. Soc. 101*:752 (1972).

23. Morel, F. M., and J. J. Morgan. "A Numerical Method for Computing Equilibria in Aqueous Chemical Systems," *Environ. Sci. and Technol. 6*:58 (1972).

24. Barsdate, R. J. "Transition Metal Binding by Large Molecules in High Latitude Waters," In: *Symposium on Organic Matter in Natural Waters,* Hood, D. W., ed. (University of Alaska, 1970), pp. 485-493.

25. Gjessing, E. T., and G. F. Lee. "Fractionation of Organic Matter in Natural Waters on Sephadex Columns," *Environ. Sci. and Technol. 1*:631 (1967).

26. Matson, W. R. "Trace Metals, Equilibria, and Kinetics of Trace Metal Complexes in Natural Media," Ph. D. Thesis, Mass. Inst. of Tech., Cambridge, Mass. (1968).

27. Shapiro, J. "Yellow Organic Acids of Lake Water: Differences in Their Composition and Behaviour," In: *Chemical Environment in the Aquatic Habitat*, Golterman, H. L., and R. S. Clymo, eds. (Amsterdam: N. V. Noord-Hollandsche Uitgevers Maatschappij, 1967), pp. 202-216.

28. Slowey, J. F., L. M. Jeffrey, and D. W. Hood. "Evidence for Organic Complexed Copper in Sea Water," *Nature 214:* 377 (1967).

29. Williams, P. M. "The Association of Copper with Dissolved Organic Matter in Sea Water," *Limnol. and Oceanogr. 14:*156 (1969).

14. DETERMINATION OF TRACE ELEMENTS IN WATER AND AQUATIC BIOTA BY NEUTRON ACTIVATION ANALYSIS

L. L. Thatcher and J. O. Johnson. U. S. Geological
Survey, Denver Federal Center, Denver, Colorado

ABSTRACT

Advantages offered by neutron activation for
water pollution analysis include positive identification of elements, high sensitivity, elimination
of contamination from reagents, minimization of
exchange between the water sample and container.
The latter point results from a special sampling
procedure using a pre-analyzed polyethylene bag.
Quartz ampoules are also used. Both containers
eliminate transfer of sample prior to irradiation.
Mercury is determined by a post-irradiation carrier
precipitation. Fourteen other elements are determined by a post-irradiation mixed carrier hydroxide-
sulfide precipitation. Cadmium, chromium, copper,
zinc, arsenic and antimony—important in pollution—
are included. Rare earths have been detected in
many water samples. Even with radiochemistry, it
is essential to use a high resolution lithium-
drifted germanium detector to eliminate rare earth
interference. The principal disadvantage of the
method is failure to determine lead by the outlined
procedure. A special procedure based on lead-207
with half life of 0.8 sec will be investigated.

INTRODUCTION

Fundamental research in the geochemistry of
the lithosphere and the hydrosphere relates
directly to pollution, because the geochemical
cycles of the elements ultimately determine the
assimilative capacity of the earth for non-
biodegradable substances, their deposition sites

277

and their concentration in the deposits. The literature of geochemistry provides abundant information on the natural concentrations of elements in specific geologic and hydrologic situations. Information of this type may often be required to determine the existence of pollution by artificial addition of materials and to evaluate the extent of the artificial increment above the natural baseline.

In geochemical investigation it is frequently necessary to determine the lowest concentration of trace elements in unenriched materials. Hence, the analytical method employed must have the greatest possible sensitivity. A great variety of materials from the lithosphere, hydrosphere and atmosphere are examined. Hence, the analytical method must have versatility and be relatively insensitive to matrix effects. Neutron activation analysis meets these requirements very satisfactorily, because it possesses fair to excellent sensitivity for the detection of approximately 70 percent of the elements and is applicable to a great variety of matrices in the liquid, solid and even gaseous states.

PRINCIPLE OF THE METHOD

Neutron activation analysis is based on the production of radioisotopes of certain elements by irradiation with neutrons. The radioisotopes so produced have unique properties of energy and half-life which permit their identification. Intensity of the radiation is a measure of the amount of isotope produced and hence of the amount of element. Thus, both qualitative and quantitative analysis may be performed.

The source of neutrons for activation may be a neutron generator, an isotopic source or a nuclear reactor. The reactor is by far the most powerful source and has the greatest component of thermal neutrons in the neutron energy spectrum. Thermal or low energy neutrons (average energy 0.025 electron volt) have greater reactivity toward atomic nuclei than higher energy neutrons. All work herein reported is based on irradiation with thermal neutrons in a reactor.

Thermal neutrons are captured by most atomic nuclei to produce an isotope one mass unit heavier than is commonly radioactive. The ability of a particular atomic nucleus to capture neutrons is expressed in terms of the effective cross sectional

area presented to the neutron beam by one atomic
nucleus. These neutron capture cross-sections are
reported quantitatively in barns (barn = 10^{-24} cm^2).
Barn values for the elements and some isotopes are
found in tables of nuclear data. Neutron capture
cross-sections for different elements cover an
enormous range from fractions of a barn to thousands
of barns.

Irradiation of a sample is followed by analysis
of the radiation spectrum of the isotopes produced.
This is carried out in a detector-multichannel
analyzer array which sorts the energies of the
photon radiation from the sample into several
hundred or thousand channels for storage in a
memory bank. The latter may be read out through
an x-y plotter, a teletype unit, high speed printer,
or may be fed directly to a computer.

A typical photon energy spectrum consists of
a continuum due to scattered radiation plus background (increases at lower energies) and a series
of sharp peaks which represent the gamma radiation
and x-radiation (collectively termed photon radiation) from the sample. The isotopes responsible
for the observed energy peaks are identified
through comprehensive tables, such as that of
Adams and Dam,[1] which report energy values of all
useful photon radiation peaks for isotopes produced
by neutron irradiation. Quantitative analysis is
performed by comparing the area under a characteristic energy peak for a particular isotope against
a standard irradiated under the same conditions as
the sample. Absolute quantitative analysis, based
on constants of the system, may also be used though
it is less reliable.

The type of radiation analysis described above
is the ideal, *i.e.*, purely instrumental with no
chemical separation of the radionuclides. The
practicality of purely instrumental analysis has
increased greatly in recent years because of developments in the high resolution lithium-drifted
germanium detectors (GeLi) and in the still newer
lithium-drifted silicon detectors (SiLi). Typical
resolution of a GeLi detector at the cobalt-60
energy, 1.33 Mev, is two to three kilo electron
volts (Kev) and for a SiLi detector the resolution
may be as fine as 0.2 Kev or lower. Unfortunately,
the SiLi detector is useful only at low energies
(below approximately 150 Kev).

Older work in neutron activation analysis was
based on the sodium iodide (NaI) detector, which

has very inferior energy resolution characteristics. Resolution is approximately 30 Kev. Hence, it was essential to use radiochemical separation procedures in the older work to break down even moderately complex mixtures of radionuclides into single nuclides or into mixtures containing a few nuclides that produced radiation with widely separated energy peaks. With the newer high-resolution detectors, radiochemical separations are required only for the analysis of unusually complex spectra or for the improvement of the ratio of peak radiation to scattered radiation (signal to noise ratio) when maximum sensitivity is desired. It is principally for this latter objective that radiochemical separations are used in the present work.

In addition to the analysis of photon radiation described above, the analysis of beta and alpha radiation may be used for isotope identification. Half-life analysis is also used to supplement energy analysis.

The sensitivity with which an element may be determined by neutron activation analysis is a function of three parameters: (a) the percent abundance of the activatable isotopes, (b) the neutron capture cross-sections of these isotopes, and (c) the intensity of the radiation produced as a result of neutron capture. Each of the three factors can vary by orders of magnitude between different elements. Hence, the detection sensitivities for different elements cover an enormous range from picograms to grams.

ADVANTAGES OF NEUTRON
ACTIVATION ANALYSIS

Neutron activation analysis holds particular interest for the hydrogeo-chemist because of the broad applicability to the analysis of the hydrosphere, *i.e.*, water, sediments, aquatic biota. A list of approximately 40 hydrogenic elements (elements transported and deposited principally by aqueous mechanisms) was compiled by Ginzburg.[2] The rate earth elements may be added to this list because aqueous transport is significant in their cycles. This results in a total of approximately 55 elements of interest to the hydrogeochemist. Approximately 30 of these elements may be determined with sensitivity to or in excess of the ppm range by neutron activation analysis.

Additional properties of neutron activation analysis of special importance to pollution investigation are:

1. The sensitivity for certain elements of particular importance in pollution such as mercury, arsenic, antimony, and selenium extends below the microgram per liter level.

2. Qualitative analysis is particularly reliable because two parameters, energy spectrum and half life, may be used for the identification of a specific element. Furthermore, the energy spectrum usually contains more than one peak which improves the certainty of identification. This property is particularly important in analytical work connected with litigation.

3. Application of the technique is not significantly affected by the physical form of the sample or by the chemical combinations of the trace elements. The total quantity of element is determined (in conventional neutron activation analysis) regardless of chemical combination. If the chemical form of the sought element is to be determined it is necessary to apply identifying separations before irradiation, because irradiation affects oxidation state, complexation and pH in solutions.

4. The ultimate sensitivity of neutron activation analysis is not limited by trace concentrations of contaminating elements in the reagents, a fundamental limitation in all analytical methods that use reagents. In activation analysis the problem is eliminated by irradiating the sample with neutrons as the first step in the analytical procedure. The elements in the sample are thus made radioactive. Elements that may be added in later chemical operations on the sample have no effect since these elements are not radioactive.

This property is especially significant because it means that the sensitivity for many elements (those that have isotopes of longer half life) may be greatly extended beyond the limits practical for routine operation by simply extending the irradiation time (for a given neutron flux) or the time given to counting of the irradiated product. This is frequently of great value with samples of special importance that warrant more than routine treatment. In conventional chemical analysis, no amount of extra effort is beneficial once the sensitivity limit imposed by impurities in the reagents has been reached.

5. The possibility of carrying out non-destructive activation analysis is valuable in hydrogeochemistry when difficult-to-obtain or irreplaceable samples are being examined. For example, in water-from-rock inclusions, the sample remains intact after neutron activation analysis to permit further analysis by other micro methods.

6. Neutron activation analysis presents the possibility of eliminating the long-standing problem of the exchange of trace elements between the water sample and container. While the usual effect is loss of trace elements from the water sample to the container by sorption, possibility of trace element impurities being leached from the container into the sample should not be overlooked. The sorption loss of trace elements is customarily minimized by the addition of acid, a practice which introduces two possible sources of contamination of the sample: impurities in the acid and increased leaching of the container.

Special sampling procedures have been developed in our laboratory which solve some of the problems associated with collecting a water sample for trace element analysis. These procedures are presently applicable only when neutron activation is used as the analytical method.

The first procedure eliminates the addition of acid and permits accurate correction for the exchange of trace elements between the water sample and container. This is accomplished by the use of a pre-analyzed container, a lightweight polyethylene bag. The trace element content of the "polybag" has been established by non-destructive neutron activation analysis before the bag is put into use as a sample container. On completion of the water sample analysis, the trace element content of the polybag is again determined. Any increase or decrease in the bag's trace element content is added to or subtracted from the trace element content of the water sample. This procedure can be precise only if the trace element content of the container is considerably lower than that of the sample. Experience shows the trace element content of the light polybag to be satisfactorily low, consisting principally of sodium, chlorine, bromine and manganese. Only the last element reaches a sufficiently high level in the polybag to occasionally cause a problem.

Unfortunately, the polybag method is not universally applicable. Mercury volatilizes under irradiation and is partially lost through the thin polyethylene film. The addition of nitric acid eliminates this problem, but also eliminates one of the attractions of the method.

The second procedure is collection of the sample in a quartz ampoule. The quartz is sealed and the water sample is irradiated in the same container. Nitric acid is added after irradiation to desorb trace elements that may have been taken up by the quartz. Impurities in the acid have no effect. Since quartz is very resistant to acid attack the possibility of leaching impurities from the container is reduced, but not eliminated.

In both of the above methods, the sample collection container is also the irradiation container. Transfer of the sample from the collection container to another container in which the analysis is to be performed (common to other analytical methods) is thus eliminated with its attendant risk of contamination.

The major disadvantage of neutron activation analysis in water pollution work is that lead is not included in the group determination of heavy metals. Determination of lead requires a separate and highly specialized analytical procedure.

While the cost of activation analysis has been somewhat higher than for other analytical methods, the difference on a cost per element basis is being reduced by technical improvements (principally in detectors) that permit the determination of a greater number of elements simultaneously. The small sample volume required for activation analysis reduces both sampling and shipping cost.

APPLICATION TO WATER CHEMISTRY

Blanchard and others[3] were the first to develop a multi-element procedure for neutron activation analysis applied to water, although the method had been used for individual elements of special interest by several others. Their analytical method predated the development of the GeLi detector which has made multi-element neutron activation analysis practical, and of necessity used the NaI detector. Consequently, they found it necessary to make relatively clean chemical separations of the irradiated water sample into six groups for energy spectrum analysis. The method was relatively laborious and could not cope

with interference by rare earths. In a later publication by a member of the same group, Leddicotte[4]
mentions the use of a small volume GeLi detector to
supplement the NaI detector.

The work reported here indicates that a high
resolution, high sensitivity GeLi detector is not
only desirable but necessary for reliable and
practical water analysis by neutron activation.
As a result of a preliminary survey of natural
waters in the United States it has become apparent
that rare earth elements are not uncommon in water.
Since these elements have high neutron capture
cross-sections and complex spectra (especially in
the low energy region) the possibility of spectral
interference in the determination of other elements
cannot be overlooked. The use of the high resolution detector is essential in water analysis to
eliminate possible rare earth interferences even
when radiochemical separations are used. The rare
earths are carried down as contaminants in many
precipitates and the detector must be capable of
quantitatively indicating their presence so that
corrections may be made if necessary. It was found,
for example, that samarium accompanies metallic
mercury carrier precipitated with stannous chloride
in the mercury determination. This would be a
serious interference using the NaI detector, but
the samarium spectrum is accurately separated from
the mercury spectrum using the GeLi detector. The
results of water analysis by neutron activation
based on the NaI detector must be critically
examined for possible unsuspected rare earth
interference.

A large volume (high efficiency) GeLi detector
is essential to avoid impractically long counting
times, particularly when isotopes emitting higher
energy photons such as cobalt-60 and iron-59 are
to be determined. Maximum sensitivity in reasonable
time is necessary to compete successfully with some
of the newer, more sensitive modifications of atomic
absorption photometry and other advanced analytical
techniques.

COLLECTION OF THE WATER SAMPLE

The customary procedure for water sampling in
preparation for trace element analysis is to collect
the sample in a chemically clean and resistant container of polyethylene, Teflon, or Pyrex, and
acidify immediately to prevent sorption loss of

trace elements. This is usually an acceptable procedure for neutron activation analysis of routine quality, even though it involves a partial sacrifice of one of the advantages of the method--avoiding introduction of reagents (with their possible impurities) into the sample prior to analysis. Here the reagent introduced, nitric acid, is one that may be easily purified. It is used in low concentration and it is the only reagent added. Hence, the possible introduction of impurities is minimized. In other analytical techniques such as atomic absorption, the introduction of reagents prior to analysis may be much greater. Nitric acid (or other acid) is always added to minimize sorption. In addition, it is usually necessary in the other methods to introduce reagents for concentration of sought elements prior to analysis. The concentration step may be extraction, precipitation, ion exchange or another technique involving significant addition of material to the sample.

The preferred methods for collection of the water sample in preparation for neutron activation analysis use pre-analyzed polybags or quartz ampoules. As stated above, it is necessary to add nitric acid to the sample if mercury is to be determined when the sample is collected in a polybag. The oxidizing acid inhibits formation of volatile metallic mercury.

The quartz ampoule technique is preferred for the highest quality of work because of the relatively low exchange of trace elements with quartz and the elimination under all conditions of the necessity to add acid before irradiation. The disadvantages of quartz are limited sample volume in the small ampoules, high cost and difficulty of handling in the field.

Use of polyethylene bags--the double polybag used for sample collection (Figure 61) comprises a lightweight inner bag (2 mil polyethylene) and a heavy outer bag (4 mil polyethylene). The trace element content of the inner bag has been determined by irradiating for one hour followed by instrumental analysis of the activated material. The inner bag is inserted into the outer bag in a clean environment using polyethylene gloves and other customary precautions to avoid contamination. After waiting approximately one week for complete decay of induced radioactivity, the bag assembly may be delivered for use. Experience indicates that trace element content of a particular batch of bags is relatively

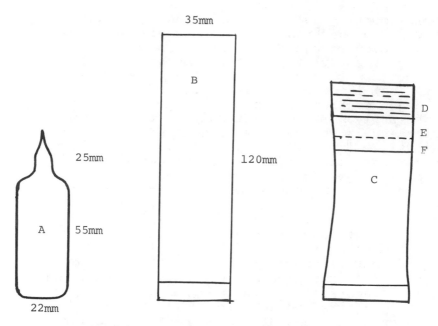

A-Quartz ampoule(1mm wall)   D-Masking tape seal
B-Polyethylene bag (empty)   E-Heat seal
C-Polyethylene bag (filled)   F-Water level (20 ml)

Figure 61.   *Containers for collection of water sample and*
                 irradiation in Triga Reactor.

consistent, and it is possible to analyze randomly-
selected bags for work of routine quality.
    The sample collector never touches the inner
bag. He rinses the inner bag with the sample and
retains approximately 20 ml of it (estimated by
means of a mark on the outer bag). Sample volume
is later determined more accurately in the
laboratory by weighing on a triple-beam balance.
The sample collector folds over the end of the
double bag three or more times and tapes it firmly
with a length of masking tape supplied with each
bag (constant length to maintain constant weight).
Two 20 ml samples are taken in polybags to give a
total of 40 ml for analysis. The water sample is
sent to the activation laboratory for weighing.
Two grams (+ 0.1 g)(the weight of the container)
are subtracted to obtain the weight of water. The
polybag package is heat-sealed, using a commercial
sealer, as a final step in preparation for

irradiation. The masking tape seal is snipped off
after heat sealing.

The quartz ampoule (Figure 61) is rinsed out
several times by immersion in the water to be
sampled. After filling, it is capped with a polyethylene finger for transportation to the nearest
field office. There it is sealed with an oxy-
hydrogen torch and is sent to the activation
analysis laboratory.

ANALYTICAL METHODS DEVELOPMENT

An analytical method for mercury has been
developed which eliminates the effect of rare earth
interference and provides sensitivity to 0.05
µg/liter. Mercury is determined with the best
sensitivity in a procedure specific for that element
using the 197 isotope. Mercury-197 has a photopeak
at 77 Kev, a more prominent gold X-ray at 69 Kev
(Figure 62) and half life of 65 hours.

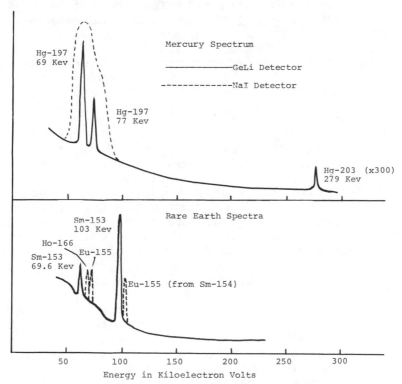

Figure 62. *Radiation spectrum of mercury and possible
interfering rare earths.*

Identification of the two energies is important for positive detection of the element. The identification may be confirmed through half-life measurement. If mercury is determined as a member of a group of elements, the sensitivity is significantly reduced by the higher background in the mercury energy region due to brehmsstrahlen and scattered radiation from the higher energy activity of the other radionuclides in the sample. Therefore, a radiochemical procedure that isolates mercury is used for work requiring the best sensitivity.

The principal interference is samarium-153, which is produced in a significant percentage of water samples. Samarium-153 has energy peaks at 69.6 and 103 Kev. The first peak interferes directly and the second peak produces a Compton smear that increases background in the mercury region (Figure 62). Addition of samarium holdback carrier was first used to minimize this interference but the addition of aluminum ion proved to be more effective. Any remaining interference can be eliminated by measuring the residual samarium at 103 Kev and applying a small correction to the measurement of mercury-197 radiation.

A possible interference in the radiochemical separations following irradiation is complexation by organic matter and inorganics, such as cyanide, that have strong affinity for mercury. It was found that the four-hour irradiation breaks up organo-mercury complexes to achieve full recovery. Cyanide is similarly destroyed, presumably through oxidation by peroxide generated in irradiation. Originally an ammonium persulfate treatment applied after irradiation eliminated organic matter, but tests showed this to be unnecessary. For example, mercury was recovered quantitatively in the analysis of solutions of the three fungicides; methyl ethyl mercuric acetate, ethyl mercuric p-toluene sulfanilimide and methyl mercuric 2,3,dihydroxypropyl mercaptide, without the addition of ammonium persulfate.

A second analytical method has been developed for the determination of a group of metallic elements including several that are very important in pollution investigations, and also including three rare earth elements. The elements determined are: copper, mercury, manganese, molybdenum, chromium, cobalt, zinc, iron, arsenic, antimony, scandium, lanthanum, europium, and samarium. The procedure involves the addition of carriers of the sought elements (except

for mercury) to the water sample after irradiation
and the addition of holdback carriers to retain
interfering radionuclides. Sodium sulfate is added
to hold back sodium-24 and sulfur-35 (in sulfate
form) and sodium bromide is added to hold back
bromine-82. The first two nuclides raise the
general background to impair sensitivity, and the
latter complicates the mid-energies of the spectrum
due to its several peaks. Mercury carrier is not
added because this would precipitate bromide. The
recovery of mercury as sulfide is complete due to
coprecipitation with the other metallic sulfides.

The procedure continues with a mixed hydroxide-
sulfide precipitation after reduction by stannous
chloride to convert metallic anions, such as chromate,
to the form that readily reacts with hydrogen sulfide.

Percent recovery in the radiochemical separation,
determined by adding known amounts of standards to
Denver city water, is shown in Table 27. Trace
element content of the city water was first determined. Counting time was extended to eliminate
the effect of counting statistics on apparent
recovery. Irradiation in quartz was used. Minimum

Table 27

*Percent Recovery in the Analysis of Water
Using Radiochemical Separation*

Element*	Percent Recovery	Element*	Percent Recovery
Antimony	98	Lanthanum	101
Arsenic	98	Manganese	86
Cadmium	95	Mercury	101
Chromium	85	Molybdenum	84
Cobalt	86	Samarium	102
Copper	102	Scandium	102
Europium	95	Zinc	92
Iron	100		

*Recovery tests were made on a solution prepared by the addition
of the following standards to one liter of pre-analyzed
Denver tap water: 500 µg of iron; 20 µg of cadmium, cobalt,
chromium, molybdenum, zinc, manganese; 5 µg of copper,
lanthanum, scandium, arsenic, antimony; 1 µg of europium and
samarium.

detection levels using normal counting times are given in Table 28. Nuclear characteristics of the radionuclides used in the analysis are given in Table 29.

Table 28

Detection Limit in the Analysis of Water Using Radiochemical Separation

Element	Detection Limit* µg/l	Element	Detection Limit* µg/l
Antimony	0.3	Lanthanum	0.1
Arsenic	0.1	Manganese	1.5
Cadmium	1.0	Mercury	0.2
Chromium[†]	5.0	Molybdunum	2.0
Cobalt[†]	2.0	Samarium	0.05
Copper	0.2	Scandium[†]	0.2
Europium	0.01	Zinc	2.0
Iron[†]	500		

*Detection limit is the concentration required to produce a signal peak with area three times greater than the mean area of noise peaks in the energy region used.

[†]Detection limit based on eight-hour irradiation of 15 ml sample at 2×10^{12} neutrons/cm^2/sec and counting four hours after seven days "cooling." Detection limit for the other elements based on the same irradiation but two-hour count after 24-hour "cooling."

The steps in the analytical procedures are outlined below. Full details for the determination of mercury are reported by Thatcher[5] and for the group trace element determination by Thatcher and Johnson.[6]

Table 29

*Radiation Characteristics of the Radionuclides
Used in the Multi-element Analysis of Water*

Element	Isotope	Half Life	Energy (Kev)
Antimony	Sb-122	67 h	564
Arsenic	As-76	26.5 h	559,657
Cadmium	Cd-115,In-115	55 h	337,492,528
Chromium	Cr-51	27.8 d	320
Cobalt	Co-60	5.27 y	1173,1332
Copper	Cu-64	12.9 h	511,1345
Europium	Eu-152m	9.3 h	122,842,963
Iron	Fe-59	45 d	1098,1291
Lanthanum	La-140	40.2 h	328,487,815,1595
Manganese	Mn-56	2.6 h	847
Mercury	Hg-197	66 h	69,77
Molybdenum	Mo-99	66 h	141
Samarium	Sm-153	47 h	69,103
Scandium	Sc-46	84 d	889,1120
Zinc	Zn-69m	13.8 h	439

EXPERIMENTAL

Reagents

   Standards - Solutions of iron, cadmium, cobalt,
chromium, molybdenum, zinc, manganese, copper,
lanthanum, and scandium are prepared from the nitrate
salts or metals dissolved in pure nitric acid.
Solutions of arsenic and antimony are prepared from
the oxides dissolved in hydrochloric acid. Solutions
of samarium and europium are prepared from the
nitrate, acetate, or other available soluble salt.
All solutions are made acid to preserve initial
concentration.

Carriers - Prepared the same as standards at tenfold higher concentration.

Holdback carriers - Aluminum sulfate, sodium sulfate, sodium bromide.

Miscellaneous - Purified nitric acid, stannous chloride, ammonia, ammonium sulphide.

Analytical Instrumentation

A block diagram of the instrumental analysis assembly is shown in Figure 63. The GeLi detector has a detection efficiency of nine percent and energy resolution of 2.0 Kev under standard test conditions. (Efficiency is measured at 1.33 Mev relative to a three-by-three inch NaI detector, 25 cm distance. Resolution is measured as width of the 1.33 Mev peak at half maximum peak height.)

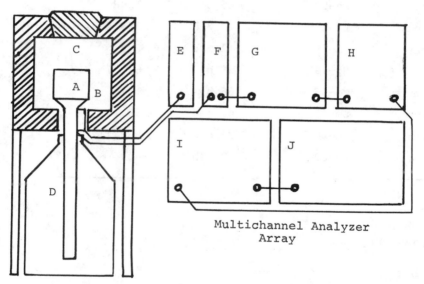

Multichannel Analyzer Array

Detector Array

A - GeLi detector
B - Preamplifier
C - Lead shield
D - Cryostat
E - Bias supply

F - Input amplifier
G - Analog to digital converter
H - Control unit
I - Memory
J - High speed plotter

Figure 63.   Instrumentation for neutron activation analysis.

Signals from the detector are amplified and shaped by an integrally mounted preamplifier and are fed into a linear amplifier for further shaping, amplification and noise rejection. The signal is fed into a multichannel analyzer having 4096 channels of energy discrimination. Readout of the 4096 channel memory is by means of a high speed x-y plotter.

Analytical Procedure for
Mercury in Water

1. Irradiate a 15-ml water sample in quartz for eight hours at flux of 2×10^{12} neutrons/cm^2/sec. If the sample was collected in polybags, add 0.5 ml of concentrated pure nitric acid to each of two 20 ml samples. Irradiate the polybag samples for four hours at flux of 2×10^{12} n/cm^2/sec. Irradiation of the samples in quartz may be prolonged beyond eight hours to improve sensitivity. Irradiation of samples in polybags should not exceed four hours at the indicated neutron flux because of possible rupture due to radiation damage and pressure. Allow to decay overnight to minimize short-lived nuclides in the sample and to reduce the silicon-31 activity of the quartz container to manageable levels. Open the quartz or polyethylene container under safe conditions. Discard the outer polybag.

2. Add five mg mercury carrier (mercuric nitrate), one ml concentrated nitric acid, aluminum sulfate and sodium sulfate holdback carriers to the sample in the irradiation container. Let stand at least one hour for complete leaching of any sorbed mercury from the container.

3. Transfer to a polyethylene test tube and adjust the pH to about 1 with sodium hydroxide. Reduce to grey-black metallic mercury with stannous chloride. Centrifuge and discard the supernate.

4. Count the precipitate in a GeLi detector of adequate resolution (as described above) coupled to a multichannel analyzer providing one Kev or better resolution per channel. Count for two hours. Examine the spectrum for traces of a samarium interference and correct if necessary.

Analytical Procedure for
Mixed Trace Elements in Water

1.   Irradiate a 15 ml sample in quartz for at
least eight hours or two 20 ml samples in polybags
for four hours at 2×10^{12} n/cm^2/sec. Add nitric
acid to the polybag sample before irradiation if
mercury is to be determined. Omit the nitric acid
if mercury is not required. Let the sample decay
overnight. Open the quartz or polyethylene container under safe conditions. Discard the outer
polybag.

2.   Add carriers of the sought elements (except for mercury) in nitric acid solution. Add
sodium sulfate and sodium bromide holdback carriers.
Add stannous chloride in hydrochloric acid to reduce
metallic anions. Let stand two hours to complete
the reaction.

3.   Transfer to a 60-ml polyethylene test tube
and precipitate the sulfides by adding excess
ammonia and dilute ammonium sulfide. Centrifuge
down the mixed hydroxide-sulfide precipitate and
discard the supernate.

4.   Count the activity of the precipitate for
at least two hours on a GeLi detector of adequate
resolution. Count the sample within 24 hours after
irradiation for the determination of copper, cadmium, mercury, arsenic, antimony, molybdenum, zinc,
samarium, lanthanum, europium, and manganese. The
sensitivity for manganese is improved by counting
as soon after irradiation as possible. Count the
polybag for sufficient time to determine whether
any significant change in trace element content has
occurred. (Usually 30 minutes is satisfactory.)
Increase of a particular trace element in the polybag indicates sorption. The amount of increase is
determined and is added to the trace element content
of the sample. Decrease of a particular trace
element indicates leaching, and the difference is
subtracted from the content of that element in the
sample.

Set the sample aside for one week to allow
shorter-lived radioisotopes to decay so that background is reduced, and then count the sample again
for at least four hours (preferably overnight).
The second count determines cobalt, chromium,
scandium, and iron under conditions of improved
sensitivity. Again, correct for trace element
content of the polybag if necessary.

Analysis of Sediments and Biota

Solid material analysis follows essentially the same procedure as that for the water sample. The solid samples are irradiated in polyethylene vials or in small quartz tubes (quartz is recommended if mercury is to be determined). Instrumental analysis of the sample is then made to identify the principal constituents and determine whether radiochemical separations are required. It is frequently possible to analyze sediments instrumentally, but radiochemical separations are usually required for the biological material because of the higher sodium and bromine content. The analysis of mercury in sediments is made by means of radiochemical separations when the best sensitivity is required.

Sediment samples are dissolved by treating with aqua regia followed by hydrofluoric acid. Mercury is not lost in the evolution of silicon tetrafluoride because of the oxidizing condition. A residue sometimes remains. The nature of this has not yet been determined. No mercury has been detected in the residues examined. Biological specimens are dissolved in aqua regia or nitric acid. The best solvent is determined by experiment before irradiation of the sample. A yellow residue containing phosphorous sometimes remains. Tests have shown that insignificant amounts of metallic trace elements are trapped in this residue.

Radiochemical separations are carried out after diluting and neutralizing the highly acid solutions. Determination of mercury is made on the solution of sedimentary or biological material by neutralizing to pH of approximately one and proceeding as for a water sample. The group hydroxide-sulfide precipitation which concentrates copper, cadmium, mercury, arsenic, antimony, molybdenum, zinc, samarium, lanthanum, europium, manganese, cobalt, chromium, scandium, and iron may be applied to the solution of biological material for improvement of sensitivity as compared to the purely instrumental analysis. This method is not generally effective for sedimentary material because of excessive concentration of the rare earths.

The principal difficulty in the radiochemical analysis of biota is a high background caused by brehmsstrahlen from long-lived phosphorous and sulfur isotopes. The radiochemical separation reduces the effect of sodium and bromine to insignificant levels but some effect from phosphorous and sulfur remains, due to significant amounts of the latter elements being carried down by the hydroxide-sulfide precipitation.

BRIEF REVIEW OF RESULTS

Distribution of trace elements between water and sediments has been investigated in streams representing different geological settings. The sediment, for purposes of this investigation, was taken as that portion that settled out of the sample after two weeks standing. Results from the Clark Fork River at Deer Lodge, Montana, are shown in Table 30.

Table 30

Trace Elements in Clark Fork River at Deer Lodge, Montana, August 12, 1971

Element	Concentration in Water-suspended Sediment Mixture, µg/l			Concentration in sediment µg/g (275 mg of sediment from one liter water)
	Dissolved Phase	*Suspended Phase*	*Dissolved Plus Suspended*	
Arsenic	0.9	1.05	1.95	3.8
Copper	29	6.2	35.2	22.5
Samarium	1.0	9.0	10.0	33.1
Mercury	< 0.2	< 0.2	< 0.2	0.35
Zinc	3.1	5.2	8.3	18.9
Bromide	1.5	< 0.02	1.5	< 0.1

It is apparent that both the sediment and water phases must be analyzed if a complete picture of the trace element distribution in the stream is to be developed.

Rare earths have been detected in many samples, occasionally at relatively high concentrations. Three samples of coal mine water from West Virginia had very high concentrations of arsenic and rare earths (Table 31).

Table 31

Determination of Arsenic and Rare Earths in Three Waters
from West Virginia Coal Mines

Mine #1	arsenic	60 µg/l
	lanthanum	40 µg/l
	samarium	9 µg/l
	scandium	100 µg/l
Mine #2	arsenic	50 µg/l
	lanthanum	51 µg/l
	samarium	13 µg/l
	scandium	66 µg/l
Mine #3	arsenic	3.3 µg/l
	lanthanum	39 µg/l
	samarium	11 µg/l
	scandium	64 µg/l

The data on rare earth elements in fresh water are very limited although there is a significant amount of data for ocean water. Oda[7] determined scandium, lanthanum, samarium, europium, and dysprosium in the hot springs at Waireki, New Zealand, using neutron activation analysis. Onuma[8] determined lanthanum, samarium, and europium in the Misasa hot springs of Japan by activation analysis. Robinson, *et al.*[9] determined yttrium, lanthanum, and traces of other rare earths in a sample of ground water collected a few miles east of Washington, D.C. Emission spectrography was used after evaporating large volumes of water. Because of the paucity of information on the rare earths in water, it is planned to actively pursue this aspect of hydrogeochemistry made practicable by neutron activation analysis.

Infiltration of arsenic into semi-confined aquifers in areas in New Jersey and Maine where lead arsenate had been used for crop-dusting over a period of years was investigated. The New Jersey sample had 0.3 µg/l of arsenic, slightly above average, and the Maine sample showed no detectable arsenic. Preliminary work indicates that the sorption capacity of soil for arsenic is very large. The effect tends to protect the ground water supply but pollutes the soil.

Because of the great sensitivity of neutron activation analysis it is possible to analyze extremely small water samples. Six samples of pore water obtained by subjecting rock cores to tremendous pressure were analyzed nondestructively. The sample volumes ranged from two to five ml. Sodium, chloride, bromide, aluminum vanadium, copper, and manganese were determined in these small volumes.

Lead determination may be possible through the use of lead-207 which has a half life of 0.8 sec. Investigation will be undertaken when a rapid transfer system becomes available to the laboratory.

Publication authorized by the Director, U.S. Geological Survey. A contribution to the US/IHD program.

REFERENCES

1. Adams, F. and R. Dam. *Jour. Radioanalytical Chem. 3:* 99 (1969).
2. Ginzburg, I. I. *Principles of Geochemical Prospecting* (translated from the Russian) (London: Pergamon Press, 1960).
3. Blanchard, R. L., G. W. Leddicotte, and D. W. Moeller. *Jour. A.W.W.A.,* 967 (Aug, 1959).
4. Ledicotte, G. W. *Modern Trends in Activation Analysis* Vol. 1 (U.S. Dept. of Commerce, 1969).
5. Thatcher, L. L. *The Determination of Mercury in Water, Sediments and Aquatic Biota by Neutron Activation Analysis,* U.S. Geological Survey Water Supply Paper (1972).
6. Thatcher, L. L. and J. O. Johnson. *The Determination of Trace Elements in Water by Neutron Activation Analysis,* U.S. Geological Survey Water Supply Paper (1972).
7. Oda, T. *Radioisotopes (Japan) 18:*39 (1969).
8. Onuma, N. *Ousen Kogakkaishi 4(3):*1933 (1967).
9. Robinson, W. O., H. Bastron, and K. J. Murata. *Geochemica Acta 14:*55 (1958).

15. CONTINUOUS FLOW NUTRIENT BIOASSAYS WITH NATURAL PHYTOPLANKTON POPULATIONS

John P. Barlow, William R. Schaffner, Frank deNoyelles, Jr.,[†] and Bruce J. Peterson.* Section of Ecology and Systematics, Division of Biological Sciences, Cornell University, Ithaca, New York

ABSTRACT

Experiments are described in which responses of natural phytoplankton populations to nutrients were studied in continuous flow cultures. Changes in metabolic, chemical, and taxonomic characteristics of these mixed populations observed under these conditions were similar to changes in the unconfined populations presumed to result from variations in nutrients. Because it is possible to investigate such changes, continuous flow systems have important advantages over usual static nutrient bioassay procedures.

INTRODUCTION

Identification of the nutrients that limit growth of phytoplankton populations is a central problem in the management of our water resources. Decisions on such matters as whether to regulate use of phosphorus-containing detergents, to build waste treatment facilities designed to remove nutrients, and so forth, almost invariably must be made with little understanding of the effect of specific nutrients on existing algal populations.

This difficulty arises not so much from a lack of understanding of the role of nutrients in the life processes of these plants as from the overwhelming difficulty in applying such understanding as we do have to mixed populations growing in nature. This is the kind of problem that seems particularly

*Present address: Department of Natural Resources, Cornell University.
[†]Present address: Zoology Department, University of Oklahoma.

appropriate for a good bioassay technique and, indeed there has been much effort in just that direction. Some techniques depend on the response of single, carefully studied organisms;[1] but applications of this approach to natural populations are limited because no single species can be expected to represent adequately the response of a natural association of some dozens of species. Much more promising, at least from an ecologist's point of view, are those techniques that use the natural plankton association in what are now usually called "nutrient enrichment experiments."[2-5]

In these experiments portions of the existing populations are confined, often in bottles in the laboratory,[2] but sometimes in much larger enclosures suspended in the open water.[5] Short period changes in rates of photosynthesis, or longer term growth responses to perturbations in specific nutrients, are measured. From these responses are drawn inferences about those nutrients limiting photosynthesis and growth of the unconfined populations.

These nutrient enrichment experiments are easy to perform and yield results that seem simple to apply directly to natural situations. This simplicity may, however, be illusory. We believe this is true because in nature phytoplankton associatesion respond to perturbations in their environment, not just by changes in rates of photosynthesis and growth but also by changes in kinds and proportions of the species that compose the populations. This process is termed species succession. It is probable that there are successional changes taking place in most nutrient enrichment experiments that last for more than a few hours, but with but few exceptions (*e.g.* 5, 6) they have neither been observed nor considered. Indeed the static conditions under which these experiments are ordinarily made allow neither for easy observation of the changes nor for rigorous analysis of their relationship to nutrients.

It is the purpose of this paper to describe an application of continuous flow culture techniques to determine the effect of nutrients on the growth and composition of natural phytoplankton associations, and to consider the implications of changes observed in these experiments for design and interpretation of nutrient bioassay experiments. Continuous flow methods are now so widely used for culture of microorganisms that there is no need here to attempt to elaborate on general principles or to describe specific applications to the study of growth kinetics of microorganisms; there are several excellent summaries.[7-9] In the most frequently used type, the chemostat, a reaction vessel containing the microorganisms is supplied with a nutrient medium at a constant slow rate, displacing via a constant level overflow a fixed proportion of the contained population. The rate of displacement is given by the ratio of flow rate to reaction vessel volume. Under these defined conditions it is possible to make certain statements about rates of cellular processes that are difficult or imprecise with the usual batch or static culture technique.[8] Moreover, the conditions of the continuous flow experiment are much more like those in nature during periods of nutrient limitation, when nearly constant levels of both plants and critical nutrients indicate that continuous processes that resupply nutrients are in equilibrium with uptake by the plants.

Continuous culture has nearly always been used with single species cultures of microorganisms, or selected combinations of a very few interacting species. Recently, however, Jannasch[10,11] has used continuous culture to study the selective effect of organic enrichment on natural populations of marine bacteria. Present studies are an extension of the Jannasch approach. Two sets of experiments will be described: the first was made in a small experimental pond and is intended to provide evidence that the responses to nutrients in continuous culture may be similar to those produced in a natural environment by similar treatment; the second was made in Cayuga Lake to demonstrate an application to a specific problem.

METHODS

Large-volume chemostats were constructed from 12-liter Pyrex bottles bored to receive an overflow at the 10-liter level (Figure 64). These were kept in a constant temperature room at environmental temperature, illuminated by 20-watt daylight fluorescent lamps either continuously or on a light-dark cycle corresponding to the seasonal day length. The number of lamps varied to approximate the mean daytime intensity in the part of the water column from which the population studied had been taken, but most experiments were run with nine lamps giving about 1,000 footcandles at the surface of the vessel. The reaction vessel was vigorously mixed with a strong stream of filtered air and a

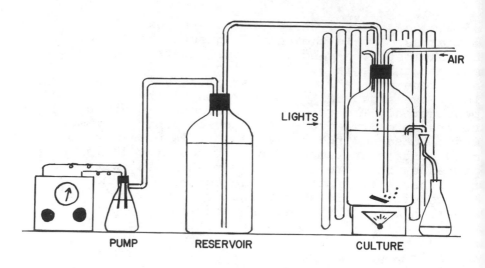

Figure 64. Diagram of chemostat.

magnetically driven stirrer. Pond or lake water
from which all algae and most bacteria had been removed by filtration through a 0.45 μ membrane filter
was supplied continuously during the illuminated
period by means of an electrolytic pump.[12]   The
displacement rate has been between 0.1 to 0.5 day^{-1}.
The water was enriched with nitrogen (NaNO$_3$ or
NH$_4$NO$_3$), phosphorus (K$_2$HPO$_4$), and/or a trace metal
mixture based on Guillard's "F" medium,[13] or used
unmodified as a control. The initial phytoplankton
association in the reaction vessel was that in pond
or lake water screened to remove most of the grazing
animals, but otherwise unmodified.

Continuous individual experiments extended for
periods of 7 to 30 days, depending on temperature,
light, and nutrient levels. Growth of visible,
attached forms was not a problem except at the
highest nutrient levels after many days, at which
time the experiment was always terminated.

Daily samples were taken from the overflow for
examination of the phytoplankton, determination of
nutrients, pigment content, and rates of ^{14}C uptake.
Nutrient assays were made on frozen samples by
methods given in Reference 14. Photosynthesis was
determined from ^{14}C uptake in 2- to 4-hour experiments
in an illuminated light tank at about 1,000 ft-c.[15]
Phytoplankton was estimated by cell counts on samples
preserved with formalin, from *in vivo* fluorescence
of chlorophyll a,[16] or by standard pigment extraction
methods.[14]

Parallel nutrient enrichment experiments have also been conducted with the Cayuga Lake samples. These were of two kinds: (1) short term experiments made by enriching 100 ml samples with nutrients NaH^{14}CO$_3$ and incubating 4 hours in an illuminated light bath, after which ^{14}C uptake was measured; (2) longer term experiments made by keeping 200 ml samples enriched with the same nutrients under lights in the constant temperature room for 3 to 6 days, and measuring changes in pigment by *in vivo* fluorescence. Enrichment was made with phosphate, silicate, and the trace metal mixture.

RESULTS

Experimental Ponds

These experiments were made with phytoplankton from a small (0.1 hectare) experimental pond, one of a series in which deNoyelles[17] and O'Brien[18] have made intensive comparative studies of the kinds and abundance of phytoplankton at various levels of fertilization. In these previous studies short period nutrient enrichment experiments failed to demonstrate nutrient limitation, although fertilization by N caused marked increases in phytoplankton biomass. The object of the present experiments has been to reproduce the changes in the phytoplankton association that were observed when the ponds were fertilized. Consequently, we used the same nutrient treatment as deNoyelles and O'Brien, *i.e.*, 40 µg at N l^{-1} (as NH$_4$NO$_3$) and 5 µg at P l^{-1} (as K$_2$HPO$_4$). The pond chosen for this study had never been fertilized, and had nearly constant levels of soluble reactive phosphorus (SRP) of about 0.5 µg at P l^{-1} and nitrate of about 1.0 µg at N l^{-1} through the winter-spring period when these experiments were made. The phytoplankton populations were small, quite stable in their composition, and entirely characteristic of what deNoyelles[17] had found previously in unfertilized ponds.

There were three experiments in all, but the one made in midwinter at low light and temperature showed no response to nutrients and will not be considered further. In the others, nutrients had marked effects that were so similar, both qualitatively and quantitatively, that they will be represented by only the last experiment run April 7 to May 10.

Major features of changes in the phytoplankton are shown in Figure 65. An initial observation to

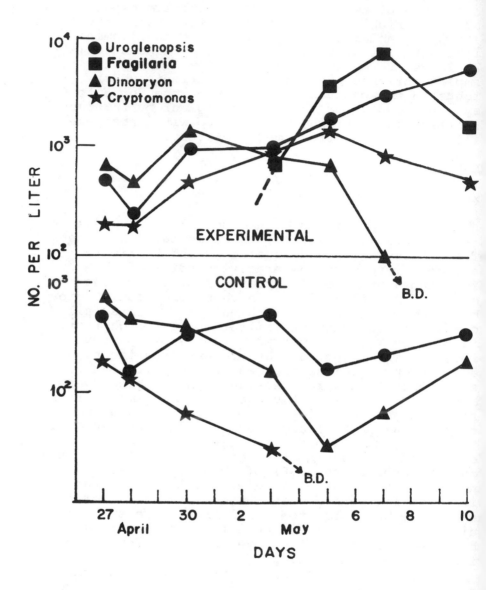

Figure 65. *Abundance of principal forms in pond chemostat experiment III. Control supplied with filtered pond water, experimental with pond water enriched with N and P. "B.D." = below detection.*

be made is that we were able to maintain the pond
association close to its original state for long
periods of time on unenriched pond water alone.
There were a few forms, initially fairly abundant,
that were evidently not dividing or were dying
slowly since their numbers declined at a rate
equivalent to or somewhat greater than the displacement rate. This is illustrated by the sharp
decline in *Cryptomonas*. By the end of the experiment
about half the original species had declined below
detection. However, in every experiment the forms
that contributed the most to the biomass of the
original population, *e.g.*, *Uroglenopsis* and *Dinobryon*,
were able to maintain themselves for relatively long
periods, in one experiment up to 30 days. Thus,
after a 2- to 3-day period of minor adjustment,
there was established a small but stable population.

In chemostats supplied with enriched water,
there was a rapid sequence of changes in both kinds
and abundance. Changes in the four initially most
abundant species is shown in Figure 65. Only a few
forms such as *Cryptomonas* did not exhibit changed
rates of growth. Some, such as *Uroglenopsis*, increased rather steadily. Others showed more abrupt
changes in abundance, *e.g.*, *Dinobryon* which initially
was most abundant and declined below detection between the 6th and 8th day, and the diatom *Fragilaria
crotonensis* which initially was undetectable and
which became the numerically dominant form as
Dinobryon disappeared. A rather similar succession
of species was seen in the first experiment, made
in early fall; then, however, *Uroglenopsis* was the
initially dominant form which gave way to the same
diatom, *F. crotonensis*.

The phytoplankton association in the enriched
chemostat had at the end of the experiment as many
species as it had originally, although not all the
original ones remained in detectable numbers. Some
of the increase in total abundance was due to growth
of some forms that had been present from the beginning, such as *Uroglenopsis*, but a large proportion
can be laid to the blooming of the diatom *Fragilaria*.
All of these forms are common components of the
experimental ponds that have been fertilized.
Although *Fragilaria* was not formerly as conspicuous
in the ponds as in the chemostats, it became more
abundant in the study pond during the summer following these experiments.[19]

The increase in total population in the enriched
chemostat resulted eventually in large increases in

photosynthetic activity (Figure 66). It will be
noted, however, that this did not occur immediately;
up to the 5th day there had been no detectable increase in photosynthesis, even though there had been
substantial increases in nutrient levels. Shortly

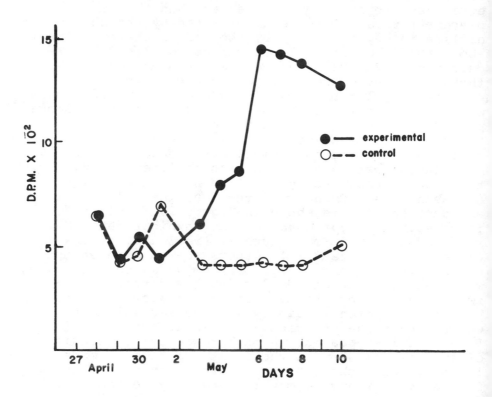

Figure 66.   *Photosynthetic activity in disintegrations per
minute (DPM) of radiocarbon in pond chemostat
experiment III. Control supplied with filtered
pond water, experimental with enriched pond water.*

thereafter there was a sharp decline in nitrate
concentration, but not phosphorus, indicating increased N uptake (Figure 67). These changes
correspond with the appearance and blooming of
Fragilaria and suggest that it utilized N and P in
substantially different proportions from its pre-
decessors. Photosynthesis reached its maximum
with the maximum abundance of this diatom and
declined as it ceased dividing on the 10th day.

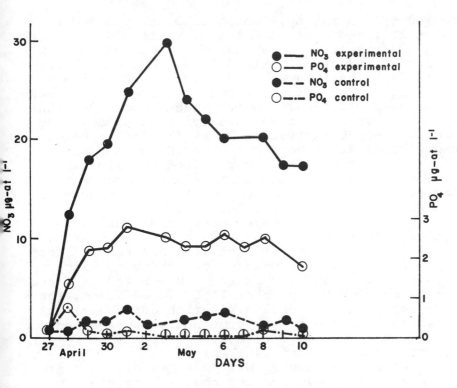

Figure 67. *Nitrate and soluble reactive phosphorus in pond chemostat experiment III. Control supplied with filtered pond water; experimental with pond water + 40 μg at N l^{-1} and 5 μg at P l^{-1}.*

In spite of growth of the population in the enriched chemostats, neither nitrate nor phosphate were reduced to levels as low as in the control. Nevertheless, it is evident from trends in both abundance and total photosynthesis that in both experiments populations had reached an equilibrium by about the 5th to 7th day and were no longer increasing. Apparently at this point populations were not limited by P or N, but we have no evidence as to what other factor may have been limiting.

Cayuga Lake

These experiments were made on phytoplankton taken during the summer from this large, deep, mesotrophic lake. Here we will consider only the latter part of the stratified summer period when all the usual limnological criteria indicate con-tinuous and acute nutrient limitation. During this period SRP in the epilimnion is usually below 0.1 µg at l^{-1}, and phytoplankton pigments a sparse 2 to 5 µg-Chl a l^{-1}.[20] In spite of the prevailing low levels of SRP short period ^{14}C nutrient enrichment experiments have shown some evidence of limitation by silicon[21],[22] and some trace metals,[23] but almost never by phosphorus. In our experiments with the Cayuga phytoplankton we have evidence of a strong and consistent limitation by P. In fact, these experiments differed from those in the ponds described above in that P remained the limiting factor almost continuously in all the chemostat experiments. This led to important differences in the responses of the populations.

There were four experiments made during the latter summer period, all with a similar design: a continuous flow experiment with one control and two experimental chemostats, one enriched with 1 µg-at P l^{-1}, and the other either with P + trace metals or with 0.5 µg-at P l^{-1}. In addition, there were parallel batch-culture enrichment experiments in which response to the same nutrient treatments was measured either by 4-hour ^{14}C uptake or by longer period pigment synthesis.

The batch-culture enrichment experiments are of interest largely because they illustrate the inconsistencies that may be met in such experiments. As illustrated in Table 32, the results of these two kinds of experiments did not agree with one another. In none of the short period ^{14}C experiments was there more than a small and inconsistent stimulation of photosynthesis by any enrichment, while in the 3- to 5-day experiments phosphorus always

Table 32

Enrichment Experiments Accompanying Lake Chemostat IV.
Photosynthesis Uptake in Counts/Minute After 4 Hours
and Pigment in Arbitrary Fluorescence Units After
5 days. Each entry is the mean of two flasks.

Treatment	Control	Trace Metals	Phosphate	Phosphate + Trace Metals
Photosynthesis counts/min X 10^{-2}	46.6	46.6	47.4	52.5
Pigment units	24	24	78	79

produced a 2- to 3-times increase in pigment synthesis.
No other single nutrient had any stimulatory effect,
although in one experiment P + trace metals did have
a greater effect than P alone.   This apparent dis-
crepancy between the two kinds of experiments can
be explained by changes in characteristics of the
phytoplankton revealed by the continuous flow experiments.   These will be illustrated with a single
experiment, Chemostat IV, run July 28 to August 6.
This was chosen because the two enriched chemostats
differed only in that one received a trace metal
mixture in addition to P.   Parallel batch-culture
experiments confirm that trace metals had no effect
on either pigment synthesis or photosynthesis, and
the two treatments can be regarded as replicates.
    Changes in abundance are shown by *in vivo*
fluorescence in Figure 68.   In the first 24 to 48
hours there was always some decline in pigment,
even in the enriched chemostats.   This initial

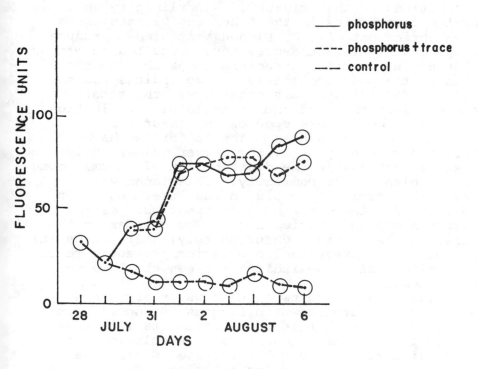

Figure 68.   Pigment in lake chemostat experiment IV. Control
            supplied with filtered lake water, experimentals
            with lake water + 1 μg at P-PO₄l⁻¹, or phosphate
            + trace metal mixture. Pigment expressed in
            arbitrary units representing relative fluorescence.

decrease is probably largely accounted for by disappearance of certain small flagellates. After the
first 2 or 3 days the control decreased more slowly,
whereas the enriched always began a rapid increase,
the rate of increase coming into equilibrium with
the displacement rate in some 5 to 8 days. The
size of the equilibrium population depended principally on the amount of additional P supplied,
differing by not more than about 15% in experiments
with the same amount of P at the same displacement
rate.

The rapid increases in abundance in the enriched
chemostats were evidently the result of approximately
equivalent increases in division rates of almost all
of the numerically important members of the community.
As shown in Figure 69, in Chemostat IV the small
diatom *Cyclotella* grew more rapidly during the first
3 days. However, the final population on the 8th
day, although more than 17 times greater in numbers,
contained most of the same species in about the same
proportions as the original. A preliminary analysis
of the communities in the other experiments shows
that enrichment with P, although it always produced
large changes in abundance, never resulted in marked
changes in community structure comparable to those
seen in the pond communities. Two indices can be
used to demonstrate this constancy: the total
number of species and the evenness of distribution
of volume among these species as measured by
Pielou's[24] index, H/Hmax. The latter can have
positive values up to 1, the larger values indicating
greater evenness in the distribution of volume among
all species in the community. This index varies
seasonally from 0.3 to 0.8 in the lake. As shown
in Table 33, there was no consistent changes in
either number of species nor Pielou's index in the
enriched chemostats. Unfortunately, differences in
methods of assessment of population prevent direct
comparisons of these indices between the lake and
pond chemostats.

The relations between changes in abundance
and nutrients are shown in Figures 70 and 71. It
should be noted that in the experiments with the
Cayuga populations a concentrated P-solution was
added directly to the culture vessel during the
first 12 hours to bring its calculated P-concentra-
tion to equal that in the supply medium. As shown
in Figure 71, this added P was so rapidly assimilated

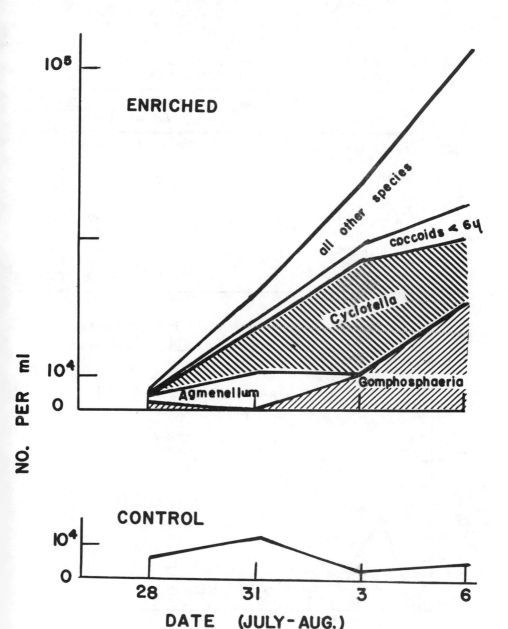

Figure 69.   *Changes in abundance of major components of population in one enriched chemostat, and of total population in the control from Run IV. 1971.*

Table 33

Comparison of Numbers of Species and Pielou's Index (H/Hmax)
Chemostat Experiments III-VI, 1971

Experiment	Date	Treatment μg-at P l^{-1}	Initial # spp.	Initial H/Hmax	Final # spp.	Final H/Hmax
III	July 14-26	1.0	21	0.49	19	0.79
IV	July 28- August 6	1.0 1.0	20	0.71	30 21	0.76 0.79
V	August 12-20	1.0 1.0	21	0.73	22 32	0.58 0.79
VI	August 25- September 3	0.5 1.0	21	0.67	24 29	0.83 0.79

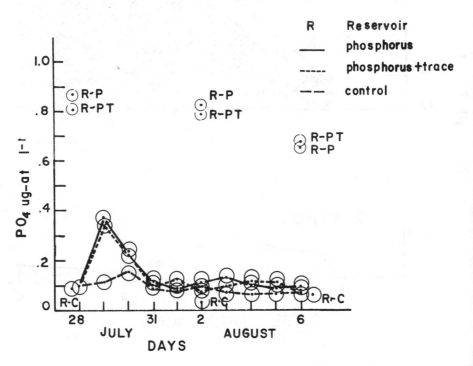

Figure 70.   Soluble reactive phosphorus in lake chemostat IV.
C-unfertilized lake water, P = lake water + 1 µg
at P l^{-1}; P+T = lake water + 1 µg at P l^{-1} and
trace metal. Concentrations in supply reservoirs
were determined on three dates and are shown on
figure.

that SRP values never reached half this calculated level and were in fact reduced to that in the control vessel within 2 to 3 days. However, in spite of this evident strong demand for P, SRP levels in neither control nor enriched chemostats were ever reduced below about 0.1 µg-at l^{-1}. This is within the limit of the analytical method used and is probably accurately measured. That the algae did not exhaust this residual SRP suggests that it may be largely in some form, possibly organic, that is not readily available to the algae, but does react with the acid molybdate reagent used in the analysis.

Changes in particulate phosphorus (PP) clearly reflect the accumulation of P by the plants, nearly all the P lost reappearing in particulate form. As a consequence, as shown in Figure 71, there were

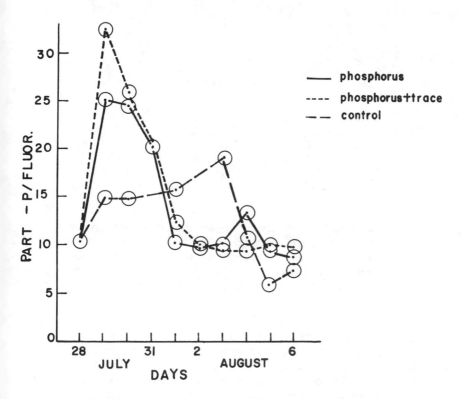

Figure 71. *Particulate phosphorus: pigment ratios in lake chemostat IV. C = unfertilized lake water, P = lake water + 1 µg at P l^{-1}, P+T = 1 µg at P l^{-1} + trace metals. Ratio has units of µg at P/ fluorescence.*

large increases in the PP:pigment ratios during the
first days corresponding to the period of maximum
P uptake. The first evidence of the effect of this
assimilated phosphorus is in the photosynthetic
rate. In the enriched chemostat the rate per unit
pigment increased immediately following the period
of rapid P-assimilation, reaching a maximum in the
2nd or 3rd day (Figure 72). It took substantially

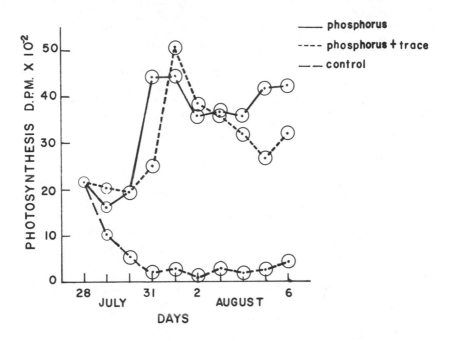

Figure 72.   *Photosynthetic activity as disintegrations per
minute of radiocarbon in lake chemostat experiment IV. Control supplied with filtered lake
water, experimentals with lake water + phosphate,
or lake water + phosphate + trace metals.*

longer for the assimilated P to affect pigment synthesis, no detectable increase occurring until at
least the 3rd day. As a result, the maximum rate
of photosynthesis per unit pigment was attained well
before the maximum population. It then declined
rapidly as with the increase in population and the
decline in cellular phosphorus as measured by PP:
fluorescence ratios. Equilibrium populations in
the enriched chemostats had photosynthesis per unit
pigment ratios as low as the controls.

DISCUSSION

Whether the changes in species composition that were observed in the experiments with the pond phytoplankton should properly be termed a species succession is a question that, although not trivial ecologically, is unimportant here. What is important is that the changes that did take place in response to added nutrients were similar to those that had been seen in the ponds when they were treated in a similar way. In these experiments there was a shift to species that could be said to be characteristic of more eutrophic environments. The experiments did not become dominated by some "laboratory" form, rarely found in abundance in nature. Moreover, in spite of sometimes very great increases in total abundance in the enriched chemostats, both pond and lake populations remained rich in species, which Hulburt[25] maintains is characteristic of most plankton blooms in nature.

That there was not a dramatic species succession in the Cayuga phytoplankton is, we believe, a consequence of the role of phosphorus as the singular limiting factor in these experiments. In several experiments the small diatom *Cyclotella*, which sometimes blooms in the lake, increased for a time more rapidly than other forms, but never came to dominate the way *Fragilaria* did in the pond studies. The reason for this is that the populations were almost continuously limited by phosphorus. Even in enriched chemostats the initially P-limited populations increased so rapidly to the point where they were again P-limited that there were only transient periods of P-sufficiency. Apparently these periods were long enough for a rapidly growing form such as *Cyclotella* to gain over other species, but not to produce marked shifts in species composition. This is actually what would be expected in chemostats in which there is a single limiting factor.[7] In the pond experiments it seems evident that neither N nor P was limiting in the enriched chemostats, and this seems to account for the marked species succession.

In more recent studies in Cayuga we have used turbidostats in order to maintain these populations continuously under P-sufficient conditions. Although analysis of the results of these experiments is still incomplete, it appears that this produces a rapid species succession somewhat like that seen in the pond studies. In one experiment analysis showed an increase in the proportion of diatoms, as well as an increase in dominance as measured by H/Hmax.

These experiments emphasize the limitations of single-species bioassays as a means of determining nutrients limiting natural populations. In both pond and lake populations several species were stimulated by increased nutrients, and although under some conditions a single species might come to dominate, this was not necessarily the one that originally had been conspicuous. Thus, it would be difficult to single out one species that could be said to characterize the response of the original population in any experiment.

It is also apparent that nutrient bioassays that use natural plankton associations are not always so easy to interpret as they seem. Immediate reactions to added nutrients, whether measured in terms of photosynthetic rate, pigment changes, or cell numbers, may be complicated by shifting meta-bolic patterns that may lead to marked time-lags in response. Similar lags have been observed in studies with pure cultures many times.[26,27] Obviously this is the reason short period ^{14}C enrichment ex-periments failed to reveal any effect of P on the plankton in the lake. An examination of the sequence of events in the enriched chemostats shows that because of the time-lag in response to added P, and the very short period of time when rate per unit plant was actually significantly elevated, it might be quite difficult to design a short-period experi-ment that would detect such a response.

Longer period experiments, although perhaps freed from these short period delays in response, may be confounded by changes in dominance that take place, possibly even more rapidly in the usual "batch" experiment than in these continuous flow experiments. Because of these changes in composi-tion, nutrient requirements, rates of growth, etc., the final population could be markedly different from the original.

Continuous flow experiments as we have made them can be cumbersome, and the plankton identifi-cation and counting that have been part of them are not only laborious but also depend on rare skills. Nonetheless, this kind of experiment has much to recommend it because conditions are closer to those in the natural environment and offer possibilities for more powerful analyses than ordinary batch experiments. It seems probable that much can be learned about nutrient relations of existing populations from changes in nutrient uptake and distribution in inorganic and organic form that take place in the initial stages of these experiments. These are observations that can be made relatively rapidly and easily using well established chemical techniques. Increases in population sizes, which are signalled by changes in pigment, should be evidence that the plant association may be changing rapidly. At this point, if not before, one must turn to biological observations in order to have real understanding.

ACKNOWLEDGMENTS

This research was supported by grants provided by the Water Resources Research Act of 1964 administered by the Office of Water Resources Research, Department of the Interior. We wish to acknowledge the examination of the Cayuga Lake phytoplankton by Mrs. Alice Savage, and analyses of pond samples provided by Gene E. Likens and John Eaton. Hugh F. Mulligan has provided valuable encouragement and assistance.

REFERENCES

1. Skulberg, O. M. "Algal Problems Related to the Eutrophication of European Water Supplies, and a Bio-assay Method to Assess Fertilizing Influences of Pollution on Inland Waters," In *Algae and Man,* Jackson, D. F., ed. (Plenum Press, 1964), pp. 262-299.
2. Ryther, J. H. and R. R. L. Guillard. "Enrichment Experiments as a Means of Studying Nutrients Limiting Phytoplankton Production," *Deep Sea Res.* 6:65 (1959).
3. Goldman, C. R. "Micronutrient Limiting Factors and Their Detection in Natural Phytoplankton Populations," In *Primary Productivity in Aquatic Environments, Mem. Inst. Ital. Idrobiol. 18 Suppl.,* Goldman, C. R., ed. (University of California Press, 1965), pp. 121-136.
4. Wetzel, R. G. "Nutritional Aspects of Algal Productivity in Marl Lakes with Particular Reference to Enrichment Bioassays and their Interpretation," In. *Primary Productivity in Aquatic Environments. Mem. Inst. Ital. Idrobiol. 18 Suppl.,* Goldman, C. R., ed. (University of California Press, 1965), pp. 137-160.
5. Schelske, C. L. and E. F. Stoermer. "Eutrophication, Silica Depletion, and Predicted Changes in Algal Quality in Lake Michigan," *Science 173*:423 (1971).
6. Menzel, D. W., E. M. Hulburt, and J. H. Ryther. "The Effect of Enriching Sargasso Sea Water on the Production and Species Composition of the Phytoplankton," *Deep Sea Res. 10*:209 (1963).

317

7. Herbert, D. "Some Principles of Continuous Culture," *Recent Progress in Microbiology, Int. Cong. of Microbiol.* 7:381 (1958).

8. Malek, I. *Theoretical and Methodological Bases of Continuous Culture of Microorganisms*. (New York: Academic Press, 1966), 655 pp.

9. Herbert, D., P. J. Phipps, and D. W. Tempest. "The Chemostat: Design, and Instrumentation," *Lab. Prac.* 14:1150 (1965).

10. Jannasch, H. W. "Continuous Culture in Microbial Ecology," *Lab Prac.* 14:1162 (1965).

11. Jannasch, H. W. "Enrichments of Aquatic Bacteria in Continuous Culture," *Archiv f. Mikro* 59:165 (1967).

12. Symons, J. M. "Simple, Continuous Flow and Variable Rate Pump," *J. Water Poll. Cont. Fed.* 35:1480 (1963).

13. Guillard, R. R. L. and J. H. Ryther. "Studies of Marine Planktonic Diatoms. I *Cyclotella nana* Hustedt, and *Detonula confervaca* (Cleve) Gran.," *Can. J. Micro.* 8: 229 (1965).

14. Strickland, J. D. H. and T. R. Parsons. "A Manual of Sea Water Analysis," *Bull. Fish. Res. Bd. Can. 125 (rev.)* (1965).

15. Strickland, J. D. H. "Measuring the Production of Marine Phytoplankton," *Bull. Fish. Res. Bd. Can. 122* (1960).

16. Lorenzen, C. J. "A Method for the Continuous Measurement of *in vivo* Chlorophyll Concentrations," *Deep Sea Res.* 13:223 (1966).

17. deNoyelles, F. "Factors Affecting the Distributions of Phytoplankton in Eight Replicated Fertilized and Unfertilized Ponds," Ph. D. Thesis, Cornell University (1971).

18. O'Brien, W. J. "The Effects of Nutrient Enrichment on the Plankton Communities in Eight Experimental Ponds," Ph. D. Thesis, Cornell University (1970).

19. deNoyelles, F. Personal communication.

20. Barlow, J. P. and A. Savage. "The Phytoplankton," In. *Ecology of Cayuga Lake and the Proposed Bell Station,* Oglesby, R. T. and D. J. Allee, eds. (Water Resources and Marine Sciences Center, Cornell University, 1969), Publ. no. 27.

21. Hamilton, D. H., Jr. "Nutrient Limitation of Summer Phytoplankton Growth in Cayuga Lake," *Limnol. and Oceanogra.* 14:579 (1969).

22. Barlow, J. P. (and associates). Unpublished observations.

23. Mills, E. L. and R. T. Oglesby. "Five Trace Metals and Vitamin B_{12} in Cayuga Lake, N.Y." *Proc. 14th Conf. Great Lakes Res.* 14:256 (1971).

24. Pielou, E. C. "Species-Diversity and Pattern Diversity in the Study of Ecological Succession," *J. Theoret. Biol. 10*:370 (1966).

25. Hulbert, E. M. "Competition for Nutrients by Marine Phytoplankton in Oceanic, Coastal and Estuarine Regions," *Ecology 51*:475 (1970).

26. Fogg, G. E. *Algal Cultures and Phytoplankton Ecology* (University of Wisconsin Press, 1966), 126 pp.

27. Fuhs, C. W. "Phosphorus Content and Rate of Growth in Diatoms *Cyclotella nana* and *Thalassiosira fluviatilis*," *J. Phycol. 5*:305 (1969).

16. THE EFFECT OF CITRATE ON PHYTOPLANKTON IN LAKE ONTARIO

Walter A. Glooschenko and James E. Moore. Fisheries Research Board Detachment, Canada Centre for Inland Waters, Burlington, Ontario, Canada

Recent concern with the role of phosphorous in eutrophication has led to the search for replacements of this detergent builder. One possible compound under consideration is sodium citrate, already used in at least one detergent, Opus, produced in Sweden.[1] Citric acid and its salts occur in nature. It has been reported in coastal marine waters,[2] and in soils.[3-5] However, such natural levels are quite low, in the ppb range. Even though organisms are already exposed to citric acid and contain the substance in their intermediary metabolism, widespread use of this compound in detergents with possible increased levels in aquatic ecosystems could possibly affect algal growth by two possible modes of action, (1) directly as a carbon source for heterotrophic growth, and (2) indirectly through its ability to form complexes, or chelates, with various trace metals in the culture media used, or aquatic ecosystems. The direct utilization of citrate as a source of carbon is dependent upon the algal species involved and concentration of citrate added. Positive effects upon algal growth have been found with species involving nearly all phyla of algae.[6-14] Other experiments have shown either no effect or very poor utilization of citrate by algae. In some experiments, citric acid was the only tricarboxylic acid cycle component that failed to be utilized by algae.[11,15-18] However, in some experiments it was impossible to ascertain whether the growth or metabolic response noted was a direct one, or an indirect one due to complex formations with metals.[14,19]

The indirect effect of citrates upon algal growth is through chelation of trace metals in culture media or aquatic ecosystems. Before the development of the synthetic chelator EDTA, citrate complexes were commonly used to maintain availability of trace metals, especially iron, in culture media.[20-23] This chelation effect can (1) lead to increased algal growth by increasing availability of trace metals, (2) increase growth by decreasing toxic levels of trace metals, (3) reduce growth by decreasing available concentrations of trace metals by formation of strong chelates, and (4) reduce growth by decreasing concentrations of metals antagonistic to toxic trace metals uptake.[24] Such effects are quite dependent on both the nature and concentration of chelators and trace metals involved.

All previously mentioned studies on the effect of citrate on algae utilized unialgal cultures, a condition atypical of that found in natural aquatic ecosystems. In order to investigate the effects of citrate upon phytoplanktonic algae, field bioassay studies were performed in Lake Ontario and Hamilton Harbour, an embayment of the lake subjected to high discharges of industrial effluents primarily from steel mills.

METHODS

Beginning in February, 1971, water samples were collected in Hamilton Harbour and in Lake Ontario approximately 2-3 miles NE of the harbour entrance. Samples were well-mixed and 100 ml volume placed in plastic foam-stoppered 125 ml Erlenmeyer flasks in a growth chamber containing fluorescent lamps and set at the temperature of collection. The photoperiod used was L : D, 12 : 12 through the month of April and L : D, 16 : 8 beginning in May. The light intensity used was approximately 900 foot-candles.

Sodium citrate was added as the disodium hydrogen citrate salt $(Na_2HC_6H_5O_7 \cdot 1\frac{1}{2} H_2O)$. For comparison with the chelating ability of EDTA, the disodium form was used. Copper was added as $CuSO_4 \cdot 5H_2O$ and zinc as $ZnSO_4$. All compounds were ACS quality.

Growth of algae in cultures in long-term studies was measured by *in vivo* fluorescence of chlorophyll with a Turner fluorometer.[25] Precision on this method was approximately +8%. All chlorophyll values in the paper are relative for the particular assignment, and absolute chlorophyll values were not determined. Measurements of photosynthesis in short-term studies (less than 5 hours) were done by uptake of $NaH^{14}CO_3$[25] followed by liquid scintillation counting. Samples for species identification and enumeration were preserved in Lugol's solution, but preliminary examination of the samples showed no definite trends in terms of species composition in this paper.

Effects of Sodium Citrate Upon Long-term Algal Growth

Long-term growth studies were performed for periods of 10-32 days. Results of these experiments showed that the concentration of added sodium citrate, temperature, and time of exposure were important variables affecting algal growth. The first experiment beginning in February, 1971 (Figure 73) with Hamilton Harbour water was done by acclimatizing the

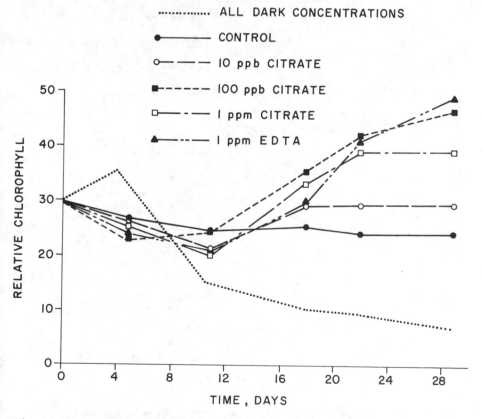

Figure 73. *Effect of Citrate and EDTA Upon Hamilton Harbour Algae. 12 hr. L-D Cycle, 15°C. Started 2 Feb. 1971.*

algae for 4 days at 3°C (ambient water temperature) and 15°C before citrate was added. At 15°C and L : D, 12 : 12 significant increases in algal growth took place with 100 ppb and 1 ppm citrate with highest growth over controls at 100 ppb. The addition of 1 ppm EDTA gave a similar growth increase indicating that chelation was probably the mode of action of citrate. At 15°C in the dark no growth occurred, while at 3°C in both the light and dark, decrease in algae was noted. A similar experiment carried out in Lake Ontario water (Figure 74) showed a decline in the algal population at all citrate concentrations besides controls both in the light and dark (not shown). No significant differences were found between different added concentrations of sodium citrate.

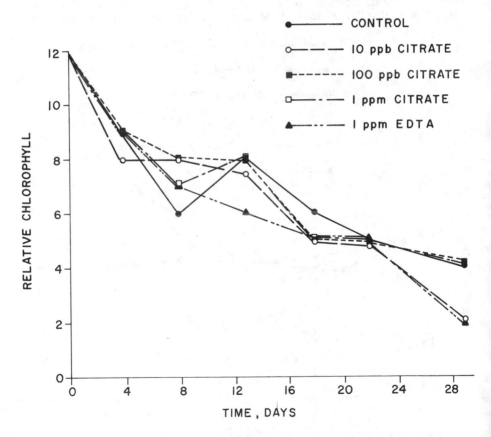

Figure 74.   Effect of Citrate and EDTA Upon Lake Ontario Algae. 12 hr. L-D Cycle, 15°C. Started 31 Mar. 1971.

The same experiments were repeated beginning
30 April, 1971, with both Hamilton Harbour and Lake
Ontario waters at 15°C in Hamilton Harbour water,
both citrate (all concentrations) and 1 ppm EDTA inhibited algal growth over a 19-day period (Figure 75).

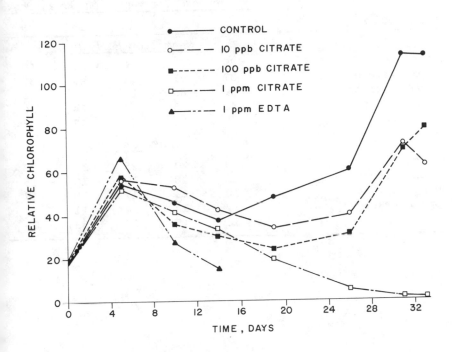

Figure 75. *Effect of Citrate and EDTA Upon Hamilton Harbour*
Algae. 16 hr L - 8 hr.D Cycle, 10°C. Started
3 May 1971.

However, in Lake Ontario, citrate again inhibited
algal growth, but EDTA stimulated growth (Table 34).
At the same time, Hamilton Harbour algae were
acclimated at 3°C for four days and citrate added
(Figure 76). After 19 days of incubation, no apparent differences between various treatments were seen.
However, after 24 days, 10 and 100 ppb citrate inhibited algal growth while 1 ppm citrate or EDTA
significantly increased growth.

In order to compare the action of citrate, a
weak chelator, with EDTA, experiments were begun in
May with Hamilton Harbour algae (Table 35). Maximum
growth over controls took place in the culture to
which 1 ppm citrate was added. This same concentration

Table 34

Effect of Sodium Citrate and EDTA Upon Growth of Algae in
Lake Ontario and Hamilton Harbour. Experiment Begun
on 31 April, 1971. Temperature = 15°C

	Relative Chlorophyll on Given Day:				
Concentration	3	7	10	14	19
	Hamilton Harbour				
Control	49	65	46	38	49
10 ppm Citrate	37	61	53	43	35
100 ppb Citrate	39	60	36	31	25
1 ppm Citrate	42	53	42	34	20
1 ppm EDTA	46	63	29	17	17
	Lake Ontario				
Control	9	20	25	30	28
10 ppb Citrate	13	18	19	16	15
100 ppb Citrate	10	23	17	18	17
1 ppm Citrate	12	14	14	11	12
1 ppm EDTA	11	15	25	53	76

Table 35

Comparison of Sodium Citrate vs EDTA. Effect Upon Growth
of Hamilton Harbour Algae. Experiment Begun on 10 May, 1971

	Relative Chlorophyll on Given Day:						
Treatment	2	3	4	5	8	10	12
Control	17	14	14	14	51	41	47
10 ppb Citrate	17	13	13	13	62	52	32
10 ppb EDTA	14	13	14	17	71	81	39
100 ppb Citrate	12	11	12	17	84	--	--
100 ppb EDTA	18	15	13	13	39	30	17
1 ppm Citrate	13	11	13	18	95	87	69
1 ppm EDTA	15	14	13	14	32	27	17

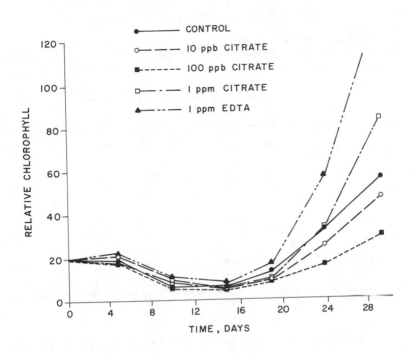

*Figure 76. Effect of Citrate and EDTA Upon Hamilton Harbour
Algae. 16 hr L - 8 hr D Cycle, 3°C. Started
3 May 1971.*

of EDTA was inhibitory along with the 100 ppb concentration of essential trace elements. With the
exception of 10 ppb citrate at 12 days, citrate was
seen to increase algal growth. The experiment was
repeated one month later (Table 36). Again, citrate
at 10 ppb decreased growth while 100 ppb and 1 ppm
increased growth over controls after 10 days. In
contrast to the previous experiment, EDTA increased
growth at 100 ppb while 10 ppb and 1 ppm showed no
significant difference over controls.

Thus, sodium citrate was seen to exhibit a
variable effect on algal growth. In general, Lake
Ontario algal growth was inhibited by additions of
this compound while Hamilton Harbour experiments
exhibited either increased or decreased growth depending on the added concentrations of citrate,
length of incubation and temperature. Hamilton
Harbour algal growth was generally increased at
higher concentrations of added citrate in contrast
to the Lake Ontario algae. This increased growth

Table 36

Comparison of Sodium Citrate vs EDTA. Effect Upon Growth
of Hamilton Harbour Algae. Experiment Begun 28 June, 1971

	Relative Chlorophyll on Given Day:						
Treatment	1	3	6	8	10	15	17
Control	60	96	180	195	192	222	189
10 ppb Citrate	59	84	162	180	153	78	66
10 ppb EDTA	59	87	165	162	162	204	189
100 ppb Citrate	60	66	135	174	192	213	195
100 ppb EDTA	55	165	195	183	175	246	258
1 ppm Citrate	61	93	174	183	221	291	261
1 ppm EDTA	56	87	126	144	183	192	186

can be explained by two possibilities: 1) a decrease
of the available concentration of a toxic trace element, or elements, or 2) maintenance of an essential
trace metal in an available state. The decreased
algal growth could be due either to 1) reduction
of availability of an essential trace metal, or 2)
removal of a trace metal antagonistic to uptake of a
toxic trace metal, a difficult effect to measure.
Since offshore water from Lake Ontario contains low
trace metal concentrations (unpublished data), decreased growth due to additions of chelators may be
explained by decreased availability of essential
trace elements due to chelation. However, Hamilton
Harbour contains high levels of potentially toxic
trace metals, mainly zinc, and a series of experiments were begun to investigate the interaction of
citrate with several toxic trace elements and the
subsequent effect on algal growth.

Interaction of Citrate with Zinc
and Copper and Effect Upon Algal Growth

In order to investigate further the interaction
between chelators and toxic trace metals, both
short-term (less than 5 hours) studies involving
measurement of $NaH^{14}CO_3$ uptake as an index of
photosynthesis, and long-term growth studies

involving chlorophyll measurement were initiated.
Zinc was chosen due to its high concentration (between 50-80 ppb dissolved zinc) found in Hamilton
Harbour and copper due to previous studies on its
algicidal properties.[26-29]
   Copper and zinc were added to Hamilton Harbour
and Lake Ontario algae at 10 and 50 ppb alone and
in combination with citrate or EDTA (Table 37).
In Hamilton Harbour, added copper at 50 ppb or 50
ppb copper + 1 ppm citrate were inhibitory to algal
growth while all other treatments involving copper
were stimulatory. Zinc was toxic at the 10 ppb

Table 37

Interaction of Copper, Zinc, Sodium Citrate (CIT) and EDTA
Upon Short-Term ^{14}C Uptake by Hamilton Harbour and
Lake Ontario Algae (5-hour exposure time).
Experiment done 12-13 May, 1971

| | % ^{14}C Uptake of Control | | | |
| | Copper | | Zinc | |
Treatment	H. Harbour	L. Ontario	H. Harbour	L. Ontario
Control	100.0	100.0	100.0	100.0
10 ppb Metal (M)	124.6	58.8	62.8	78.2
50 ppb M	82.2	22.8	48.2	57.8
10 ppb M + 1 ppm CIT	124.4	66.7	112.0	60.0
10 ppb M + 1 ppm EDTA	141.5	93.5	109.5	106.3
50 ppb M + CIT	92.2	50.4	48.3	72.1
50 ppb M + EDTA	131.9	81.6	101.1	103.4
10 ppb M + CIT + EDTA	132.4	96.4	101.1	102.1
50 ppb M + CIT + EDTA	146.2	83.9	112.2	104.6
1 ppm CIT	110.2	97.5		
1 ppm EDTA	120.3	90.4		

addition and the combination of zinc with either
citrate or EDTA reduced toxicity except where 50 ppb
zinc was added with citrate. This combination caused
a slight inhibition which may not be significant.
Experiments with Lake Ontario algae showed copper
to be toxic to growth at 10 ppb. Citrate alone was
unable to reverse the copper toxicity in contrast
to the stronger chelator EDTA which did. Zinc

329

additions produced similar results to those found with copper. Additions of citrate alone slightly increased photosynthesis in water from Hamilton Harbour, while in water from Lake Ontario, a slight inhibition was seen. The addition of EDTA alone behaved similarly.

Further studies were performed in early June to determine long-term effects of zinc and copper in combination with citrate or EDTA (Table 38). In Hamilton Harbour, copper was toxic to algal growth, with the chelated forms of copper being slightly more toxic than copper alone up to day 15. Zinc was slightly more toxic than copper. The zinc-EDTA complex was less toxic up to 10 days. After that time, toxicity of this complex increased, indicating a possible breakdown of the chelate with release of

Table 38

Copper, Zinc, Citrate, EDTA
Effect Upon Growth of Hamilton Harbour and
Lake Ontario Algae

Treatment	*Relative Chlorophyll on Given Day*					
	2	6	8	10	15	20
Hamilton Harbour						
Control	52	54	78	81	126	168
+ 50 ppb Cu	34	43	67	117	87	60
+ 50 ppb Cu + 1 ppm CIT	31	38	45	54	66	60
+ 50 ppb Cu + 1 ppm EDTA	44	29	39	39	54	45
+ 50 ppb Zn	36	33	41	54	60	90
+ 50 ppb Zn + 1 ppm CIT	34	27	26	27	45	66
+ 50 ppb Zn + 1 ppm EDTA	45	46	62	63	27	21
Lake Ontario						
Control	17	21	19	26	20	18
+ 50 ppb Cu	6	4	5	5	6	14
+ 50 ppb Cu + 1 ppm CIT	8	4	5	4	11	11
+ 50 ppb Cu + 1 ppm EDTA	22	12	14	19	21	12
+ 50 ppb Zn	17	21	19	26	20	18
+ 50 ppb Zn + 1 ppm CIT	9	5	7	4	5	8
+ 50 ppb Zn + 1 ppm EDTA	22	57	59	45	35	43

zinc. The zinc-citrate complex was intermediate between zinc and zinc-EDTA in its toxicity. Lake Ontario experiments exhibited zinc and copper toxicity, but EDTA was able to overcome this toxicity and stimulate growth over controls. Both the zinc and copper citrate complexes were more toxic than the metal alone indicating increased availability of this complex to algae.

DISCUSSION

The influence of chelators on algae have been reviewed by several authors.[22,24,30] This influence can be quite complicated as chelators can either inhibit or stimulate growth depending upon such factors as the concentration and nature of the chelator and trace metal (or metals) involved, presence of antagonistic metals, other chemical properties of the culture media used such as pH and oxygen,[28,31] physical factors such as temperature and light,[29] and effects of microbial activity on the chelator.[26,27] Other complicating factors in such studies in the case of citrate is the possibility of direct utilization by algae of citrate as a source of carbon in heterotrophic growth, although no evidence in this study was found to indicate such usage. The chemical nature of the growth medium used is also quite important, and conflicting work by different authors may be explained by use of different growth media.[26] The presence of naturally occurring chelators may also cause complications in interpreting experimental results.

The experiments performed showed that citrate caused a different response by the algae in two water bodies tested, 1) Lake Ontario--a slightly eutrophic, low trace metal content water, and 2) Hamilton Harbour--a highly eutrophic, high trace element content water body. Generally, citrate additions to Lake Ontario caused decreases in algal growth compared with controls. A similar effect has been noted with the chelator nitrilotriacetic acid, NTA, in sea water media with the marine diatom, *Cyclotella nana*.[32] With increasing concentrations of NTA between 1 and 5 ppm, photosynthesis by this organism was reduced due to induction of trace metal deficiency when the ratio of NTA to trace metal concentration exceeded three. This study also involved copper toxicity to algae and found that while 50 ppb copper inhibited photosynthesis, 500 ppb of NTA reversed this copper toxicity to varying degrees.

Our experiments also demonstrated this effect. Sodium citrate, a weak chelator, either increased toxicity or did not reduce trace metal toxicity as much as EDTA in short-term ^{14}C uptake experiments (Table 37, 38). This may be explained by low concentrations of a weak chelator such as sodium citrate causing increased availability of toxic metals, and higher concentrations inducing decreased availability of a toxic element. However, too high a concentration of a strong chelator such as 1 ppm EDTA (Table 35) might decrease growth by decreasing available trace metals essential for algal growth as seen in Table 38 where EDTA in combination with toxic levels of copper decreased growth compared to copper complexed with citrate.

Therefore, it appears that chelators such as sodium citrate can exhibit variable effects on algal growth, with the trace metal concentration of the culture medium (or natural water in which the algae is present) being a most important variable. Large-scale additions of sodium citrate to natural aquatic systems could either increase or decrease algal growth depending on the concentration added and the chemical composition of the natural water involved. Due to the complexity of this interaction, bioassays under field or simulated field conditions using a particular water body appear to be the only methods at present to predict what will happen if sodium citrate is added in large quantities. Such bioassays must take into account seasonal variations in species composition of algae and in the chemical composition of the water tested, especially trace metals (both essential and toxic). The effect of sodium citrate also may be much more significant in waters subjected to high levels of trace metals discharge due to the chelating ability of this compound. If such a water body contains near toxic levels of metals to algae, citrate may possibly accelerate eutrophication by decreasing availability of such toxic substances. On the other hand, citrate may possibly cause some algal growth inhibition or have no discernible effects on algae in culture media or in natural waters.

Sodium citrate is readily degradable by microbial activity.[33] Therefore, one might expect very low levels of the substance discharged to receiving waters from sewage treatment plants. The variable effects of stimulation and inhibition of algae would probably be transient at most.

REFERENCES

1. Anon. *Chemical Week, 108*:19 (1971).
2. Creach, P. *C. R. Acad. Sci. Paris 240*:2551 (1955).
3. Kononova, M. M. *Soil Organic Matter* (New York: Pergamon Press, 1961).
4. Muir, J. W., R. I. Morrison, C. J. Brown, and J. J. Logan. *J. Soil Sci. 15*:220 (1964).
5. Stevenson, F. J. In *Soil Biochemistry*, McLaren, A. D. and G. H. Peterson, eds. (New York: Marcel Dekker, 1967), p. 119.
6. Belcher, J. H. and G. E. Fogg. *Archiv für Mikrobiol. 30*:517 (1958).
7. Belcher, J. H. and J. D. A. Miller. *Archiv für Mikrobiol. 36*:219 (1960).
8. Bunt, J. *J. Phycol. 5*:37 (1969).
9. Carefoot, J. R. *J. Phycol. 4*:129 (1968).
10. Droop, M. R. *J. Mar. Biol. Assoc. U.K. 46*:673 (1966).
11. Eny, D. M. *Plant Physiol. 25*:478 (1950).
12. Eny, D. M. *Plant Physiol. 26*:268 (1951).
13. Lewin, J. C. *J. Gen. Microbiol. 9*:305 (1953).
14. Wong, P. T. S., Canada Centre for Inland Waters, Burlington, Ontario. Personal communication.
15. Barker, H. A. *J. Cell. Comp. Physiol. 7*:73 (1935).
16. Cosgrove, W. B. and B. K. Swanson. *Physiol. Zool. 25*: 287 (1952).
17. Cramer, M., and J. Meyers. *Archiv. für Mikrobiol. 17*: 384 (1952).
18. Hoare, D. S., S. L. Hoare, and R. B. Moore. *J. Gen. Microbiol. 49*:351 (1967).
19. Hutner, S. H., L. Provasoli, and L. Filfus. *J. Annals. N.Y. Acad. Sci. 56*:852 (1953).
20. Droop, M. R. *J. Mar. Biol. Assoc. U.K. 38*:605 (1959).
21. Goldberg, E. D. *Biol. Bull. 102*:243 (1952).
22. Johnston, R. *J. Mar. Biol. Assoc. U.K. 44*:87 (1964).
23. Rodhe, W. *Symb. Bot. Uppsalienses 10*:1 (1948).
24. Saunder, G. W. *Bot. Rev. 23*:389 (1957).
25. Strickland, J. D. H. and T. R. Parsons. *A Practical Handbook of Seawater Analysis*, Bull. No. 167, Fisheries Research Board of Canada, 1968.
26. Fitzgerald, G. P. and S. L. Faust. *Appl. Microbiol. 11*: 345 (1963).
27. Fitzgerald, G. P. and S. L. Faust. *Water Sewage Works 112*:271 (1955).
28. McBrien, D. C. H. and K. A. Hassall. *Physiol. Plant. 20*: 113 (1967).
29. Steeman-Nielsen, E., L. Kamp-Neilsen, and S. Wium-Andersen. *Physiol. Plant. 22*:1121 (1968).
30. Spencer, C. P. *J. Gen. Microbiol. 16*:282 (1957).

31. Jackson, P. C., J. M. Taylor, and S. B. Hendricks. *Proc. Nat. Acad. Sci.* 65:176 (1970).
32. Erickson, S. J., T. E. Maloney, and J. H. Gentile. *Jour. W. P. C. F.* 42:R329 (1970).
33. Wolfe, M. *Ann. Bot. (London)* 18:309 (1954).

7.   A DESCRIPTION OF BIOASSAY FLOW-THROUGH
     TECHNIQUES, AND THE USE OF BIOASSAY TO
     MEASURE THE EFFECTS OF LOW OXYGEN AT THE
     WHOLE-ANIMAL AND THE MOLECULAR LEVEL

Arthur Scheier and Dennis T. Burton. Department
of Limnology, Academy of Natural Sciences of
Philadelphia, Philadelphia, Pennsylvania

ABSTRACT

     This discussion of the advantages of continuous
flow bioassay as opposed to static batch methods
includes the removal of metabolic products which
might affect bioassay results, the creation of
stress effects upon the test organisms much like
those found in the normal environment, the maintenance of more constant concentrations of volatile
materials, and the control of dilution factors where
toxic materials cannot be chemically analyzed. The
factors which were considered in the design of a
bioassay system are enumerated and a description
of this system is presented.
     The use of bioassay as an important tool in
understanding gross and molecular effects of toxic
materials is illustrated by a description of increases in toxic effect of zinc, napthenic acids,
and potassium cyanide under low oxygen conditions
(*i.e.*, 2mg O_2 per liter for two hours) as compared
to normal oxygen (*i.e.*, 5-9mg O_2 per liter). The
stress of low oxygen conditions is illustrated on
the molecular level by following the shift from
aerobic metabolism to partial anaerobic metabolism
at 2 mg O_2 per liter through analyses of increases
of muscle lactic acid with concomitant decreases
in muscle pyruvic acid. Liver lactic acid concentration also increased significantly but with no
significant changes in liver pyruvic acid.

INTRODUCTION

It is recognized that flow-through bioassay techniques are more valid than batch methods. Recommendations are now commonly made which require the use of flow-through bioassays to determine the application factor for potential pollutants.[1] It is obvious that static batch methods do not take into account the accumulation of metabolic wastes, nor the changes in the environment made by the test organisms in an attempt to create a more comfortable environment. Probably most important, static bioassay methods do not create stress effects on the test organism similar to those that would be found in a continuously flowing environment. Also note that volatile materials are difficult to maintain at a constant concentration in a batch static test, but can be metered continuously in a flow-through system.

Because of the great variety of new products being manufactured, the number of potential toxic pollutants is rapidly increasing. The increase in these pollutants often far outstrips our ability to accurately determine their concentrations in the environment--or even their presence--but flow-through bioassay techniques allow us to accurately meter whatever pollutants we are concerned with, through the test system, so that we can be reasonably sure of our dilution factors.

SYSTEM DESIGN

The design of a flow-through bioassay system for use in the laboratories at the Academy of Natural Sciences required that careful analysis of the types and uses of our test procedures be made. To meet our research requirements, the system had to be extremely accurate. It had to be portable and simple, in order to permit bioassay testing at plant sites. It had to require minimum space, so that many dilutions could be tested concurrently to satisfy the needs of our consulting work. It had to be flexible enough to be modified for use with many types of aquatic organisms--including fish and invertebrates. Its reliability had to be assured since a great many of our test procedures, both in research and in consulting, are chronic long-term tests extending over 90-day, or longer, periods. After consideration of various types of serial diluters and other systems, we decided on a two-pump system with a mixing chamber. This system uses two metering pumps, capable of pumping from 5 mls to 100 mls per minute per pump, with some variability from pump to pump.

Test sample and dilution water are pumped separately at pre-set rates in accordance with the test required into a mixing chamber, where mixing is accomplished by a magnetic stirrer. The mixed dilution is then gravity-fed into the bottom of the test container. At the same time and at the same rate the used mixture of test sample and dilution water overflows the top of the test chamber.

Because the dilution water is well aerated, test chamber aeration is not necessary, thus providing greater stability for volatile test material. The flow velocity can be regulated, depending on the requirements of the experiment. The usual flow velocity for fish bioassays is one gallon of mixed sample per hour flowing through the test container, a 15-gallon glass aquarium framed with stainless steel.

Snail test containers are 1500 ml glass jars with perforated plastic caps. Screening is placed in the perforations to allow for gas exchange and yet prevent the test snails from escaping. The flow velocity through the snail tests is calculated so that it will evacuate and replace the mixture at the same rate as the fish bioassay, in order that equal stress is placed on all test organisms. Thus, when the fish bioassay flow velocity is one gallon per hour, the snail flow velocity is set at 100 mls per hour. Equalization of the stress of adaptation to the toxic environment is carefully considered during multiple organism bioassay procedures.

Figures 77 and 78 illustrate fish and snail bioassay apparatus.

Most bioassay investigation has concerned itself with the measurement of gross effects of toxic substances as expressed by a TLm, or 50% survival concentration. Certainly, death is as crude a measure of any effect as can be devised, and it is obvious that the use of probes or other devices to assess changes in function before they become irreversible is a much more meaningful approach to bioassay investigation. Experiments designed to investigate the gross effects of low oxygen upon the bluegill sunfish[2] and some molecular effects of low oxygen on the bluegill sunfish[3] show relationships

Figure 77.   Flow-through bioassay apparatus for studying fish.

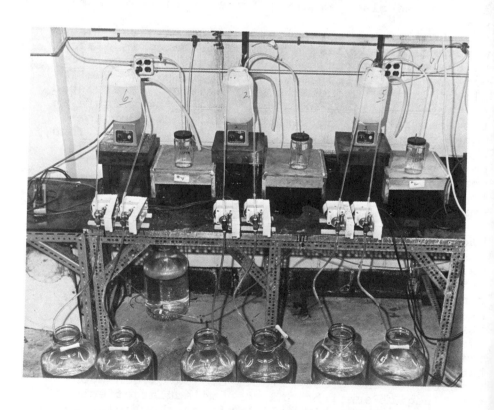

Figure 78.   Flow-through bioassay apparatus for studying snails.

f the gross to the molecular effects and also point
toward more precise measurement of toxic effects.

RESULTS

The gross effects of periodic low oxygen on
the toxicity of various chemicals to the bluegill
sunfish were studied and compared with the toxicity
of these chemicals at a constant 5 to 9 mg O_2 per
liter, using as test chemicals zinc, napthenic acids,
and potassium cyanide. Over an eight-hour period
the oxygen content of the dilution water was lowered
slowly to a minimum of 2 mg O_2 per liter, maintained
there for two hours, and then allowed to increase
slowly to saturation.

Preliminary investigation indicated that 2 mg
O_2 per liter was the minimum periodic dissolved
oxygen concentration that could be endured by the
bluegill over a 96-hour period without gross injury
or death of any test fish.

Four basic sets of conditions were tested:
(1) 5 to 9 mg O_2 per liter plus the test chemical;
(2) 5 to 9 mg O_2 per liter with no test chemical;
(3) periodic low dissolved oxygen plus test chemical;
(4) periodic low dissolved oxygen with no test
chemical.

Test temperature was 18°C ± 1°C and the dilution
water was soft water with a total hardness of 60 ppm
as $CaCO_3$. Table 39 gives the tolerance to various
test chemicals under "normal" dissolved oxygen
conditions of 5 to 9 mg O_2 per liter. In addition
to the 96-hour TLm, an attempt was made to include
both the highest concentration which permitted some
survival and the lowest concentration at which some
death occurred. When this was not possible because
of a sharp threshold, data on the lowest concentration
producing no survival and the highest concentration
allowing complete survival are presented.

Table 40 provides the same information with
periodic low oxygen conditions. There was complete
survival of all test fish under "normal" dissolved
oxygen with no chemical added, and also under low
periodic oxygen with no chemical added.

Comparisons of the critical 96-hour TLm concentrations illustrate the increase in stress leading
to greater lethality from these test chemicals under
periodic low oxygen conditions.

The basis for these stresses can also be seen
at the molecular level. In this study changes in
muscle and liver metabolism as indicators of low

Table 39

Results with "Normal" Oxygen, Fish

Chemical	High Threshold Concn. (ppm)	Concentration % Survival	96-Hour TL_m* (ppm)	Low Threshold Concn. (ppm)	Concentration % Survival
Zinc Chloride (as Zn^{++})	12.5	5	8.02	6.0	75
Naphthenic acid	7.0	0	5.6	3.5	100
Potassium cyanide	0.62	0	0.45	0.175	100

*The 96-Hour TL_m (50% survival) is the critical value for comparative purposes. The threshold concentrations were included in order to delimit the range between complete survival and complete death.

Table 40

Results with Low Periodic Oxygen, Fish

Chemical	High Threshold Concn. (ppm)	Concentration % Survival	96-Hour TL_m* (ppm)	Low Threshold Concn. (ppm)	Concentration % Survival
Zinc Chloride (as Zn^{++})	6.5	20	4.9	3.7	90
Naphthenic acid	3.2	10	2.0	1.0	80
Potassium cyanide	0.2	10	0.12	0.087	80

*The 96-Hour TL_m (50% survival) is the critical value for comparative purposes. The threshold concentrations were included in order to delimit the range between complete survival and complete death.

oxygen stress were measured. The environmental oxygen concentrations at which the bluegill sunfish started shifting from aerobic to anaerobic metabolism were determined, because a shift from aerobic metabolism to anaerobic metabolism indicates a severe lack of oxygen in most vertebrates. This shift can be demonstrated by an increase in the cellular metabolite lactic acid. Lactic acid is unique for this type of study because it accumulates during anaerobic metabolism and is not converted to any other form until sufficient oxygen is present in the system. In order to gain more insight into the possible biochemical processes occurring during low oxygen stress, pyruvic acid, the metabolic precursor of lactic acid, was also examined.

Bluegill sunfish were acclimated in the laboratory to 20°C for a period of at least two weeks in a continuous flow of dechlorinated tap water with an oxygen concentration near saturation. All fish were maintained under a constant photoperiod of 12 hours light and 12 hours dark, and were exposed to a gradual reduction in oxygen concentration in the apparatus described in detail by Burton.[3] Briefly, this involved placing the fish in individual plexiglass cylinders supplied by a continuous flow of water of approximately 75 ml per minute at all times before (15 to 24 hours) and during an experiment. Water was not recirculated through the experimental chambers. The lowering of oxygen was achieved by bubbling nitrogen gas through a displacing column which supplied six cylinders housing the fish.

Hypoxia was induced by reducing the oxygen concentration of the water in steps of approximately 0.5 mg O_2 per liter at half-hour intervals from 8.6 mg O_2 per liter down to 0.5 mg O_2 per liter. Over this 7.5 hour stress period, approximately 10 fish were sampled at each of six different oxygen concentrations. Skeletal muscle and liver tissues were analyzed for lactic acid and pyruvic acid as described by Burton.[3] All data were analyzed statistically by an extension of Duncan's multiple range tests to group means with unequal numbers of replications.[4]

The results of this study are summarized in Figure 79. A highly significant increase ($P < 0.01$) in skeletal muscle lactic acid concentrations was found in bluegills exposed to oxygen concentrations below approximately 2 mg O_2 per liter. Muscle pyruvic acid concentrations decreased significantly

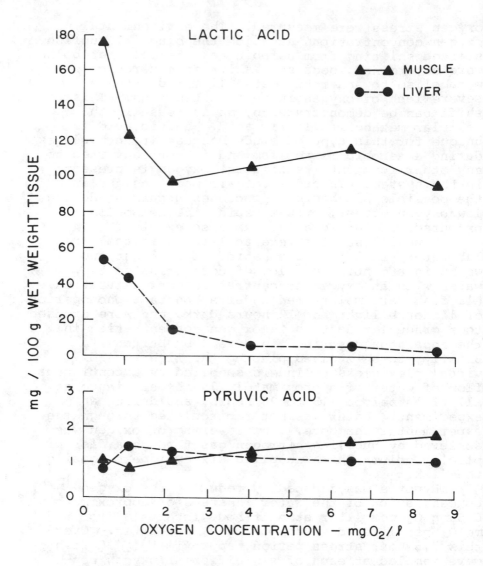

Figure 79. *Changes in muscle and liver lactic acid and pyruvic acid concentrations during gradual environmental hypoxia in bluegill sunfish. Each point on the curve is the mean of approximately 10 fish.*

(P< 0.005) as lactic acid concentrations increased. This decrease probably represents a shunting of pyruvic acid to lactic acid is a means of anaerobic energy production for the cell while the organism is subjected to low oxygen conditions.

342

Liver lactic acid concentrations increased significantly ($P < 0.01$) at approximately 2.2 mg O_2 per liter. In contrast to pyruvic acid changes in muscle, no significant changes ($P < 0.05$) occurred in liver pyruvic acid concentrations. Because no apparent shunting of pyruvic acid to lactic acid occurred in liver, this raises an interesting question. Is the increase in liver lactic acid a result of anaerobic metabolism in that tissue; or is the increase simply a reflection of an equilibrium between muscle-produced lactic acid and blood levels of lactic acid; or is it a combination of both factors? Until the contribution of anaerobic metabolism in the liver is related to total lactic acid accumulation in that tissue, we cannot definitely state that liver tissue shifts from aerobic to anaerobic metabolism under extreme hypoxia.

SUMMARY

In summary, significant increases in muscle lactic acid with concomitant decreases in muscle pyruvic acid indicated that a shift in muscle metabolism from aerobic to anaerobic conditions occurred in bluegills below approximately 2 mg O_2 per liter. However, this shifting to anaerobic metabolism should not be interpreted as a complete shift from aerobic to anaerobic metabolism at approximately 2 mg O_2 per liter. At this oxygen concentration the animal is most likely utilizing a combination of both aerobic and anaerobic metabolism. As the oxygen concentration is lowered further, the bluegill has to rely more and more on anaerobic metabolism for energy production. Anaerobic metabolism appears to supply some metabolic energy for short term exposure to extremely low oxygen concentrations.

It thus appears that 2 mg O_2 per liter causes sufficient metabolic stress so that any increase in stress--caused, for instance by some toxic substance--significantly decreases the organism's tolerance to the toxic substance. Bioassay should and can be used not only to study gross effects, but also to provide the knowledge necessary to understand the gross effects and perhaps to alter them.

REFERENCES

1. *Report of the Committee on Water Quality Criteria,* F.W.P.C.A., U.S. Department of the Interior, April, 1968.
2. Cairns, John Jr., and Arthur Scheier. "The Effects of Periodic Low Oxygen Upon the Toxicity of Various Chemicals to Aquatic Organisms," *Purdue University Engineering Bull., No. 94,* pp. 155-176.
3. Burton, Dennis T. "A Reevaluation of the Anaerobic Endproducts of Freshwater Fish Exposed to Environmental Hypoxia," *Comp. Biochem. Physiol.,* in press.
4. Kramer, C. Y. "Extension of Multiple Range Tests to Group Means with Unequal Numbers of Replications," *Biometrics, 12:*307 (1956).

18. DESCRIPTIONS OF A CONTINUOUS-FLOW BIOASSAY LABORATORY TRAILER AND THE MICHIGAN DILUTER

Mark Wuerthele, John Zillich, Mike Newton, Carlos Fetterolf.* Bureau of Water Management, Water Resources Commission, Michigan Department of Natural Resources, Lansing, Michigan

ABSTRACT

Michigan's mobile bioassay laboratory and diluter system provides on-site, continuous flows of any desired mixture of wastewater discharge and receiving water. One-horsepower pumps deliver wastewater (effluent) and receiving water (diluent) to supply tanks located above several one liter plastic bioassay containers, each capable of holding five small fish and/or other test organisms. The Michigan diluter system supplies a mixed flow of effluent and diluent to each container by gravity. A water quality monitor mounted in the trailer continuously measures and records the pH, temperature, conductivity, and dissolved oxygen in both the effluent and diluent water. A grab water sample can be taken automatically for later analysis if any of these parameters varies abruptly. Temperature control for off-site experimental bioassays is provided.

This technique exposes test animals to the effects of actual multi-element wastes, varying mixtures of these elements, slug discharges, changes in receiving water quality, and the same toxicity relationship they would face in the receiving system. The continuous-flow system is superior to short-term static bioassays for recognizing noxious effects and supporting enforcement decisions.

---

*Present affiliation, Dow Chemical Co., Midland, Michigan.

INTRODUCTION

Enforcement agencies and dischargers find reaching a decision on how much of a toxic waste effluent can be introduced safely to a stream extremely difficult. The literature provides much data from static bioassays conducted with single toxic materials on specific organisms. However, a discharge seldom contains only one toxic material. Usually, a unique combination of toxicants is encountered whose elements may have accumulative, additive or synergistic biological effects. A further complicating factor is that these discharges enter receiving waters of various qualities which support diverse aquatic biota in various life stages. The chances of finding a written description of a bioassay that would meaningfully predict a safe discharge for a particular stream are extremely small. Such information often has been all that decision makers have had to work with.

Off-site continuous flow bioassays using a waste sample as the toxicant eliminate some of the guess work of interpreting the static bioassay. Since the solution is continuously renewed there is little danger of the test organisms using up the toxic material or the available oxygen. On-site continuous flow bioassays provide further advantages. If the actual discharge and receiving water are continuously renewed in the test solutions, all the toxic materials are present, they react and have similar toxicity relationships as they would in the receiving system, and the fish are subjected to varying toxicant concentrations, slug discharges and variations in receiving water quality.

Floor Plan

To provide these advantages to the enforcer and discharger alike, we developed an on-site, continuous flow bioassay laboratory housed in a trailer twenty by eight feet. This size couples easy mobility with adequate laboratory facilities and storage space (Figure 80). A two-man desk which may also serve as additional counter space to the central counter and cupboards is situated in the front of the trailer. Five stainless steel sinks, one large and four smaller ones, are located in the rear of the trailer. With the addition of standpipes, they may be used as fish holding facilities. A 254-liter holding tank located above the sinks is used for storage of water or chemical solutions. Both interior side walls are

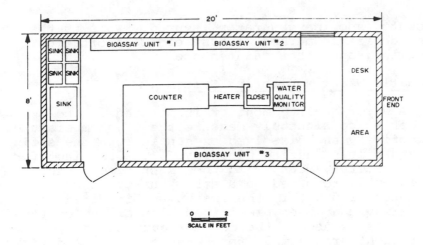

Figure 80. Bioassay Trailer--Floorplan

of peg board and combine easy accessibility to the mounting surface with provisions for alteration of design of the bioassay arrangement. There are three ceiling exhaust fans for ventilation. Other equipment includes: air conditioning, fuel oil furnace, and dehumidifier. Each of these items control conditions in the trailer during a study. The trailer is also equipped with a small refrigerator for storage of samples and reagents.

Water Supply

Water has been supplied to the trailer by either placing one horsepower submersible pumps in the effluent and diluent waters and stringing flexible piping or hooking directly to an effluent or diluent which is under pressure. Major problems may arise when it becomes necessary to obtain wastewater from a submerged discharge, but normally an industry can provide adequate access.

Electrical Supply

The trailer's electrical system requires 220 volts of single phase AC at 30 amps which can be provided in three ways. Most desirable is an in-plant connection to the commercial power supply. If unavailable, field hook-ups directly into a power supply normally can be arranged through the utility. When no electrical hook-up is available, a 10 kilowatt gasoline generator is used. This is the least desirable method because of transportation, refueling, and other maintenance.

Automatic Water Monitoring System

A water quality monitor is mounted in the trailer to continuously measure the pH, temperature, conductivity and dissolved oxygen in both the effluent and diluent during bioassays. The readings for each parameter, along with the exact times, are continuously recorded on a strip chart. Each parameter card within the instrument is wired in such a way that when prescribed high or low critical levels are exceeded, the monitor automatically takes an effluent grab sample for later analysis and marks the sampling time on the chart paper. Instrument and wet chemistry analyses performed in the trailer depend on the wastes being bioassayed and the difficulty of the procedure. The most difficult analyses are performed in our main laboratory in Lansing. On-site analyses confirming operation of the automatic system are routine.

Modifications of Wastewater

A waste discharge may have a characteristic which masks the effects of the parameter to be bio-assayed, *i.e.*, excessive temperature masking a biocidal cleaning agent, low dissolved oxygen masking a toxic component. The waste may be adjusted in the supply tank, or a sink or other holding facility, by insertion of such a reservoir into the waste intake line. Chlorine added to municipal wastewater treatment plant discharges as a disinfectant is toxic to aquatic organisms at concentrations below 0.1 mg/l for 96 hours.[1,2] When conducting on-site bioassays with chlorinated wastes addition of a weak sodium thiosulfate solution will remove total residual chlorine. Removal of chlorine permits evaluation of other constituents. Heating, cooling, pH alteration and other modifications have been made.

Supply Tanks and Bioassay Containers

The supply tanks or "header boxes" (Figure 81) are constructed of clear 3/16" cast acrylic plastic. Their size and capacity is dictated by the number of supply tubes desired.

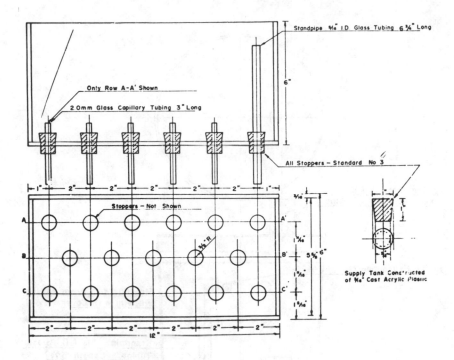

Figure 81.   Supply Tank.

Bioassay Containers

The bioassay containers presently used in the
trailer (Figure 82) apparently are no longer commer-
cially available, and we are constructing our own.
The new containers are made of cast acrylite resin
tube 5" I.D. cut 6" long and have a 1.5 liter
capacity. Other changes include standpipe design,
container support, etc. Each container is equipped
with one mixing chamber for the influent(s), slotted
guards over the effluent tube and standpipe, and
free-fall chambers with air relief holes to permit
constant flow to the next container.

The Michigan Diluter

Planned uses of our trailer facility required
a simple, continuous flow dilution apparatus that
could be rearranged easily on-site to deliver dilution ratios to accommodate effluents of varying
degrees of toxicity and provide a wide range of
concentrations. Our original design was refined to
that described below which meets the criteria above
and is inexpensive to build, reliable, durable, and
requires minimum upkeep.

349

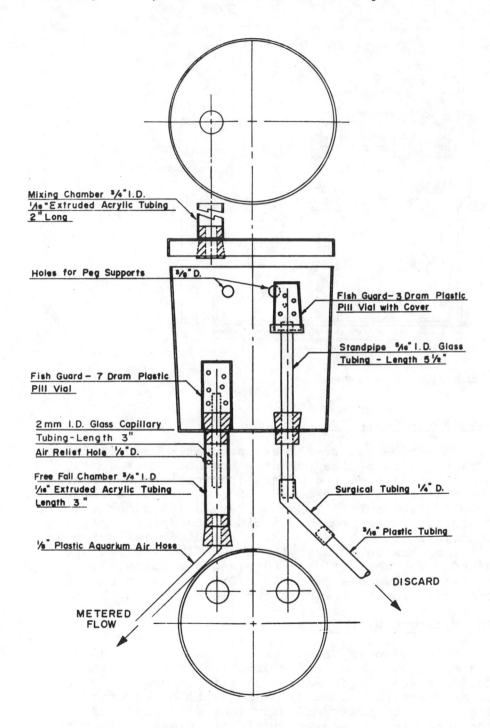

Mixing Chamber ³/₄" I.D.
¹/₁₆" Extruded Acrylic Tubing
2" Long

Holes for Peg Supports ³/₈" D.

Fish Guard— 3 Dram Plastic
Pill Vial with Cover

Standpipe ⁵/₁₆" I.D. Glass
Tubing - Length 5½"

Fish Guard— 7 Dram Plastic
Pill Vial

2 mm I.D. Glass Capillary
Tubing-Length 3"

Air Relief Hole ¹/₈" D.

Free Fall Chamber ³/₄" I.D
¹/₁₆" Extruded Acrylic Tubing
Length 3"

Surgical Tubing ¹/₄" D.

³/₁₆" Plastic Tubing

¹/₈" Plastic Aquarium Air Hose

DISCARD

METERED
FLOW

Figure 82.   Bioassay Container.

Desired test dilutions are made by mixing the appropriate number of equal flows of effluent and diluent water. The flow is regulated by maintaining a constant head of water in the supply tanks and passing this water through pieces of 2.0 mm I.D. glass capillary tubing 3" in length. These flows enter 1/8" aquarium air hoses and flow by gravity to appropriate bioassay containers where they enter a single mixing chamber. In Figure 83 container "X" constantly receives 100% effluent, all supply tubes originate from the effluent supply tank. Container

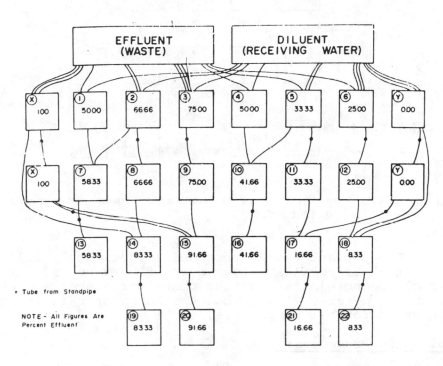

Figure 83.   *A serial diluter delivering effluent concentrations of 8.33 to 100.00%.*

"1" constantly receives 50% effluent, one tube each from the effluent and diluent supply tanks. With this set-up the following concentrations in percent effluent are achieved: 100, 92, 83, 75, 67, 58, 50, 42, 33, 25, 17, 8 and 0. By starting with a 5% waste concentration in the supply tank (Figure 84) the following concentrations in percent effluent are achieved: 5.0, 4.6, 4.2, 3.7, 3.3, 2.9, 2.5, 2.1, 1.7, 1.3, 0.8, 0.4 and 0. If extreme accuracy is desired the flows from each tube must be measured.

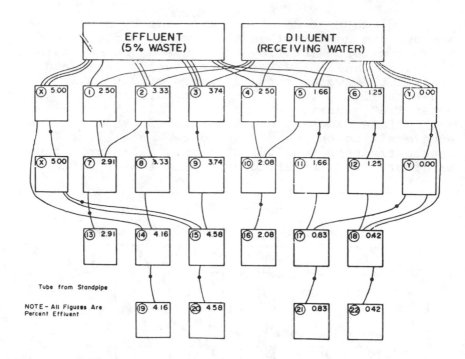

Figure 84.   *A serial diluter delivering effluent concentrations of 0.42 to 5.00%.*

The present system provides approximately 1 liter of water per capillary tube every five minutes. It should be noted that various containers receive more than one metered flow. Thus the turnover rate is not the same for all containers.

Bioassay Procedures

Each bioassay container maintains a constant liquid volume and supports five fish, 2 to 4 inches in length, or other test animals. Each bioassay concentration is replicated so that ten fish are exposed to each concentration. The test fish are usually acclimated to the dilution water for seven days prior to each test. This is accomplished by holding the fish in live cages placed at the time of site reconnaissance.

At the start of the test, both waste and diluent water are run through the bioassay unit for at least one-half hour to stabilize the test concentrations. Fish are then placed in each container and their

initial response is noted. Dead fish are counted
at the end of 24, 48, 72 and 96 hours. Lethal and
sublethal effects on the fish are noted during the
study. Once started, the bioassay does not require
continual attention but is tended intermittently by
a staff member performing routine laboratory and
receiving system investigations.

Advantages

On-site, continuous-flow bioassays using wastewater discharges as the tested material and receiving
system water as the diluent exposes the test animals
to all the changing characteristics of the effluent
and receiving water body. Intermittent and slug
discharges may be evaluated and reactions of test
fish in water quality conditions similar to those
encountered in the natural environment may be observed.
This technique has been used successfully to
bioassay many industrial and municipal discharges.
The approach permits both the discharger and the
enforcement agency an opportunity for on-site assessment of toxicity without total dependence on static
bioassays in remote laboratories, or literature cita-
tions of bioassays conducted with far different
quality diluent water, wastes faintly similar to the
one in question, and using test organisms foreign to
the receiving system of concern.

Materials Needed for Michigan Diluter

The following materials are necessary to construct a Michigan diluter and bioassay system on a
peg-board wall:

Quantity	Description
36	2.0 mm I.D. glass capillary tubing 3" long
36	3/4" I.D. extruded acrylic tubing 1/16" thick, 3" long
12	3/4" I.D. extruded acrylic tubing 1/16" thick, 2" long
28	Standpipes 5/16" I.D. glass tubing 6" long
26	3 dram plastic pill vials with snap on tops
12	7 dram plastic pill vials without tops
28	1/4" I.D. surgical tubing 1-3/4" long

26	One liter plastic containers with removable tops
2	Supply tanks
roll (approx.)	1/8" plastic aquarium tubing
roll (approx.)	3/16" plastic tubing
48	No. 3 rubber stoppers with 3/8" hole
72	No. 3 rubber stoppers with 1/4" hole
26	6" two prong peg supports
4	8" two prong peg supports
1	Tube of Sealastic or similar material

REFERENCES

1. Basch, Robert E., Michael E. Newton, and Carlos M. Fetterolf. "*In Situ* Investigations of the Toxicity of Chlorinated Municipal Wastewater Treatment Plant Effluents to Rainbow Trout (*Salmo gairdneri*)," Symposium on Bioassay Methods, American Chemical Society (1972).
2. Zillich, John A. "Toxicity of Combined Chlorine Residuals to Freshwater Fish," *Water Pollution Control Fed.* 44:212 (1972).

SECTION V

APPLICATIONS

CONTRIBUTORS TO SECTION V

John Cairns, Jr. Department of Biology and Center for Environmental Studies, Virginia Polytechnic Institute and State
University, Blacksburg, Virginia.

Nicholas L. Clesceri. Rensselaer Fresh Water Institute,
Rensselaer Polytechnic Institute, Troy, New York.

E. M. Colon. Rensselaer Fresh Water Institute, Rensselaer
Polytechnic Institute, Troy, New York.

David L. Correll. Radiation Biology Laboratory, Smithsonian
Institution, Rockville, Maryland.

M. Cranmer. Environmental Protection Agency, Perrine Primate
Laboratory, Perrine, Florida.

Darrell L. King. Department of Civil Engineering, University
of Missouri-Columbia, Columbia, Missouri.

R. Kohberger. Rensselaer Fresh Water Institute, Rensselaer
Polytechnic Institute, Troy, New York.

Leif L. Marking. Fish Control Laboratory, Fish and Wildlife
Service, United States Department of the Interior,
La Crosse, Wisconsin.

Gerald C. McDonald. Albany County Sewer District, Albany,
New York.

A. G. Payne. Environmental Water Quality Research Department,
Proctor & Gamble Company, Cincinnati, Ohio.

A. Peoples. Environmental Protection Agency, Perrine Primate
Laboratory, Perrine, Florida.

Dennis M. Sievers. Department of Civil Engineering, University
of Missouri-Columbia, Columbia, Missouri.

Richard E. Sparks. Department of Biology and Center for
Environmental Studies, Virginia Polytechnic Institute
and State University, Blacksburg, Virginia.

R. G. Stross. Department of Biological Sciences, State
University of New York at Albany, Albany, New York.

R. N. Sturm. Environmental Water Quality Research Department,
Proctor & Gamble Company, Cincinnati, Ohio

Charles R. Walker. Branch of Pest Control Research, Fish and
Wildlife Service, U.S. Department of the Interior,
Washington, D.C.

William T. Waller. Department of Biology and Center for
Environmental Studies, Virginia Polytechnic Institute
and State University, Blacksburg, Virginia.

S. L. Williams. Rensselaer Fresh Water Institute, Rensselaer
Polytechnic Institute, Troy, New York.

19. THE USE OF FISH BIOASSAYS TO DETERMINE THE RATE OF DEACTIVATION OF PESTICIDES

Leif L. Marking. Fish Control Laboratory, Fish and Wildlife Service, U.S. Department of the Interior, La Crosse, Wisconsin, and
Charles R. Walker. Branch of Pest Control Research, Fish and Wildlife Service, U.S. Department of the Interior, Washington, D.C.

ABSTRACT

The biological half-life of a pesticide is the time required to decrease its toxicity by one-half. Half-lives can be established by determining the tolerance of fish or other organisms in bioassays of aged solutions containing unknown residual concentrations, and in concurrent reference tests containing known concentrations. The half-life of biological activity is calculated in two procedures by plotting (1) the percent concentration remaining in aged solutions or (2) the deactivation indices against aging time on cyclic semi-logarithmic graph paper. The half-life of biological activity of antimycin, a powerful fish toxicant, was determined for rainbow trout, goldfish, channel catfish, black bullhead, green sunfish, and bluegill. Antimycin was deactivated rapidly in high pH waters, and the half-lives were as follows: pH 10 = 1.5 hours, pH 9.5 = 4.6 hours, pH 9.0 = 9.7 hours, pH 8.5 - 46 hours, pH 8 = 100 hours, pH 7.5 = 120 hours, and pH's 6.0 and 6.5 = 310 hours.

INTRODUCTION

The biological activity of a pesticide is the killing power, and the activity generally decreases as a chemical is deactivated biologically and chemically with time. Included in biological activity

357

are the toxic effects of the original molecule, chemical degradation products, and metabolites of the pesticide.

The half-life of a pesticide's biological activity is the time necessary for that activity to be reduced by one-half. Regulatory agencies now require data on the persistence of pesticides in the environment and the half-life or rate of disappearance is one measure of persistence.

Other than bioassay, analytical methods to detect and measure very small concentrations of some fish toxicants in water are unavailable. For example, antimycin kills rainbow trout at 30 parts per trillion,[1] but analytical methods aren't available to quantitate this concentration. Investigators have observed that antimycin degrades rapidly in water, especially under alkaline conditions.[2] Other studies show that the persistence of most fish toxicants in water is influenced by pH, temperature, sunlight, metabolism by aquatic organisms, and other environmental variables.

The purpose of the study reported here was to develop a bioassay method for determining the rate of deactivation of toxic substances such as antimycin, and to apply the method to a study of effects of pH on the deactivation of antimycin.

METHODS

Standard laboratory techniques were employed for the static bioassays.[3] Technical grade antimycin was dissolved in acetone at the rate of 10 mg/100 ml. Dilutions of the stock solution were prepared as necessary just prior to administering the toxicant. Aliquots of the diluted stock solution corresponding to the desired concentration were delivered to glass jars containing 15 liters of reconstituted water and 10 fish.

The test water was prepared by adding reagent grade salts to deionized water in the following quantities: $NaHCO_3$--48 mg, $CaSO_4 \cdot 2H_2O$--30 mg, $MgSO_4$--30 mg, and KCl--2 mg per liter. The water was saturated with oxygen at 12°C prior to the bioassay.

Different pH solutions were prepared by adding buffering chemicals to the standard test water used in all of the bioassays.[4] The pH was checked daily and corrected with additional buffering when necessary to keep the pH within 0.2 units of the desired level.

The bioassays were observed daily, and the mortalities were removed and recorded. They were terminated after the fish had been exposed to the antimycin for 96 hours.

Six fish species of different sensitivites to antimycin were selected to assess the degree of deactivation at selected aging periods. These fish were obtained from state and federal fish hatcheries, held and cared for by a qualified fish culturist, and acclimated to the test conditions. The species tested were rainbow trout (*Salmo gairdneri*), goldfish (*Carassius auratus*), channel catfish (*Ictalurus punctatus*), black bullhead (*Ictalurus melas*), green sunfish (*Lepomis cyanellus*), and bluegill (*Lepomis macrochirus*).

Two procedures for determining the deactivation rate of antimycin were evaluated as follows.

PROCEDURE ONE

The concentrations of biologically active chemicals remaining in solutions and permitted to age for a selected period of time are determined by introducing fish or other susceptible aquatic organisms for bioassay. More resistant species are exposed to lethal solutions of toxicants that have been aged for short periods of time while more sensitive species are exposed to older solutions which have become deactivated to lower concentrations. Concurrent with the tests in aged media, the same organism must be employed in bioassays of fresh solutions under identical test conditions and exposure periods at concentrations expected to produce mortality. The mortality data from the tests in aged and fresh solutions are analyzed to determine the statistical LC50 (concentration producing 50% mortality).[5] The LC50's are calculated for both tests on the basis of applied concentrations even though the concentration has decreased in the aging tests. The ratio between the two LC50 values is expressed as the percent concentration remaining and represents the total biological activity (toxicity) of the test solutions. The percent concentration remaining is plotted against aging-time on semi-logarithmic paper for each test and the half-lives are read directly from the graph.

Results

Four series of uniform concentrations of antimycin were set up using standard test water at pH 7.5 and

12°C and permitted to age prior to introducing the test fish. Each series included concentrations of 20, 40, 60, 80, and 100 ppb of antimycin and a control with no toxicant. After 7 days of aging, 10 channel catfish were introduced into each concentration and into a control vessel of a single series. On the same day, antimycin was added to five vessels containing 10 channel catfish each at concentrations ranging from 1 to 40 ppb. The calculated 96-hour LC50 was 47.3 ppb in 7-day old solutions, and 20.4 ppb in fresh solutions (Table 41). Thus, the percent concentration remaining after 7 days was 43.13.

Table 41

Toxicity of Antimycin to Fish in Aged and Unaged Solutions and Residual Activity in Aged Solutions at 12°C

Species	Aging Time (days)	96-hour LC50 (ppb) of Aged Solutions	Unaged Solutions	Percent Concentration Remaining
Channel catfish	7	47.3	20.4	43.1
Goldfish	30	60.5	0.650	1.07
Green sunfish	35	52.0	0.308	0.59
Rainbow trout	44	34.4	0.049	0.14

Goldfish, green sunfish, and rainbow trout were exposed similarly to solutions of toxicant that had been aged for longer periods, and the percent concentration remaining was calculated for each aging period (Table 41). The percent concentration remaining and aging-time from Table 41 are plotted on cyclic semi-logarithmic graph paper (Figure 85). The curve approximates a first order decay curve and describes the rate of disappearance for antimycin in soft water of pH 7.5. From the curve, the percent concentration remaining is 50 after 5 days and 25 after 10 days; therefore, the half-life of biological activity is 5 days.

The same procedure was repeated for antimycin solutions at different pH's. Different species of fish were used depending on their availability. The aging time ranged from 2,808 hours at pH 6 to 1 hour

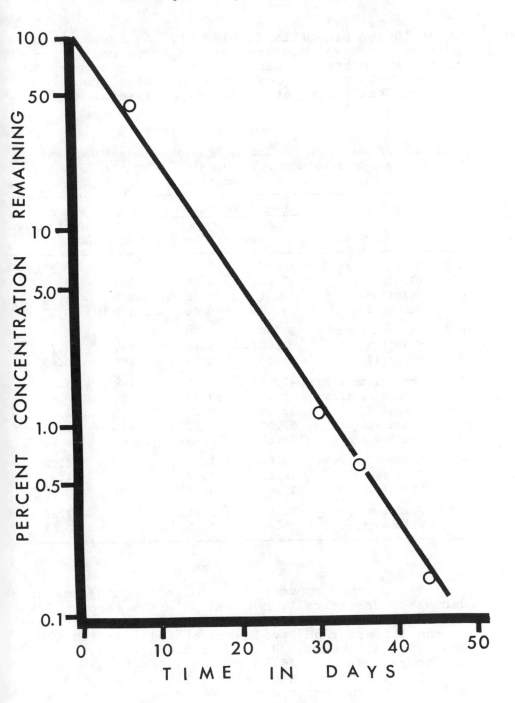

Figure 85.   The rate of inactivation of antimycin at pH 7.5
and 12°C.

at pH 10 for bluegills (Table 42). At these aging periods, survival occurred at the lower concentrations, and mortality was produced at the higher concentrations. The results were plotted to determine the half-lives at different pH levels.

Table 42

Toxicity of Antimycin to Fish in Aged and Unaged Solutions and Residual Antimycin in Aged Solutions at Different pH's

pH	Species	96-hour LC50 (ppb) of Aged Solutions	Unaged Solutions	Aging Time (hours)	Percent Concentration Remaining
6.0	Black bullhead	55.00	12.18	432	22.15
	Bluegill	0.53	0.08	864	15.09
	Goldfish	28.44	0.42	1,800	1.48
	Bluegill	29.90	0.08	2,808	0.27
6.5	Channel catfish	86.50	10.30	1,008	11.91
	Goldfish	12.00	0.45	1,440	3.75
	Bluegill	34.70	0.14	2,640	0.40
7.5	Channel catfish	47.30	20.40	168	43.13
	Channel catfish	60.00	17.30	336	28.83
	Bluegill	13.30	0.17	672	1.28
	Green sunfish	52.00	0.31	840	0.60
	Rainbow trout	28.00	0.05	1,056	0.21
8.0	Channel catfish	22.60	21.70	24	96.02
	Goldfish	28.20	1.22	552	4.33
	Bluegill	23.00	0.22	672	0.96
8.5	Goldfish	21.80	1.61	192	7.39
	Rainbow trout	59.40	0.06	504	0.10

The percent concentration of antimycin remaining decreases dramatically between pH's 6.5 through 9.0 as the aging time increases (Figure 86). Half-lives at the different pH levels are as follows: pH 6.0 and 6.5 = 310 hours, pH 7.5 = 120 hours, pH 8.0 = 100 hours, pH 8.5 = 46 hours, pH 9.0 = 9.7 hours, pH 9.5 = 4.6 hours, and pH 10 = 1.5 hours. The critical pH for antimycin in water is 9.0. Above this its persistence is definitely limited. These data are essential for safe, efficacious, and economical control of target organisms.

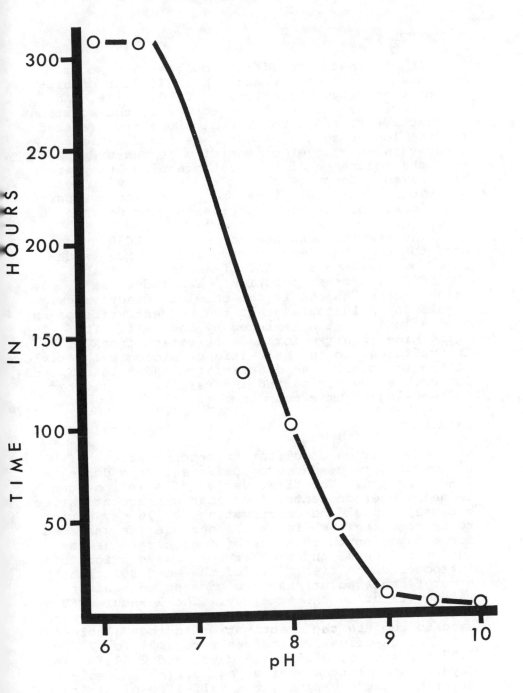

Figure 86. Summary curve for the half-life of antimycin in solutions having different pH's at 12°C.

PROCEDURE TWO

An alternate procedure requires the use of only one species to estimate the half-life of biological activity. Durations of 2, 4, 8, 16, and 32 or more days are selected to permit the solutions to deactivate, and stronger solutions are prepared for the longer aging periods. As the solutions deactivate to a level of toxicity commensurate with the tolerance of the test organism, they are bioassayed. Concurrent with the bioassay of aged solutions, fresh solutions are bioassayed against individuals of the same species to provide reference data.

The results obtained in terms of LC50's are used to compute a deactivation index which is derived by dividing the LC50 of an aged solution by the LC50 of a fresh solution. The index has a value greater than 1 if the chemical deactivates. A value of 2 indicates that the concentration of an aged solution has diminished by one-half. The aging time required for a deactivation index of 2 is equivalent to the half-life of biological activity of the toxicant. The deactivation index is computed for each aging period and the results are plotted on semi-logarithmic coordinates.

Results

Five series of different concentrations of antimycin were prepared on different days during a 16-day period. The first series of solutions contained higher concentrations than subsequent series because it would be deactivating for 16 days prior to introducing the fish. The next series was aged for 8 days, another series of concentrations was aged for 4 days, and another for 2 days prior to introducing the fish. All of the aged solutions and reference solutions were bioassayed against rainbow trout on the same day. The 96-hour LC50's for solutions aged for different time periods were used to compute the deactivation indices (Table 43). The deactivation indices were 1.25 for 2 days, 1.97 for 4 days, 3.25 for 8 days, and 7.50 for 16 days. When plotted against aging time on semilogarithmic graph paper, the half-life of biological activity for antimycin in soft water at pH 7.5 and 12°C is approximately 5 days (Figure 87). This value agrees with that determined using the previous method.

Table 43

Toxicity of Antimycin to Rainbow Trout in Aged and Unaged
Solutions in 96-hour Exposures at pH 7.5 at 12°C

| Aged Solutions | | Unaged | |
Aging Time (days)	96-hour LC50	Solutions LC50	Deactivation Index
2	0.0498	0.040	1.25
4	0.0787	0.040	1.97
8	0.130	0.040	3.25
16	0.300	0.040	7.50

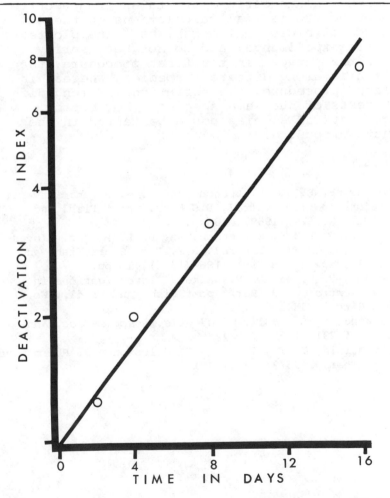

Figure 87.   The inactivation of antimycin at pH 7.5
and 12°C by the alternate procedure.

DISCUSSION

The first procedure is limited by the number of differentially sensitive organisms available. In fact, to more accurately derive the half-life by this method, three or more values are desirable to establish the slope of the curve. Several of these values may be obtained using one species, although more than one species should be used in deriving the half-life curve.

Advantages of the alternate procedure are that only one species is required, fewer reference tests are necessary, and the total time and facility requirements are less. Both methods are estimates because the values derived are taken from a curve representing statistical calculations on the resistance of biological organisms. The plotted values are experimental and do not necessarily occur on the curve. In the first procedure, the variation is among different species, whereas in the second procedure, the variation is among different tests of the same species. For better results, several species could be tested in the alternate procedure.

REFERENCES

1. Berger, B. L., R. E. Lennon, and J. W. Hogan. *Investigations in Fish Control*, U.S. Bur. Sport Fish. and Wildl., No. 26 (1969), 19 pp.
2. Walker, C. R., R. E. Lennon, and B. L. Berger. *Investigations in Fish Control*, U.S. Bur. Sport Fish. and Wildl., No. 2 and Circ. 186 (1964), 18 pp.
3. Lennon, R. E. and C. R. Walker. *Investigations in Fish Control*, U.S. Bur. Sport Fish. and Wildl., No. 1 and Circ. 185 (1964), 15 pp.
4. Marking, L. L. *Bull. Wildl. Dis. Assoc. -Proc. Ann. Conf.* 5:291 (1969).
5. Litchfield, J. T., Jr., and F. Wilcoxon. *J. Pharm. and Exp. Therap.* 96(2):99 (1949).

20. DETERMINATION OF TRACE QUANTITIES OF ANTICHOLINESTERASE PESTICIDES IN AIR AND WATER

M. Cranmer and A. Peoples. Environmental Protection Agency, Perrine Primate Laboratory, Perrine, Florida

There is currently a great emphasis on establishing the quality of our environment. Two compartments of the environment which became contaminated with a seemingly endless number of potentially toxic agents are air and water. Since the need for air and water is for the most part excluded from the realm of freedom of personal selection and since any contamination is widely dispersed, it is imperative that appropriate standards, enforced, insure quality.

Organophosphate and carbamate insecticides are two classes of toxic chemicals which find their way into both air and water. Since these compounds are extremely toxic even in low concentrations and since absorption from air and water is efficient, even trace amounts may be of biological significance. There is currently no single multi-residue method which is applicable to all organophosphate or carbamate insecticides, and many of the available procedures lack specificity.[1-3] Several extremely sensitive methods for certain organophosphate insecticides utilize gas liquid chromatography and selective detectors. The most sensitive requires 10^{-10} g of parent compound for accurate determination.[4-6]

If extraction and partition, selective detection or derivative formation are considered as part of the procedure, a multi-residue method cannot be based on a common denominator, for both the organophosphates and carbamates structural types represented are too diverse. An obvious common denominator

is the ability to inhibit cholinesterase. Quality
standards usually are set to limit toxicity--and
the toxicity of various anticholinesterase insecticides differs markedly. It is reasonable that
screening of environmental samples would be
accomplished most expeditiously by indexing the
total anticholinesterase activity of a sample,
rather than by determining the absolute amounts of
the several components present, since it is the
anticholinesterase potential of these compounds that
determines their toxicity. Samples with unacceptable
levels of anticholinesterase activity would then be
subjected to additional, more specific, analytical
manipulation.

The following method is a modification of a
sensitive procedure for the gas chromatographic
determination of acetylcholinesterase activity in
μl volumes of blood.[7] It is applicable to the
screening of large numbers of samples with the
confidence that if anticholinesterase components
are present in toxicologically significant form,
they will be detected, and their collective anticholinesterase activity determined.

METHODS

Apparatus

A MicroTek MT-220 gas chromatograph equipped
with a single hydrogen flame ionization detector
and a Westronics dual pen recorder were used in
this study.

The incubation of the reaction was carried out
with a Thermolyne Dri-Bath permanently set at 37°.
Components of the assay were added with Oxford
pipettors, the reaction was carried out in 5-ml
glass-stoppered test tubes and a clinical centrifuge
was used to separate the CS_2 and the aqueous layers
in the final reaction mixtures.

Reagents

Reagents were obtained from the following
sources: 3,3-dimethylbutyl acetate (DMBA) and 3,3-
dimethylbutanol (DMB), K & K Laboratories; true
cholinesterase, Type I, Sigma Chemical Works;
paraoxon, Farben Fabriken-Bayer-A.G; malaoxon,
American Cyanamid; carbaryl, Union Carbide; Tween
20[R] emulsifier, Biscayne Chemical Laboratories.
All other reagents were ACS grade.

368

Preparation of Reagents

A stock solution of buffer was prepared by dissolving 44.73 g of potassium chloride, 4.12 g of sodium barbital, and 0.55 g of potassium monobasic phosphate in 200 ml of water.[8] The working buffer solution was prepared by adding 20 ml of stock buffer solution to 75 ml of water, adjusting the pH to 8.0 with 0.1 N HCl, and diluting the solution to a final volume of 100 ml with water.

To remove butanol impurities from DMBA, 5 parts of DMBA were mixed with 1 part of acetic anhydride. The mixture was kept at 37°C for 24 hours and then washed once with water to remove the acetic anhydride and impurities. A 0.2 M solution was prepared by diluting the DMBA with working buffer and adding emulsifier to a concentration of 0.2%. The pH was then adjusted to pH 8.0 with 5% sodium hydroxide.

For gas chromatographic standards, DMB solutions containing 50, 100, 150, and 200 ng/μl were prepared in water containing 0.1% emulsifier. Two ml of CS_2 were added to 2 ml of each standard solution, and the mixtures were shaken well and centrifuged. The top layer was discarded and an appropriate volume of the CS_2 layer was injected into the gas chromatograph. Standard curves were prepared by plotting DMB concentration versus peak height or peak area.

Concentrated formic acid (88%) was diluted 1:1 with distilled water.

A stock enzyme solution was prepared by adding 3 ml of working buffer solution to 50 μM units of true cholinesterase. Dilutions of 16.6 μM units/ml(A), 3.3 μM units/ml(B), 0.66 μM units/ml(C) and 0.13 μM units/ml(D) were prepared in working buffer solution.

Three solutions of the inhibitor of choice *i.e.* carbaryl (2.0 μg/μl, 201 ng/μl and 20.1 ng/μl), paraoxon (27.5 pg/μl, 2.75 pg/μl and 0.275 pg/μl) and malaoxon (3.15 ng/μl, 315 pg/μl and 31.5 pg/μl) are prepared in ethylene glycol and served as standards for comparison of unknown samples.

A 0.2% bromine water solution is prepared by diluting 0.2 ml bromine with 100 ml of distilled water.

Preparation of Sample

Samples of anticholinesterase compounds in water or ethylene glycol have been found to be compatible with the enzymatic system. Alcohol,

369

ether, acetone and other organic solvents are not
acceptable, unless compared to standard curves prepared in a similar way, since they alter the activity
of the enzyme solutions.

Samples of anticholinesterase compounds which
require activation, *e.g.*, parathion to paraoxon,
must be treated with one part of 0.2% bromine water
solution to 10 parts of sample solution for one
hour at 37°C. The bromine water residue does not
interfere with the acetylcholinesterase reaction.

Chromatography

Column: a glass U-tube, 1/4 in. x 6 ft., was
packed with Johns-Manville Chromosorb 101, mesh
size 80-100, and conditioned overnight at 250°C.
Instrument conditions: column oven, 215°C; detector,
265°C; inlet, 245°C; 300 ml/min; hydrogen, 20 ml/
min; nitrogen. 40 ml/min.

Procedure

Standard Curves: Inhibition curves were prepared by incubating for one hour at 37°C the reaction
mixtures given in Step I below:

	Composition of Reaction Mixture			
	A	B	C	D
I Enzyme Solution plus Inhibitor Solution	100 µl of A + 10	200 µl of B + 20	500 µl of C + 50	2,000 µl of D + 200
II Reaction Mixture used for GLC Analysis	10 µl	50 µl	250 µl	1,250 µl

To 1 ml of working buffer in a 5 ml glass-
stoppered test tube add the amounts of incubated
reaction mixtures designated in II and warm to 37°C.
Add 0.2 ml of 0.2 M DMBA and incubate at 37°C for
30 minutes. After 30 min., stop the reaction with
0.1 ml of 44% formic acid and add 2 ml CS₂. Shake
the tubes vigorously, centrifuge and remove aqueous
layer from top by aspiration. Inject an appropriate
volume of CS₂ into the gas chromatograph for DMB

determination. Carry a blank, containing all reagents except enzyme solution, throughout the
procedure.

Unknown: In Step I above substitute the unknown
material for the Inhibitor Solution and follow the
same procedure.

Calculation of Inhibition:

$$\% \text{ inhibition} = \frac{a - b}{a} \times 100$$

a = chart units of DMB in the uninhibited sample.
b = chart units of DMB in the inhibited sample.
Standard curves are prepared by plotting per cent
inhibition versus inhibitor concentration on semi-
log paper. Per cent inhibition of samples are taken
from these standard curves.

Calculation of Quantity:

Quantification is accomplished by comparing
the inhibition obtained with an unknown sample to
that obtained with known concentrations of a reference inhibitor. Results would then be expressed
in moles of paraoxon, carbaryl, malaoxon, or any
inhibitor of choice.

The procedure is summarized by the flow diagram
in Table 44.

RESULTS AND DISCUSSION

This procedure was based on a GLC method for
the determination of the cholinesterase activity
of small (μl) amounts of human blood.[7] The selection of 3,3-dimethylbutyl acetate for use as a
substrate rather than acetylcholine was determined
by several preferred characteristics: (1) decreased
non-enzymatic hydrolysis, (2) solubility in a solvent
of low hydrogen flame response (CS_2), (3) increased
response of 3,3,-dimethyl-1-butanol with respect to
choline or acetic acid, (4) satisfactory chromatographic characteristics of both 3,3-dimethylbutyl
acetate and 3,3-dimethyl-1-butanol, and (5) similarity to acetylcholine in substrate to enzyme
affinity.

Figure 88 illustrates the similarity of the
enzyme-substrate fit of 3,3-dimethylbutyl acetate
and that of acetylcholine.

Figure 89 illustrates the gas chromatographic
characteristics of dimethylbutyl acetate and
dimethyl butanol.

Table 44

Procedure Flow Sheet

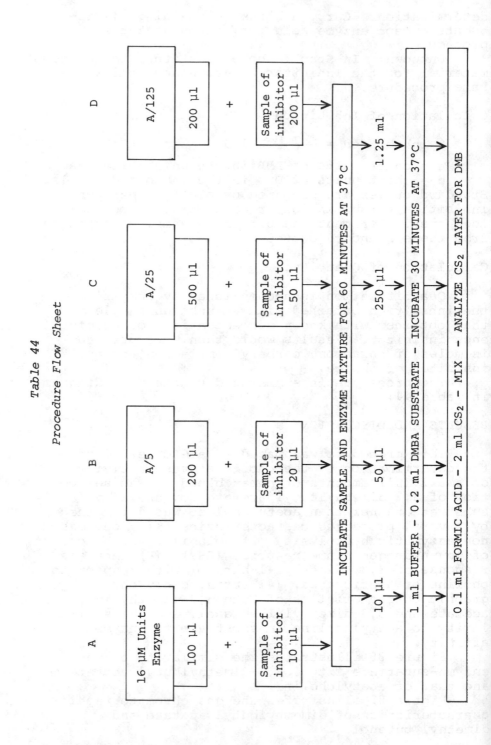

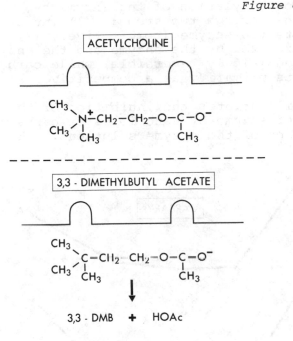

Figure 88. *Illustration of similarity of enzyme-substrate fit of 3,3-dimethylbutyl acetate and acetylcholine.*

ACETYLCHOLINE

3,3 - DIMETHYLBUTYL ACETATE

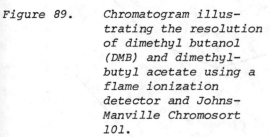

3,3 - DMB **+** HOAc

Figure 89. *Chromatogram illustrating the resolution of dimethyl butanol (DMB) and dimethylbutyl acetate using a flame ionization detector and Johns-Manville Chromosort 101.*

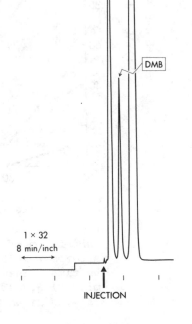

DMBA

DMB

1 × 32
8 min/inch

INJECTION

373

It was anticipated that competitive and non-competitive inhibitors might require different incubation periods with cholinesterase since the mechanism of inhibition is quite different. For example, the phosphorylation of the enzyme by malaoxon proceeds through two steps: (1) intact malaoxon inhibits the enzyme (competitive); (2) malaoxon is hydrolyzed by the enzyme and the enzyme is phosphorylated (slowly reversible) while carbaryl, a carbamate, acts primarily as a competitive inhibitor.

Figure 90 illustrates the inhibition of the four dilutions of enzyme by nano moles of carbaryl per 0.1 ml of each of the enzyme solutions. The

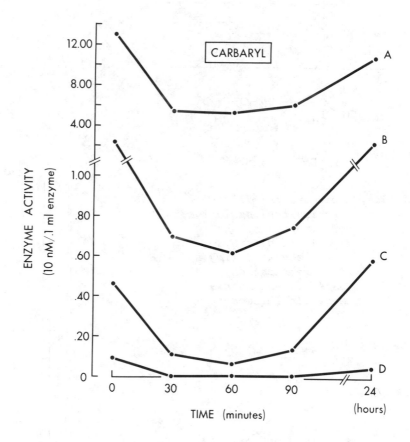

Figure 90.   Inhibition of four activities of true cholin-esterase by 10 nmoles of carbaryl/.1 ml enzyme vs time. Enzyme activity: (A) 16.6 μM units/ml, (B) 3.3 μM units/ml, (C) .66 μM units/ml, (D) .13 μM units/ml.

reduction in activity approaches a maximum in 60
minutes and begins a slow trend toward reactivation
after 90 minutes. Malaoxon demonstrated the same
trend as illustrated by Figure 91. Similar results

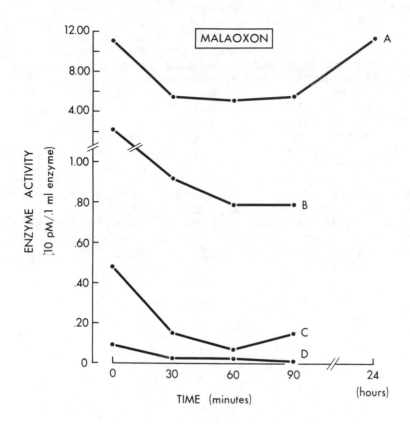

Figure 91. Inhibition of four activities of true cholin-
este ase by 10 p moles of malaoxon/.1 ml enzyme
vs time. Enzyme activity: (A) 16.6 µM units/
ml, (B) 3.3 µM units/ml, (C) .66 µM units/ml,
(D) .13 µM units/ml.

are obtained for other inhibitors including the very
potent inhibitor paraoxon. These observations led
to the conclusion that a 60-minute incubation period
was sufficient and that there were no problems in
analyzing either competitive or non-competitive
inhibitors with this assay system.

The conditions of this assay were arranged to
screen a wide range of concentrations of cholinesterase inhibitors. Four dilutions differing by

a factor of 5 provide a 125 fold difference in enzyme concentration available for inhibition. By varying the ratios of buffer to enzyme, plus buffer mixture, the dilutions are adjusted to provide an equivalent amount of activity present in each reaction. For example, only 10 µl of the most active enzyme solution (A) is added to 1 ml of buffer while 1.25 ml of the least active (D) solution is used. This compensation results in the same range of reaction product being produced for all initial enzyme concentrations (A, B, C, and D) and greatly simplifies the GLC determinative step. However, since the kinetics of the inhibition of this enzyme depend on the concentration of the enzyme and inhibitor, the different enzyme concentrations do not produce four identical reaction systems. There is, however, a clear relationship between the different enzyme dilutions and the amount of inhibition which results at each inhibitor concentration.

Figure 92 illustrates the inhibition of each of the four enzyme dilutions produced by three paraoxon concentrations differing by a factor of 10.

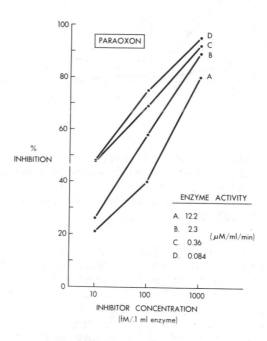

Figure 92. *Per cent inhibition of four true cholinesterase activities by three concentrations of paraoxon.*

The paraoxon concentrations used are expressed in femto moles (10^{-15}). This illustration demonstrates that for paraoxon the sensitivity of the assay is in the femto gram region.

Figure 93 illustrates the inhibition characteristics of malaoxon in the pico mole range on the various enzyme dilutions. There is roughly a 100 fold difference in the inhibitory effectiveness of paraoxon and malaoxon in this assay procedure. This difference roughly parallels the difference in the intravenously administered LD_{50} of the two compounds.

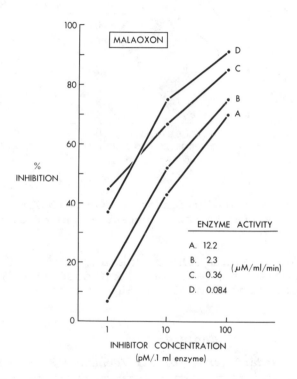

ENZYME ACTIVITY

A. 12.2
B. 2.3
C. 0.36   (μM/ml/min)
D. 0.084

INHIBITOR CONCENTRATION
(pM/.1 ml enzyme)

Figure 93. *Per cent inhibition of four true cholinesterase activities by three concentrations of malaoxon.*

Figure 94 illustrates the inhibition characteristics of carbaryl in the nano mole range. There is roughly a 100,000 fold difference in the inhibitory effectiveness of paraoxon and carbaryl and again this indicates the order of magnitude separating the toxicities of the two compounds.

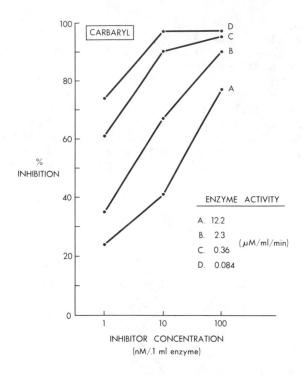

Figure 94.   Per cent inhibition of four true cholinesterase activities by three concentrations of carbaryl.

Table 45 presents the means and standard errors of six replicate samples. In each case the inhibitor concentrations produced about 50% inhibition of the enzyme.

Table 45

Mean and Standard Error for Six Identical Samples

Enzyme Activity	A	B	C	D
Carbaryl				
Mean µM/ml/min	5.4	0.64	0.065	0.0046
S. E.	0.11	0.01	0.001	0.0004
Malaoxon				
Mean µM/ml/min	6.10	1.0	0.154	0.0235
S. E.	0.11	0.02	0.001	0.0005

Table 46 shows mean and standard error of day to day variation. The values are of seven determinations made on a given sample of enzyme in a two month period.

Table 46

Mean and Standard Error of Day to Day Variation of Seven Determinations

		A	B	C	D
Mean	M/ml/min	11.2	2.2	0.41	0.083
S. E.		0.4	0.07	0.02	0.003

The data from which Table 47 was derived support the observation that the assay procedure is sufficiently precise to distinguish at the 95% confidence level the difference between 1 and 3 pico moles of malaoxon.

Table 47

Illustration of Inhibitor Concentration and % Inhibition

Enzyme Activity	Sample	% Inhibition	
µM/ml/min.	Inhibitor conc. in moles	Expected*	Observed†
A 12.2	$1 - 1 \times 10^{-6}$	44	41
	$2 - 3 \times 10^{-6}$	59	59
	$3 - 5 \times 10^{-5}$	66	66
B 2.3	$1 - 1 \times 10^{-6}$	51	48
	$2 - 3 \times 10^{-6}$	64	67
	$3 - 5 \times 10^{-5}$	70	74
C 0.36	$1 - 1 \times 10^{-6}$	61	54
	$2 - 3 \times 10^{-6}$	72	73
	$3 - 5 \times 10^{-5}$	77	82
D 0.084	$1 - 1 \times 10^{-6}$	70	62
	$2 - 3 \times 10^{-6}$	81	80
	$3 - 5 \times 10^{-5}$	86	86

* Results expected from standard curve.
† Results obtained experimentally.

Samples of malathion, parathion, carbaryl and other anticholinesterase compounds have been analyzed in both ethylene glycol, a common solvent used in air samplers, and in potable water. Quantities of these compounds were determined both by standard analytical procedures and by the enzyme assay technique, which produced results that were in agreement.

The anticholinesterase activities of samples containing direct inhibitors and components which must be activated can be divided by determining duplicate samples of which only one is treated with bromine water.

This assay technique shows much promise for further development in the area of environmental sample evaluation.

For most screening procedures, the oxidation of all samples and the use of a single, probably B, enzyme concentration would be sufficient to identify samples which might require further analytical evaluation.

SUMMARY

A procedure has been described which is applicable to the screening of large numbers of samples for anticholinesterase activity. The method is sufficiently sensitive to detect cholinesterase depression due to as little as 30 ppt paraoxon; 3 ppb malaoxon or 2 ppm of carbaryl in either ethylene glycol or water. While the method does not allow the absolute differentiation between compounds which might be present, some classification is possible, *i.e.*, for compounds which must be activated before they possess anticholinesterase activity. The technique can be made more specific through manipulations of types of cholinesterase used and substrates employed. The major usefulness of the technique at this time is for screening.

REFERENCES

1. Storherr, R. W., P. Ott, and R. R. Watts. *J.A.O.A.C.* *54*:513 (1971).
2. Beckman, Herman, and Dennis Garber. *J.A.O.A.C.* 52:286 (1969).
3. Porter, Mildred L., Raymond J. Gojan, and Jerry A. Burke. *J.A.O.A.C.* 52:177 (1969).
4. Butler, Lillian I., and Leslie M. McDonough. *J.A.O.A.C.* *53*:495 (1970).

5. Shafik, M. T., Diane Bradway, and Henry F. Enos. *Bull. Environ. Contam. & Tox. 6*:55 (1971).
6. Bowman, M. C., and Morton Beroza. *J.A.O.A.C. 53*:499 (1970).
7. Cranmer, M. F., and A. Peoples. *J. Chrom. 57*:365 (1971).
8. Henry, R. J. *Clinical Chemistry: Principles and Techniques.* (New York: Hoeher Medical Division, 1964), p. 494.

21.   THE RELATIONSHIP BETWEEN CONTINUOUS BIOLOGICAL
      MONITORING AND WATER QUALITY STANDARDS FOR
      CHRONIC EXPOSURE

John Cairns, Jr., Richard E. Sparks, William T.
Waller.*[†]* Department of Biology and Center for
Environmental Studies, Virginia Polytechnic Institute
and State University, Blacksburg, Virginia.

ABSTRACT

   A rapid biological monitoring system was developed
which measured changes in activity and breathing rates
of fish in order to provide an early warning of developing toxicity in the wastes of an industrial
plant. The system has been tested in the laboratory
with zinc. Fish detect potentially lethal zinc
concentrations (8 mg/l) rapidly enough so that they
survive if zinc is removed at the time of detection,
and they also detect sublethal concentrations (2-3
mg/l). Further tests are being carried out to
determine the general suitability of this method.

INTRODUCTION

   As the pollution problems associated with an
expanding industrial base and increasing population
size become more acute, the need for an overall
program of ecosystem management to minimize the
adverse effects of man's activities on the environment becomes essential.
   The recent report of the Council on Environmental Quality[1] repeatedly stressed the need for
predictive, simulative and management capabilities

―――――――――――――
*Present address: Illinois Natural History Survey, Havana,
 Illinois.
[†]Present address: New York University, Tuxedo, New York.

to combat air and water pollution. These techniques must be developed if the concept of multiple use set forth in the Water Quality Act of 1965 (Public Law 89-234) is to be achieved and if the present practice of alteration of the environment without comprehension is to be eliminated.

A great deal of progress has been made in determining biologically safe concentrations of toxicants for fish. Along with the standards for the chronic exposure of fish to toxicants, there must be safeguards against the development of acutely toxic conditions resulting from either industrial or municipal "accidental spills" or from changes in the total environment of the organism due to a combination of normal environmental variation and industrial processes.

Continuous physical and chemical monitoring systems can be used to detect, almost instantaneously, changes in individual environmental factors. However, aquatic organisms respond to the collective impact of these environmental factors, and this cannot be predicted from physical and chemical data alone.

Techniques have been developed in our laboratory which permit the continuous monitoring of either fish movement patterns or breathing.[2] Both techniques detected effects when test species were exposed to lethal and sublethal concentrations of zinc. In addition, reproduction studies were conducted in an attempt to relate the detection limits of the techniques to biologically safe concentrations for chronic exposure.

METHODS AND MATERIALS

Fish Movement Patterns

Fish movement patterns were monitored using the technique of light beam interruption described in detail by Cairns, *et al.*[2] Dawn and dusk were simulated in the experimental room by a motor-driven dimming unit which turned on and gradually increased the intensity of the room lights (0.5 to 100.0 foot-candles at the surface of the aquaria) over a half-hour period starting at 6:30 a.m. and gradually decreased the intensity (100.0 to 0.5 foot-candles) over a half-hour period before shutting the lights off at 7 p.m. The cumulative movement of each of six bluegill sunfish, a single fish per tank, was recorded every hour throughout a test except during the simulated sunrise and sunset when an additional record was made on the half hour. Each day was divided into four intervals: first half-day (7:30 a.m.-1 p.m.), second half-day (2 p.m.-7:30 p.m.), first half-night (8 p.m.-1 a.m.), second half-night (2 a.m.-7 a.m.) (Table 48). Before any statistical analysis could be performed, recordings for day 1 had to be completed. After the cumulative movement for day 1 was recorded, statistical analyses were performed after the completion of each designated time interval. For example, the cumulative movement values recorded hourly for each fish during day 1, first half-day were compared to the cumulative movement values recorded hourly for each fish during day 2, first half-day. The statistical test used was a two sample test for homogeneous variance ($\alpha = .002$).[3] If the statistical test indicated homogeneous variance, a zero was scored (Table 48) and the fish was considered to be exhibiting a normal movement pattern. If the statistical test indicated heterogeneous variance the fish was considered to be showing abnormal movement and an asterisk was recorded. As can be seen from the results of the experiment presented in Table 48, the statistical test occasionally indicated abnormal movement (day 1 *vs*. day 2, second half-day values, fish 6) during a period in which no toxicant was being added to the system. However, a positive test for abnormal movement was never recorded for more than a single fish during a given time interval unless toxicant was being added to the system. Whenever a positive test for abnormal movement was scored, the cumulative movement recorded for the most recent time interval was dropped and the values recorded during the preceding interval were compared to the next recorded interval. For example, when the positive test for abnormal movement was recorded for day 1 *vs*. day 2, fish 6, second half-day values, the cumulative movement for day 2 was dropped and where the table shows day 2 *vs*. day 3 for this fish during this interval, the actual comparison was between the movement recorded for day 1 and day 3.

Based on the results of 20 laboratory experiments, "stress detection" was defined as the presence of two or more abnormal movement patterns for two or more fish recorded during the same half day.

A series of experiments at progressively lower zinc ($ZnSO_4 \cdot 7H_2O$, reagent grade) concentrations was used to determine the lowest concentration detectable by the movement apparatus.

Table 48

Statistical Analysis of Light Beam Interruptions Recorded During Days 1-20 of Experiment 20. Bluegill exposed to a zinc concentration increasing to a maximum of 13.32 mg/1 (11.39 to 13.32) over a 6 1/2 hour period on day 7 showed no positive test for stress detection. The same fish were exposed to a maximum zinc concentration of 7.52 mg/1 (7.49-7.52) on day 13 and 14 until a positive test for stress detection occurred. Detection occurred between 24 and 29 1/2 hours.

Fish	Day 1 vs Day 2	Day 2 vs Day 3	Day 3 vs Day 4	Day 4 vs Day 5	Day 5 vs Day 6	Day 6 vs Day 7	Day 7 vs Day 8	Day 8 vs Day 9	Day 9 vs Day 10	Day 10 vs Day 11	Day 11 vs Day 12	Day 12 vs Day 13	Day 13 vs Day 14	Day 14 vs Day 15	Day 15 vs Day 16	Day 16 vs Day 17	Day 17 vs Day 18	Day 18 vs Day 19	Day 19 vs Day 20
									First Half-Day Values										
1	o	o	o	o	o	o	o	o	o	o	o	o	*	o	o	o	o	o	o
2-C	o	o	o	o	o	o	o	o	o	o	o	o	o	o	o	o	o	o	o
3	o	o	o	o	o	o	o	o	o	o	o	o	*	o	o	o	o	o	o
4	o	o	o	*	o	o	o	o	o	o	o	o	o	o	o	o	o	o	o
5	o	o	o	o	o	*	*	o	o	*	*	o	*	*	*	*	o	o	o
6	*	o	o	o	o	o	o	o	o	o	o	o	o	o	o	o	o	o	o
									Second Half-Day Values										
1	o	o	o	o	o	o	o	o	o	o	*	*	*	*	*	*	o	o	o
2-C	o	o	o	o	o	o	o	o	o	*	o	o	o	o	o	o	o	o	o
3	o	o	o	o	o	o	o	o	o	o	o	*	o	o	o	o	o	o	o
4	o	o	o	*	o	o	o	o	o	o	o	o	o	o	o	o	o	o	o
5	o	o	o	o	o	o	o	o	o	o	o	o	o	o	o	o	o	o	o
6	*	o	o	o	o	o	*	o	o	*	o	o	o	o	o	o	o	o	o
									First Half-Night Values										
1	o	o	o	o	o	o	o	o	o	o	o	*	o	o	o	o	o	o	o
2-C	o	o	o	o	o	o	o	o	o	o	o	o	o	o	o	o	o	o	o
3	o	o	o	o	o	o	o	o	o	o	o	*	o	o	o	o	o	o	o
4	o	o	o	o	o	o	o	o	o	o	o	o	o	o	o	o	o	o	o
5	o	o	o	o	o	o	o	o	o	o	o	o	o	o	o	o	o	o	o
6	o	o	o	o	o	o	o	o	o	o	o	o	o	o	o	o	o	o	o
									Second Half-Night Values										
1	o	o	o	o	o	o	o	o	o	o	o	o	o	o	o	o	o	o	o
2-C	o	o	o	o	o	o	o	o	o	o	o	o	o	o	o	o	o	o	o
3	o	o	o	o	o	*	o	o	o	o	o	*	o	o	o	o	o	o	o
4	o	o	o	*	o	o	o	o	o	o	o	o	o	o	o	o	o	o	o
5	o	o	o	o	o	o	*	o	o	o	o	*	*	*	*	*	*	o	o
6	o	o	o	o	o	o	o	o	o	o	o	o	o	o	o	o	o	o	o

[Zn++ Added] in First Half-Day Values / Second Half-Day Values region at Day 6 vs Day 7; Zn++ Added in First Half-Night Values / Second Half-Night Values region at Day 12 vs Day 13.

o = Difference in variances not significant at α = .002.   * = Difference in variances significant at α = .002.

Fish Breathing

Breathing rates were determined from polygraph recordings of breathing signals from 52 bluegill sunfish used in nine experiments. The fish were tested in plexiglas tubes through which dechlorinated tap water or zinc solutions were metered at a flow rate of approximately 100 ml/min. Breathing signals were detected by three platinum wire electrodes placed in the water; an active electrode, an indifferent electrode, and a ground. The test chambers and methods of acclimating the fish are described in more detail by Cairns, *et al*.[2] The photoperiod was the same as that for the fish movement study.

The fish were placed in test chambers by 6:00 p.m. and the recordings began at 6:00 a.m. the next day. Zinc solutions were first introduced one to six days after the experimental fish had been exposed to tap water containing no added zinc, and were continuously introduced until the experiment terminated. Each experimental fish thus served as its own control. In the experiment reported in Table 49, one fish was not exposed to zinc, and served as control throughout the experiment. In Table 50, the results of two experiments are reported together. In Experiment 1, six fish were exposed to water containing no added zinc for four days, to determine how many responses would be obtained under control conditions. In experiment 7, three fish were exposed to tap water containing no added zinc for one day, then to 2.55 mg/l zinc for 67 hours.

Preliminary evidence suggested that the data could be analyzed by separating the experimental day into four periods: a period from 6:00 to 8:00 a.m. when the breathing rates changed markedly; a period from 9:00 a.m. to 5:00 p.m. when the rates were comparatively high; another period of rapid change from 6:00 to 8:00 p.m.; and a night period from 9:00 p.m. to 5:00 a.m. when the rates were comparatively low.[4]

Bluegills increase their breathing rates when exposed to zinc.[2] Since a response is commonly considered to be a reaction to a stimulus, it seemed justifiable to arbitrarily define a "response" in these experiments as an increase in breathing rate. An individual fish was considered to have shown a response each time its breathing rate during a time period exceeded the maximum breathing rate observed during the corresponding period of the first day, before any zinc was added. The maximum breathing

Table 49

Number of Fish Showing Breathing Responses, Before and During Exposure to 4.16 mg/1 zinc

Day		6am	7	8	9	10	11	12	1pm	2	3	4	5	6	7	8	9	10	11	12	1am	2	3	4	5
2	Ex	0	0	0	0	0	0	1	0	0	0	0	0	0	0	0	0	0	0	1	0	0	1	0	0
	Con	0	0	0	0	0	0	0	0	0	0	0	0	0	0	0	0	0	0	0	0	0	0	0	0
3	Ex	0	0	0	0	0	0	0	0	0	0	0	0	1	0	0	0	0	0	0	0	0	1	0	0
	Con	0	0	0	0	0	0	0	0	0	0	0	0	0	0	0	0	0	0	0	0	0	0	0	0
4	Ex	0	0	0	0	0	0	0	0	0	0	0	1	0	0	0	0	Recorder off							
	Con	0	0	0	0	0	0	0	0	0	0	0	0	0	0	0	1	–	–	–	–	–	–	–	–
5	Ex	–	–	–	0	0	0	0	0	0	0	0	0	0	1	1	1	1	2	2	1	1	1	1	0
	Con	–	–	–	0	0	0	0	0	0	0	0	0	0	0	0	0	0	0	0	0	0	0	0	0
6	Ex	Recorder off	–	–	0	0	1	0	0	0	0	0	0	1	0	0	1	1	0	1	2	0	1	2
	Con	off	–	–	0	0	0	0	0	0	0	0	0	0	0	0	0	0	0	0	0	0	0	0
7	Ex	0	0	0	1	↓1	2	3	0	1	1	1	2	1	0	0	3	3	4	3	3	3	3	2	2
	Con	0	0	0	0	0	0	0	0	0	0	0	0	0	0	0	0	0	0	0	0	0	0	0	0
8	Ex	2	2	0	1	4	3	3	0	1	1	1	2	1	1	0	4	4	3	3	4	3	4	3	1
	Con	0	0	0	0	0	0	0	0	0	0	0	0	0	0	0	0	0	0	0	0	0	0	0	0
9	Ex	0	0	1	1	2	1	3	1	1	2	1	1	0	2	2	2	3	3	3	3	3	2	2	2
	Con	0	0	0	0	0	0	0	0	0	0	0	0	0	0	0	0	0	0	0	0	0	0	0	0
10	Ex	0	2	0	0	0	0	2	0	1	0	1	1	0	0	1	2	3	2	2	3	3	1	2	2
	Con	0	0	0	0	0	0	0	0	0	0	0	0	0	0	0	0	0	0	0	0	0	0	0	0
11	Ex	0	0	0	0	0	0	0	0	end of experiment															
	Con	0	0	0	0	0	0	0	0																

⁺Measured zinc concentration of 4.16 mg/l introduced. Duration of zinc exposure shown by underline.

NOTE: There were 4 experimental fish (Ex) from Experiment 8 and 1 control fish (Con) from Experiment 8.

Table 50

Number of Fish Showing Breathing Responses, Before and During Exposure to 2.55 mg/l zinc

Day	6am	7	8	9	10	11	12	1pm	2	3	4	5	6	7	8	9	10	11	12	1am	2	3	4	5
Ex 2	0	0	1	1	1↓	1	1	1	1	1	1	1	2	1	2	1	1	1	1	0	1	1	1	1
Con	0	0	0	0	0	0	0	0	0	0	0	0	0	0	0	0	1	0	0	0	0	0	0	0
Ex 3	1	2	1	1	1	1	1	1	1	2	1	1	1	1	0	1	0	0	0	1	1	2	1	—
Con	0	0	0	1	1	1	1	0	0	0	0	0	0	1	0	0	0	0	1	0	0	0	1	0
Ex 4	recorder off			1	0	0	0	0	2	3	3	2	1	0	0	1	1	1	1	2	2	2	2	3
Con	0	0	0	0	0	0	1	1	0	0	0	0	0	0	0	0	0	0	0	0	0	0	0	0

↓Measured zinc concentration of 2.55 mg/l introduced. Duration of zinc exposure shown by underline.

Note: There were 3 experimental fish (Ex) from Experiment 7 and 6 control fish (Con) from Experiment 1.

rate of a particular fish during the dawn period of
the first day (6:00 to 8:00 a.m.) was compared to
the breathing rates recorded for that fish during
the dawn period of the second day. A response was
scored for each value on the second day higher than
the first day maximum. The same procedure was followed in comparing the maximum rate recorded during
the 9:00 a.m. to 5:00 p.m. period of the first day
to the breathing rates recorded during the same
period of the second day. The dusk and night periods
were compared in the same way. The same procedure
was followed in comparing the maxima of the first
day to values recorded on the third day, the fourth
day, *etc.*

The rationale for this method of analysis is as
follows. The breathing rates during all periods of
the first day were generally slightly higher than
rates during comparable periods of subsequent days,
perhaps due to incomplete recovery from the stress
of handling. Any increase in breathing rate after
the first day which exceeded the maxima observed on
the first day could thus reasonably be ascribed to
some sort of stress. The sensitivity of detection
could be increased by using not the first day's
results, but the results of the second or third day
to establish the maximal breathing rates. However,
when the second or third day was used as a baseline,
many responses were obtained on subsequent days even
when no zinc was added to the water. The control
periods (before any zinc was added) and the experiment where no zinc was added at all were used to
determine how many false detections this method of
analysis would produce. A false detection is one
occurring before any zinc is added to the water.
The experimental periods (after zinc was added)
determined how quickly zinc could be detected using
this method of analysis.

Fish Reproduction

Bluegill sunfish were seined from a local pond
and held for several months in the laboratory in the
same dilution water, with the same photoperiod and
water temperatures as for the monitoring experiments.
The chemical characteristics of the dilution water
are shown in Table 51.

Starting April 13, approximately 200 fish
(approximate total lengths 8-15 cm; weights 10-80
gms) were brought into breeding condition by exposing them to a photoperiod of 16 1/2 hours of

Table 51

Routinely Determined Characteristics of Dilution Water

Water Characteristic	Number of Analyses	Mean	Standard Deviation	Range
Total Hardness (mg/l as CaCO)	394	51	10	34-68
Phenolphthalein alkalinity (mg/l as CaCO)	393	0.0	0.0	Always 0.0
Total alkalinity (mg/l as CaCO)	393	41.3	8.8	29-68
pH	397	7.8	0.3	7.0-8.5
Temperature (°C)	396	19.7	1.8	16.3-23.5
Dissolved oxygen (mg/l)	Maintained at air saturation			
Chlorine	None			
Zinc, copper	Zinc and copper were less than .03 mg/l in random analyses of the dilution water by atomic absorption photospectrometry.			

light, water temperatures of 31-32°C, and by feeding them twice daily with frozen Gordon Formula[5] and once daily with live mealworms. The dimming system described earlier simulated a 1/2 hour dawn starting at 6 a.m. and a 1/2 hour dusk starting at 10 p.m.

On May 4 most of the fish could be sexed by gently squeezing the sides and observing whether eggs or milt was extruded, and three females and one male were placed in each of twenty 20-gallon tanks (standard aquaria, long type, Ramfab Aquarium Products cat. no. RA-20L).

One standard clay flowerpot (upper rim-to-rim diameter 6 in.) was placed on its side in each tank for the females to hide from the aggressive attacks of the males. An artificial nest, described by Eaton, was also placed in each tank and five river pebbles, 2-3 cm in diameter, were scattered on the bottom of each nest.

One toxicant delivery apparatus was used for each set of five tanks receiving one concentration, and one water delivery apparatus was used for five control tanks which received no added zinc. The toxicant delivery apparatus consisted of a toxicant dipper which dipped out 5 ml of stock solution,[7] a chamber where the stock solution and dilution water were mixed thoroughly by the turbulence of incoming dilution water and five metering cells[8] which received the mixed solution and delivered replicate 300 ml quantities to each tank once every three minutes. The zinc concentrations for the reproduction study were based on the lowest concentration used in the breathing experiments, 2.55 mg/l. Tanks 6-10 received .235 mg/l zinc (approximately 1/10 of 2.5 mg/l), tanks 16-20 received .035 mg/l zinc (1/75 of 2.5), and tanks 11-15 received no added zinc and served as controls. The 96-hour median tolerance limit (TL50) for adult bluegill sunfish exposed to zinc in Blacksburg tap water is 7.5 mg/l. Tanks 1-5 received approximately 1/100 this concentration (.076 mg/l). The water entered at the top and front of the tank and was removed from the bottom, carrying some detritus with it, by means of a sheathed standpipe at the rear of the tank.

A plastic egg hatching box (20.5 cm long, 7.0 cm wide, and 15.5 cm deep) hung on the front of each aquarium and was large enough to accept three egg cups (Figure 95). The egg cups were made from Turtox plastic jars (5.5 cm o.d., 6.8 cm tall), with the bottoms sawed off. In use, each end of the cup was covered with a piece of ladies' woven nylon support hose held in place with rubber bands. Each cup rested on an airstone cemented to the bottom of the hatching box. A piece of plexiglas (20 cm long, 6.3 cm wide, and 0.6 cm thick) fit into grooves in the hatching box and rested on top of the cups to keep them from floating.

Water siphoned from each tank into the hatching boxes was returned to the tank by an air lift. Another air lift delivered water from the hatching box to a plastic dishpan (38.0 cm long, 30.5 cm wide, and 17.5 cm deep) for rearing newly-hatched fry. Five 1.1 cm holes were drilled in each dishpan on centers 11.5 cm above the bottom, and were covered with a piece of woven nylon support hose. Thus the eggs were hatched and the fry reared in the same water in which their parents lived.

The fish in each tank were fed two grams of frozen Gordon Formula[5] twice a day and eight live

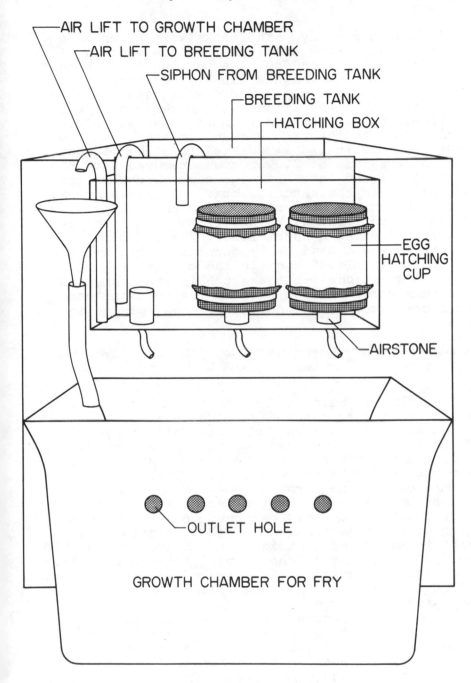

AIR LIFT TO GROWTH CHAMBER
AIR LIFT TO BREEDING TANK
SIPHON FROM BREEDING TANK
BREEDING TANK
HATCHING BOX

EGG
HATCHING
CUP

AIRSTONE

OUTLET HOLE

GROWTH CHAMBER FOR FRY

Figure 95.   *Chambers for growth and reproduction study. The
plastic piece used to keep the egg cups from
floating is not shown. One egg cup is removed
to show the air lifts. The air tubing to the
air lifts is not shown.*

393

mealworms once a day. The tanks were cleaned once
a week by siphoning detritus from the bottom.

At 1 p.m. every day, the pebbles in each nest
were removed and examined closely for eggs. If
eggs were present, a plastic dishpan (same dimen-
sions as above) was filled with water from the tank
and the nest was removed from the tank and placed
upside down in the pan over an airstone. A new
nest was substituted immediately for the old one.

A subsample of 200 eggs was removed from the
nest and placed in a hatching cup, which in turn
was placed in the hatching box. After 48 hours,
the number of fry in both the egg cup and the
dishpan were counted, by pipetting them into petri
dishes and using a Dazor model M209 fluorescent
magnifier and a hand tally counter. The hatch
in the subsample of eggs in the cup was assumed
to be proportional to the hatch in the nest, and
the numbers of fry and eggs in the cup were used
to back-calculate the number of eggs spawned in the
nest:

$$\begin{array}{l} \text{Total No. Eggs} \\ \text{in Nest} \end{array} = \begin{array}{l} \text{No. Fry} \\ \text{in Nest} \end{array} \left(\begin{array}{l} \text{No. Eggs} \\ \text{in Cup} \end{array} \div \begin{array}{l} \text{No. Fry} \\ \text{in Cup} \end{array} \right) + \begin{array}{l} \text{No. Eggs} \\ \text{in Cup} \end{array}$$

Fish that were dead or that had lost their
equilibrium were removed as soon as they were
noticed. In addition, six fish had an eye
disease that started as a white spot and gradually
consumed the entire eye, and these fish were also
removed. Fish that were removed before the end
of the experiment were weighed, measured, and
sexed, unless they were too decomposed. Fish that
were removed and could be sexed were replaced by a
fish of the same sex from a stock of ripe fish
kept in tap water containing no added zinc, until
July 20, when no further replacements were made.
The breeding experiment terminated August 19, when
all the remaining fish were killed, weighed,
measured, and sexed.

Eggs were hatched and fry reared for 90 days
in water drawn from the aquarium where the eggs
were spawned. Also, some newly-hatched fry were
transferred from high zinc concentrations to low
concentrations and vice-versa. Fry were initially
fed plankton obtained from local ponds, then
newly-hatched brine shrimp.

RESULTS

Fish Movement Patterns

Table 48 shows the results of one continuous flow experiment carried out for 20 days. During this experiment fish were exposed to zinc on day 7 from 1:00 p.m. until 7:00 p.m. at which time the flow was returned to normal dilution water. The zinc concentrations reached their maximum at 7:00 p.m. and atomic absorption analyses on effluent samples collected at this time showed the following concentrations: tank one, 13.32; tank two, less than 0.08; tank three, 11.39; tank four, 12.72; tank five, 13.32; and tank six, 12.59 mg/l Zn^{++}. The results show that these concentrations of zinc developing over the six and one half hour interval of exposure were insufficient to cause a detectable change in the movement patterns of the fish. Because the maximum concentration of zinc reached in this experiment was considerably higher than the minimum concentration detected it is felt that the time of exposure was insufficient for development of detectable stress conditions. By 8:30 a.m. of day 8 the effluent zinc concentrations were less than 0.30 in all cases. To determine the percent survival and recovery patterns of the fish once stress detection occurred, zinc flow was re-initiated at 1:00 p.m. on day 13 of this experiment. Between 8:00 and 9:00 p.m. on day 13 the zinc concentration in the effluent reached a maximum of 7.51 for tank one, less than 0.05 for tank two, 7.49 for tank three, 7.52 for tank four, 7.49 for tank five, and 7.54 mg/l for tank six. The concentrations remained near the above values until the statistical analyses showed "stress detection" which occurred during the first half-night on day 14 (Table 48). As soon as stress detection occurred, the flow was returned to normal dilution water. At 10:00 a.m. on day 15 zinc analyses showed all effluent concentrations to be less than 0.70 mg/l Zn^{++}. Stress detection continued to be registered for two consecutive time intervals following the initial detection, but after that no stress detection was registered. The frequency of abnormal patterns returned to pre-stress levels within 48 hours. In this experiment, as with all others in which dilution water containing zinc was replaced with dilution water minus zinc at the time of stress detection, *all fish survived!*

The results from the series of experiments at progressively lower zinc concentrations indicate that the lowest detectable concentration is between 3.65 mg/l zinc (Table 52) and 2.93 mg/l zinc (Table 53) for a 96-hour exposure.

Fish Breathing

Table 49 shows the number of breathing responses from four fish exposed for four days to a mean zinc concentration of 4.16 mg/l (based on daily determinations by atomic absorption photospectrometry) and one control fish. During the control period before any zinc was added there were 15 occasions when a single experimental fish responded, and 3 occasions when two experimental fish responded at the same time. At no time during the control period did more than two fish show responses together. After the zinc was introduced, all four of the exposed fish showed responses simultaneously on 5 occasions, and three fish showed responses during the same time interval on 19 occasions. If the criterion for detection of water conditions potentially harmful to fish were two or more responses during the same time period, then three false detections would have occurred before any zinc was added, and 4.16 mg/l zinc would have been correctly detected eight hours after it was introduced. If the detection criterion were three or more responses during the same time period, then no false detections would have occurred and the zinc would still have been correctly detected after eight hours.

The lowest zinc concentration tested was 2.55 mg/l. Using a detection criterion of simultaneous responses by three fish, this concentration was detected 52 hours after the zinc was added, with no false detections occurring during the four hours before zinc was added (Table 50). The responses of six control fish that were exposed to tap water containing no added zinc are also shown for comparison. Note that there was no tendency toward increased breathing rates through time in the control fish, and that no more than one control fish showed an increased breathing rate during one time period.

Table 54 summarizes information on three experiments and indicates the effectiveness of the method of analysis when different criteria for detection are used. Changing the criterion for detection from one to three responses per time

Table 52

Statistical Analysis of Light Beam Interruptions Recorded During Experiment 16. Bluegill exposed to 3.65 mg/l Zn^{++} showed a positive test for stress detection between 42 and 48 hours exposure.

Fish	*Day 1 vs Day 2*	*Day 2 vs Day 3*	*Day 3 vs Day 4*		*Day 4 vs Day 5*	*Day 5 vs Day 6*	*Day 6 vs Day 7*	*Day 7 vs Day 8*
				First Half-Day Values				
1	0	0	0		0	0	0	0
2	0	0	0		0	0	0	0
3	0	0	0		0	*	0 (5vs7)	*
4-C	0	0	0		0	0	0	0
5	0	0	*		0 (3vs5)	0	0	0
6	0	0	0		0	0	0	0
				Second Half-Day Values				
1	0	0	0		0	0	0	0
2	0	0	0		0	0	0	0
3	0	0	0	-Zn^{++} Added-	0	0	0	0
4-C	0	0	0		*	* (4vs6)	* (4vs7)	* (4vs8)
5	0	0	0		0	0	0	0
6	0	0	0		0	0	0	0
				First Half-Night Values				
1	0	0	0		0	0	0	0
2	0	0	*		0 (3vs5)	*	* (5vs7)	* (5vs8)
3	0	0	0		0	0	0	0
4-C	0	0	0		0	0	0	0
5	0	0	0		0	0	*	0 (6vs8)
6	0	0	0		0	0	0	0
				Second Half-Night Values				
1	0	0	0		0	0	0	0
2	0	0	0		*	0 (4vs6)	0	0
3	0	0	0		0	0	0	0
4-C	0	0	0		0	0	0	0
5	0	0	0		0	*	0 (5vs7)	0
6	0	0	0		0	*	* (5vs7)	* (5vs8)

0 = Differences in variances not significant at $\alpha = .002$.
* = Differences in variances significant at $\alpha = .002$.

Table 53

Statistical Analysis of Light Beam Interruptions Recorded
During Experiment 17. Bluegill exposed to 2.93 mg/l Zn^{++}
for 96 hours showed no positive test
for stress detection.

Fish	Day 1 vs Day 2	Day 2 vs Day 3	Day 3 vs Day 4	Day 4 vs Day 5	Day 5 vs Day 6	Day 6 vs Day 7	Day 7 vs Day 8
First Half-Day Values							
1	0	0	0	0	0	0	0
2	0	0	0	0	0	0	0
3-C	0	0	0	0	0	0	0
4	0	0	0	0	*	0 (5vs7)	0
5	0	0	0	0	0	0	0
6	0	0	*	0 (3vs5)	0	0	0
Second Half-Day Values							
1	0	0	0	0	0	0	0
2	0	0	0	0	0	0	0
3-C	0	0	0	0	0	0	0
4	0	0	0	0	0	0	0
5	0	0	0	0	0	0	0
6	0	0	0	0	0	0	0
First Half-Night Values							
1	0	0	0	0	0	0	0
2	0	0	0	0	0	0	0
3-C	*	0 (1vs3)	0	0	0	0	0
4	0	0	0	0	0	0	0
5	0	0	0	0	0	0	0
6	0	0	0	0	0	0	0
Second Half-Night Values							
1	0	0	0	0	0	0	0
2	0	0	0	*	0 (4vs6)	0	0
3-C	0	0	0	0	0	0	0
4	0	0	0	0	0	0	0
5	0	0	0	0	0	0	0
6	0	0	0	0	0	0	0

(In the Second Half-Day section, "$-Zn^{++}$ Added—" is noted between the Day 3 vs Day 4 and Day 4 vs Day 5 columns.)

0 = Differences in variances not significant at $\alpha = .002$.
* = Differences in variances significant at $\alpha = .002$.

Table 54

Effectiveness of Zinc Detection Using Increases in
Fish Breathing Rates

Zinc Concentration (mg/l)	No. of Fish Exposed	Detection Criterion: Minimum No. of Fish Showing Response At One Time	Lag Time (Hours from Addition of Zinc)	No. of False Detections
5.22	3	1	0	12 in 100 hours
		2	4	1 in 100 hours
		3	not detected after 45 hours	0 in 100 hours
4.16	4	1	0	19 in 123 hours
		2	11	3 in 123 hours
		3	11	0 in 123 hours
2.55	3	1	0	2 in 4 hours
		2	8	0 in 4 hours
		3	52	0 in 4 hours

period generally increases the lag time and decreases
the number of false detections. The lag time is the
time from the addition of zinc to the first detection.

Fish Reproduction

We do not wish to give the impression that the
growth and reproduction experiments were adequate to
establish a biologically safe concentration for bluegills exposed to zinc in water of the same quality
as Blacksburg tap water. The fish in these experiments were not exposed to zinc for a whole life cycle--
the adults were brought into breeding condition before
being exposed to zinc. Twenty-gallon tanks do not
appear to be suitable for bluegill reproduction
studies because they are too small, even when provided with hiding places, to allow females to escape
the aggressive attacks of males, and 49 dead or dying
females had to be replaced in these experiments.
The survival of the fry was poor, even in the control
tanks (5-8% survival after 30 days), because of

problems in obtaining plankton for the fry to eat.
Nevertheless, the results (Table 55) did show that
1/10 the lowest concentration used in monitoring
fish breathing was certainly not safe for chronic
exposure because only one spawning occurred and
the fry died within three days in this concentration (.235 mg/l). Fry taken from the control tanks
and the other experimental tanks also died within
three days when placed in .235 mg/l zinc. Reproduction, survival and growth of bluegills at the
other zinc concentrations (.076 and .035 mg/l)
were comparable to that in the control tanks
receiving no added zinc (Table 55).

Table 55

Bluegill Reproduction and Growth in Four Zinc Concentrations

(mg/l)	Zn, as Fraction of 2.55 mg/l	Total Eggs Spawned	Spawnings	Eggs per Spawning	% Hatch[a]	Length of fry (mm) Days 30	42	60	90
.021	no added Zn	13,264	13	1,046	65	8	-[b]	18	30
.035	1/75	14,476	8	2,037	80	-	14	19	31
.076	1/34	23,797	16	1,367	64	-	15	22	32
.235	1/10	1,009	1	1,009	43	*	*	*	*

[a]No. of eggs and percent hatch not determined for all spawnings
because of premature hatching, fungus infestation, etc.

[b]Dashes indicate that fry were not measured.

*None of the fry exposed to .235 mg/l zinc survived 30 days.

DISCUSSION

The experiments described above show that it is
probably feasible to use fish in continuous monitoring
systems at industrial sites to warn of developing
acutely toxic conditions in time to forestall acute
damage to the fish populations in the stream. The
movements and breathing of bluegill sunfish can be
continuously monitored and used to detect sublethal
concentrations of zinc. The criterion for detection

is a certain number of fish showing an arbitrarily
defined response in breathing rate or activity during
one time period. In conjunction with stream water
quality standards for chronic exposure, such biological monitoring systems should make it possible
for healthy fish populations to coexist with
industrial water use.

In choosing a specific criterion for detection,
the risk of not detecting stress soon enough must
be weighed against the risk of false detections and
the choice would probably be determined by the nature
of the pollutant. If a pollutant is easily detected
by the biological monitoring system, and if the toxic
effects are reversible, the criterion for detection
might be responses by 3/4 of the test fish, to avoid
the false detections that would necessitate expensive
remedial action or a temporary shut-down. On the
other hand, an effluent containing a fast-acting
toxicant whose effects are irreversible would require a criterion that leads to rapid detection
(responses by 1/4 to 1/2 of the test fish), and
would necessitate the expense of installing holding
ponds or recycling facilities to accommodate a
relatively high number of false detections. Alternatively, the lag time could be reduced by metering
proportionally more waste into the dilution water
delivered to the test fish than is delivered to
the stream. In general the higher the toxicant
concentration, the faster the toxicant is taken up
by the fish and the sooner the response appears.

The detection limits of the monitoring system
could be related to standards for chronic exposure
by additional growth and reproduction experiments.
The growth and reproduction experiments in the
present study indicated that 1/10 the lowest zinc
concentration used in the monitoring experiments
was almost certainly not a safe level for chronic
exposure.

Whether or not the detection limits of the
monitoring techniques are related to stream standards
for chronic exposure, the techniques could be useful
in the prevention of accidental spills or environmental changes that produce acutely toxic conditions,
since the above experiments with zinc show that
detection occurs soon enough for fish to recover
if zinc addition is promptly stopped.

In an actual industrial situation, water and
waste qualities are apt to vary unpredictably, and
it would certainly be desirable to monitor both
breathing and activity. It is conceivable that some

harmful combination of environmental conditions and waste quality would be detected by monitoring one biological function but not by monitoring another. It is also possible that excessive turbidity would disrupt the light beams of the movement monitor and not affect the breathing monitor, or that an excessive concentration of electrolytes would affect the electrodes of the breathing monitor but not affect the activity monitor. Therefore, the activity monitor and the breathing monitor have been combined in our laboratory for further experiments.

ACKNOWLEDGMENT

This research was supported by grants 18050 EDP and 18050 EDQ from the Water Quality Office, Environmental Protection Agency.

REFERENCES

1. Train, R.E. (ed.). *The First Annual Report of the Council on Environmental Quality* (Washington, D.C.: U.S. Govt. Printing Office, 1970), 326 pp.
2. Cairns, J., Jr., K. L. Dickson, R. E. Sparks, and W. T. Waller. "A Preliminary Report on Rapid Biological Information Systems for Water Pollution Control," *Jour. Water Poll. Contr. Fed. 42(5)*:685 (1970).
3. Sokal, R. R. and F. J. Rohlf. *Biometry* (W.H. Freeman and Co., 1969), 776 pp.
4. Sparks, R. E., W. T. Waller, J. Cairns, Jr., and A. G. Heath. "Diurnal Variation in the Behavior and Physiology of Bluegills (*Lepomis macrochirus* Rafinesque)," *The ASB Bull. 17(3)*:90 (1970) (abstract).
5. Axelrod, H. R. *Tropical Fish as a Hobby.* (London: George Allen and Unwin, Ltd., 1952), 264 pp.
6. Eaton, S. G. "Chronic Malathion Toxicity to the Bluegill (*Lepomis macrochirus* Rafinesque)," *Water Research 4*:673 (1970).
7. Mount, D. I. and W. A. Brungs. "A Simplified Dosing Apparatus for Fish Toxicology Studies," *Water Research 1*:21 (1967).
8. Brungs, W. A. and D. I. Mount. "A Water Delivery System for Small Fish-Holding Tanks," *Trans. Amer. Fish. Soc. 99(4)*:799 (1970).

22. ENVIRONMENTAL TESTING OF TRISODIUM NITRILOTRIACETATE: BIOASSAYS FOR AQUATIC SAFETY AND ALGAL STIMULATION

R. N. Sturm and A. G. Payne. Environmental Water
Quality Research Department, Procter & Gamble
Company, Cincinnati, Ohio

INTRODUCTION

The environmental safety evaluations of raw
materials considered for use in consumer products
can be of critical importance in many industries.
Since new detergent materials will be disposed of
as part of household sewage, evaluations are conducted with the objective of insuring that their
projected environmental levels would have no significant effect on the flora and fauna of the
aquatic environment.

NTA (Figure 96), as a leading candidate for
use as a phosphate substitute in detergent products,
underwent comprehensive testing, and this overall
program furnishes useful examples for future testing
programs. Trisodium nitrilotriacetate (NTA) is not
really a new chemical. First synthesized in the
mid-1930's by I. G. Farben chemists,[1] this amino-
polycarboxylate has seen many agricultural and
industrial applications.

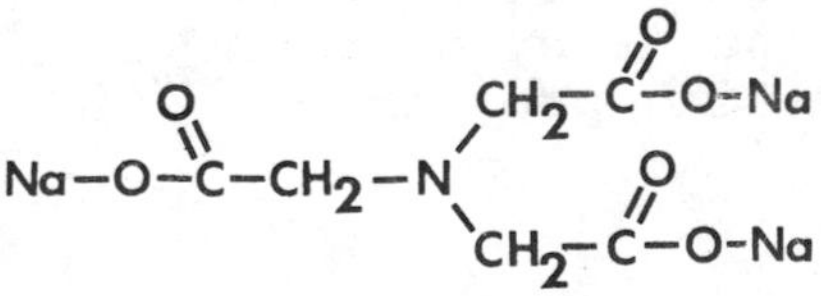

Figure 96. *Trisodium
Nitrilotriacetate
MW = 257*

403

On a performance basis, NTA is considered to be an effective substitute, but a complete assessment requires a wide range of both laboratory and field studies to qualify it as safe for human exposure and as safe for the environment. Initial stages of an environmental water quality evaluation indicated this material to be removable by laboratory-scale biological sewage treatment processes, and its degradation to be essentially complete to the end-products of CO_2, H_2O, and inorganic nitrogen.[2] Swisher *et al.*[3] reported that degradation of 20-500 mg/l NTA was essentially complete in laboratory activated sludge units and that NTA added at levels up to 200 mg/l failed to disturb the normal functioning of the activated sludge process. Shumate *et al.*,[4] conducting a field study in an operating activated sludge plant, found NTA to be readily removable and the plant's normal performance unaffected by the presence of NTA. NTA degradation continued in the receiving stream.

Degradation of NTA in natural waters has been reported by Forsberg and Lindquist[5] in which one NTA-degrading species was *Pseudomonas*. Duthie and Thompson[2] demonstrated NTA degradation in Ohio River water following a short acclimatization period. At low NTA levels, such as might be found after dilution below a wastewater outfall, it was estimated that NTA degradation would be essentially complete in one day. Based on levels that might be used in detergents, and the fact that NTA is degraded during biological waste treatment and by microorganisms present in soil and surface waters, the average of expected concentrations of NTA in surface waters is less than 0.05 mg/l.

Scope of the Paper

While a wide range of both human and environmental safety testing has been performed on NTA by industry, university, and government researchers, this paper is limited to a discussion of but two phases of the Procter and Gamble Company's environmental evaluation of NTA--Bioassays for Aquatic Safety and Algal Stimulation--and is by no means a review of the extensive NTA testing program.

Aquatic Safety

Initial acute toxicity bioassays (Table 56) had shown the acute toxicity (96 hr. TL_{50}) of NTA

Table 56

Static Acute Toxicity* of NTA to Bluegills,
Snails, and Diatoms

Organism	Chemical	Water Hardness (mg/l CaCO₃)	96 Hr. TL₅₀ (mg/l)**
Bluegill	NTA	60	252
Bluegill	NTA	170	487
Snails	NTA	60	373
Snails	NTA	170	522
Diatoms	NTA	60	185
Diatoms	NTA	170	477

*Bioassays performed by the Academy of Natural Sciences of
  Phildelphia.
**As Trisodium NTA

to bluegill (*Lepomis macrochirus*), snails (*Physa
heterastropha*), and the diatom (*Navicula seminulum*)
to be more than two orders of magnitude (200X)
higher than the average expected environmental
levels. The toxic effect of NTA varied, as expected, with the hardness of the diluent water.
96 hr. TL₅₀'s for bluegill, snails, and diatoms
were 252 mg/l, 373 mg/l, and 185 mg/l, respectively,
in soft water (60 mg/l as CaCO₃) and 478 mg/l, 522
mg/l, and 477 mg/l, respectively, in hard water
(170 mg/l as CaCO₃) in a static test. Dynamic
bioassays, which were performed in parallel with
the static test, yielded results not significantly
different from those obtained with the static bioassay. In published studies with NTA-metal complexes, Sprague[6] has shown NTA to reduce the toxicity
of copper and zinc to brook trout. In addition,
two previous studies showed no inhibition by NTA of
growth of green algae at levels far above the projected environmental concentrations. Christie[7]
showed NTA Na₃·H₂O to be non-toxic to *Chlorella
pyrenoidosa* at a concentration of 275 mg/l, and
Eyster[8] reported that NTA is non-toxic to *Chlorella*
at 10⁻³ molar (257 mg/l), causes slight inhibition

at 2570 mg/l, and inhibits growth completely at 25700 mg/l.

The present aquatic safety work has extended these initial acute fish toxicity observations to rainbow trout (*Salmo gardnerii*) and fathead minnows (*Pimephales promelas*). Additionally, the toxicity of NTA to bluegill and fathead minnows was evaluated under conditions of 28 days continuous exposure to 8 levels of NTA ranging as mean values from 3.4 to 172.8 mg/l.

The extended exposure studies were conducted in a continuous flow proportional dilution apparatus, as described by Mount and Brungs,[9] which provides for the automatic intermittent introduction of the test material in diluent water into the test chambers. Flow rate to each 30 liter test chamber was 6 liters per hour. Test fish were maintained and observed in the laboratory for a minimum of 21 days prior to testing. During that period, mortality in the fish stocks was less than 2%, and the fish were judged to be in excellent physical condition. Bioassays on fathead minnows and bluegills were conducted at 19°C ($\pm$ 0.5 C°), while those performed on trout were conducted at 15°C. The bioassay water was prepared by adding to deionized water of at least 1 megohm resistivity, 3 mg potassium chloride, 30 mg calcium sulfate, 48 mg sodium bicarbonate, and 30 mg magnesium sulfate per liter. The total hardness of the test diluent was 35 ppm as $CaCO_3$, with an initial pH of 7.1. The mean dissolved oxygen during the course of the test was 5.2 mg/l (range 4.3 - 5.7 mg/l). During the 28-day study, the fish were fed a dry pelleted fish ration daily. The desired chemical concentrations were established by adding appropriate amounts of NTA stock solution to each test chamber. Thereafter, the proportional dilution apparatus was used to maintain the desired chemical concentration. Twenty bluegills (mean weight 1.5 grams) and 20 minnows (mean weight 0.8 grams) were tested at each concentration. During the course of the 28-day study, water samples were taken at 96 hr. intervals from three tanks containing nominal concentrations of NTA (200 mg/l, 72 mg/l, and 5.6 mg/l) and analyzed to monitor the operation of the proportional dilution apparatus. Samples were not taken from tanks containing the five additional levels of NTA. NTA was analyzed using the Zinc-Zincon method for sequestrant in waste and sewage as described by Thompson and

Duthie.[2] All NTA values are reported as mg/l
anhydrous trisodium NTA (MW = 257).

Dynamic acute toxicity bioassays in soft water
(Table 57) indicated the 96 hr. median tolerance
limit (95% confidence interval) for NTA to be 98
(72-133) mg/l for rainbow trout, and 127 (93-170)
mg/l for fathead minnows. A static acute toxicity

Table 57

Acute Fish Toxicity* of NTA in Soft Water
(35 mg/l as CaCO₃)

Fish	96 Hr. TL_{50} mg/l**
Bluegill	198 (175-225)
Fathead Minnow	127 (93-170)
Rainbow Trout	98 (72-133)

*Bioassays performed by Bionomics, Inc., Wareham, Mass.
**As Trisodium NTA.

bioassay on bluegills indicated the 96 hr. TL_{50} of
NTA to be 198 (175-255) mg/l. The results of the
28-day continuous exposure bioassay are consistent
with the acute bioassays (Table 58). During the
first 48 hours exposure to 172.8 mg/l of NTA, none
of the minnows survived, while 60% of the bluegills
survived; mortality among bluegills did not increase significantly after that time. No mortality
was observed with either species at any of the
other exposure levels (3.7 - 96.0 mg/l) tested
during the first 9 days of the test; and no significant mortality occurred at any of these levels
throughout the duration of the 28 days exposure to
NTA.

Both species' accumulative mortality due to
continuous exposure to NTA is conspicuously absent,
and the fish appeared healthy and normal. In order
to evaluate the possible NTA-induced damage to gill
structure, fish were sacrificed after 28 days exposure and fixed in Bouin's fixative. After
dehydration, imbedding in paraffin, and sectioning,
a minimum of five sections from each gill were

Table 58

Results of 28-Day Continuous Exposure Bioassay*

Nominal Concentration mg/l	Actual Concentration mg/l**	Mortalities at 28 Days		NTA Induced Gill Pathology
		Fathead Minnow	Bluegill	
0 (Control)	0	2/20	2/20	−
200	172.8	20/20	11/20	+
120	96.0	0/20	2/20	−
72	56.4	0/20	0/20	−
43.2	29.4	0/20	1/20	−
25.9	17.1	0/20	0/20	−
15.5	10.0	0/20	1/20	−
9.3	6.0	0/20	3/20	−
5.6	3.4	1/20	3/20	−

*Bioassays performed by Bionomics, Inc., Wareham, Mass.
**As Trisodium NTA

stained with Heidenhahn's Azan Technique. Microscopic examination of the mounted gill sections
indicated no change in gill histology occurred due
to 28 days of continuous exposure to concentrations
of NTA up to and including 96 mg/l. Additionally,
fish exposed to NTA for 28 days were transferred
to clean water for 7 days and gill sections were
then prepared for histological examinations.
Examinations of stained gill sections indicated
NTA induced gill pathology to be absent at concentrations up to and including 96 mg/l. It appears
that 28 days continuous exposure to levels of NTA
at concentrations more than 200 times higher than
maximum expected environmental levels had no significant effect on gill structure or survival of
bluegill and minnows in soft water, where the toxic
effect would be expected to be the greatest
(Figure 97).

Algal Stimulation Bioassays

The effect of trisodium nitrilotriacetate (NTA)
on eutrophication was studied throughout 1970 by

408

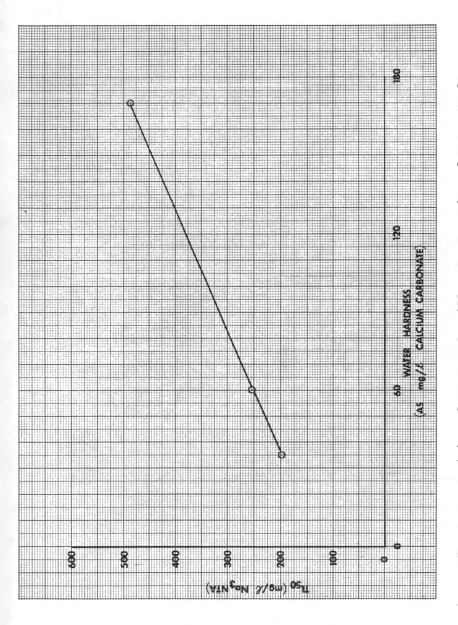

Figure 97. Acute Toxicity of NTA to Bluegills as a Function of Water Hardness

laboratory testing which, in general, paralleled that of the National Eutrophication Research Program (NERP) of EPA. The bioassay procedure followed was that of the Algal Assay Procedure (AAP): Bottle Test,[10] formerly known as the Provisional Algal Assay Procedure (PAAP). Secondary and tertiary treated sewages, with and without NTA, were added to filtered waters from lakes of different geographical locations across the U.S.; these waters varied in trophic levels from oligotrophic to highly eutrophic. All filtered waters were inoculated with a bacteria-associated green alga, *Selenastrum capricornutum*. Some waters were inoculated, in addition, with a blue-green alga, *Microcystis aeruginosa*, and/or a nitrogen fixing blue-green alga, *Anabaena flos-aquae*. Algal growth was followed by cell counts with hemocytometer and microscope for a 12-21 day period through the end of the log phase of growth.

The effects of NTA added alone (without sewage) and of other potential nutrients, P, N, and C, added alone, were also studied. Na_3 NTA was added at concentrations to yield 0.05, 0.5, and 5.0 mg/l H_3 NTA (M.W. 191). N and C were added at levels equivalent to that contained in 5 mg/l H_3 NTA. C was also added at a higher level, 50 mg $KHCO_3$/l, to test for possible C limitation.

The following test conditions of the AAP Bottle Test[10] were maintained:

Illumination - Continuous "Cool White" Fluorescent lighting - 400 ft. c. $\pm$ 10% for green algae; 200 ft. c. $\pm$ 10% for blue-green algae.

Temperature - 24° $\pm$ 2°C.

Shaking - Gyrotory, 110 oscillations/min.

Culture Flasks - Erlenmeyer, Pyrex-type. Culture volume normally was 60 ml in a 250 ml flask.

Culture Flask Closures - Aluminum foil, loosely capped.

Culture Medium - Final concentration of nutrients

Macronutrients	mg/l	*Micronutrients*	µg/l
$NaNO_3$	25.500	H_3BO_3	185.520
K_2HPO_4	1.044	$MnCl_2$	264.264
$MgCl_2$	5.700	$ZnCl_2$	32.709
$MgSO_4 \cdot 7H_2O$	14.700	$CoCl_2$	0.780
$CaCl_2 \cdot 2H_2O$	4.410	$CuCl_2$	0.009
$NaHCO_3$	15.000	$Na_2MoO_4 \cdot 2H_2O$	7.260
		$FeCl_3$	96.000
		$Na_2EDTA \cdot 2H_2O$	300.00

Inoculum - Cultures were obtained from NERP, EPA at Corvallis, Oregon and maintained through at least two

transfers in the medium described above. Initial
inoculum concentrations in the test waters were 10^3
cells/ml for *S. capricornutum* and 50×10^3 cells/ml
for *M. aeruginosa* and *A. flos-aquae*.
Replication - Three replicates were run on each variable.

All sewages used were provided by the NERP Lab of
EPA (see Table 59). Primary effluent from the nearby
town of Philomath, Oregon, sewage treatment plant
was split; one-half was spiked with 5 mg/l NTA and
both portions were separately fed to a laboratory
activated sludge secondary treatment unit. The
secondary treated effluents with and without NTA
then received tertiary treatment in the laboratory
using alum. Secondary and tertiary treated sewages,
with and without NTA, were added to the test waters
at a concentration of 2% by volume.

The rationale for testing NTA in the presence
of sewage was simply one of using the expected
pathway of NTA into receiving waters. Since NTA
would be used primarily in household detergent products, it would normally be just one added component
in household wastes. It would reach surface waters
only as part of a sewage effluent and be subject to
the same treatment as the balance of the waste. The
levels tested were based upon the dilution factor
for waste streams carrying the material. Typical
dilution from detergent product to municipal sewage
influent is approximately 1/20,000. This means
that if 10% NTA were used in all detergent products,
approximately 5 ppm could be anticipated at treatment
plants.

The major part of this study was conducted using
waters from three lakes in North Carolina, three in
Oregon, and two in California; water from a marl
lake in Michigan was also tested once. Most lakes
were sampled three times from July through October,
1970. Earlier tests of three Minnesota waters with
a somewhat different protocol are also reported.
These waters were sampled from August 1969 to May
1970. All test waters were single grab samples
collected in polyethylene containers, generally
from the top three meters of the water column.
Chemical data on the test waters are shown in
Table 60.

Figure 98, a study with *Selenastrum* using water
from Lake James, a mesotrophic lake located near
Marion, N.C., illustrates results representative of
those seen throughout the algal stimulation bioassay
work. As shown by the curves, the presence of NTA
had no significant effect on the stimulation of

Table 59

Chemical Analysis of Sewages

Two batches of secondary and tertiary sewages, supplied by NERP, were used in the testing. The first batch was used from July 22 to September 16, 1970. The second batch was used with tests from September 16 through November, 1970. The following analyses were supplied by NERP

Batch #1 (mg/l)

	NTA	Iron Total	Phosphate Tot.	Phosphate Ortho	Nitrogen NH₃	NO₂	NO₃	KJEL
Sec. Effluent	.1	.05	13.9	12.6	.33	.12	26.4	1.0
Sec. Effluent Filtered	.1	.04	12.6	12.6	.36	.12	27.3	0.1
Sec. Effluent + NTA	.1	.05	12.9	12.6	6.0	.48	25.2	10.1
Sec. Effluent + NTA Filtered	.1	.05	12.6	12.6	6.0	.52	25.2	5.3
Tert. Effluent	.4	.025	5.8	5.8	.6	.12	27.3	.1
Tert. Effluent + NTA	1.1	.025	6.0	6.0	3.0	.30	27.3	1.4

Batch #2 (Unfiltered) (mg/l)

	NTA	Iron Soluble	Phosphate Tot.	Phosphate Ortho	Nitrogen NH₃	NO₂	NO₃	KJEL	Carbon Tot. Org.	Tot. Inorg.	Tot.	Alkalinity Tot.	HCO₃	pH
Sec. Effluent	<.1	.26	–	7.56	<1	<.1	23.1	3 (?)	21	2	19	59	59	7.3
Sec. Effluent + NTA	<.1	.16	–	7.20	<1	<.1	23.1	.3	24	3	21	64	64	7.3
Tert. Effluent	<.1	.10	–	0.84	<1	<.1	24	.3	22	2	20	82	82	7.0
Tert. Effluent + NTA	<.1	.10	–	0.60	<1	<.1	24.2	.1	–	2	–	82	82	7.3

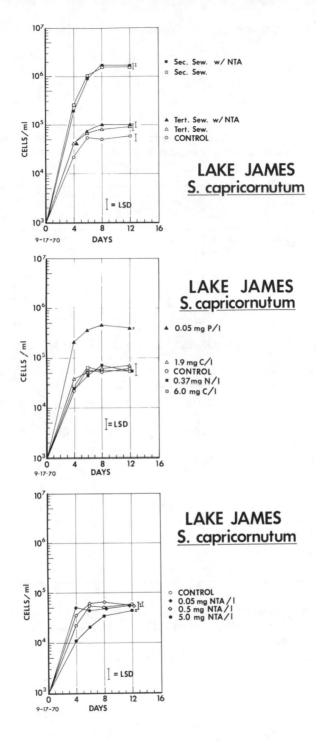

Figure 98.   Effect of spikes with sewages, NTA, and other
             potential nutrients on the growth of *Selenastrum
             capricornutum* in water from Lake James, North
             Carolina.

Table 60

Range of Chemical Concentrations (mg/l) in Unfiltered
Lake Waters During Sampling Period (6/70-10/70)*

Chemical	Waldo Lake	Triangle Lake	Cline's Pond
Total Inorganic Carbon	1.0	5.0-6.0	11.0-14.0
Total Organic Carbon	1.0	1.0-2.0	7.0-9.0
NH_3 Nitrogen	0.010	0.001-0.030	0.01-0.07
NO_3 Nitrogen	0.001-0.005	0.001-0.010	0.001-0.030
Kjeldahl-Nitrogen	0.100	0.100-0.300	1.3-5.2
Ortho-Phosphorus	0.001	0.001-0.008	0.01-0.27
Total-Phosphorus	0.001	0.003-0.008	0.04-0.34
Total Iron	0.050	0.100-0.300	0.30-0.50
Alkalinity (as $CaCO_3$)	-	-	-
Hardness (as $CaCO_3$)	-	-	-
pH	-	-	-

*Samples and chemical data supplied by Pacific Northwest
 Water Laboratory, EPA, Corvallis, Oregon, and PNWL grantees.
**Single samples.

Table 60, Continued

University Lake	Lake Michie	Lake James	Clear Lake**	Castle Lake**
-	-	-	25.3	3.4
-	-	-	-	-
.015-.22	.01-.07	.01-.01	-	-
.010-.030	.001-.02	.003-.015	.015	.001
.20-.42	.14-.30	.15-.61	-	-
.001-.045	.002-.025	.002-.038	.390	.0005
.005-.096	.003-.038	.006-.05	.431	.0031
-	-	-	2.40	.037
24.0-25.5	9.5-13.5	15.0-18.0	-	-
13.36-18.76	9.41-14.04	6.34-9.56	-	-
6.8-7.7	6.6-7.2	8.6-9.0	8.75	-

algal growth by secondary or tertiary treated
sewages. While in the absence of sewage or bottom
muds, stimulation was seen with the addition of
0.05 mg P/l, NTA at levels as high as 5.0 mg/l
caused no biostimulatory effect.

Similar results are illustrated in Figure 99
with *Anabaena* in water from (a) University Lake, a
eutrophic lake located near Chapel Hill, N.C., and
(b) Lawrence Lake, an oligotrophic marl lake located in southwestern Michigan. Stimulation by
sewages with and without NTA was found, but no
stimulation occurred with NTA alone. Water from
eutrophic Cline's Pond near Corvallis, Oregon,
however, was so highly productive that any stimulation of *Microcystis* growth was difficult to
detect, even with addition of sewages (Figure 99);
NTA again was not stimulatory.

Water taken from highly eutrophic Clear Lake,
California, in October 1970 was toxic to *Microcystis*
(Figure 100). NTA prevented *Microcystis* die-off for
several days, possibly by chelation of trace metals
present in toxic quantities.

High levels of NTA have been reported to inhibit growth of estuarine phytoplankton in sea
water due to the high ratio of chelator to trace
metals.[11] Algal growth generally was not inhibited
with NTA additions up to 5.0 mg/l in the fresh
waters reported here. A very slight inhibition was
observed in Lake James water spiked with 5.0 mg
NTA/l (Figure 98).

Results on all test waters are summarized in
Table 101A-B. NTA generally was not stimulatory,
either alone or in the presence of sewage; its
major effect was to prevent or slow the rate of
Microcystis die-off, possibly by chelation of trace
metals sometimes present at toxic levels in the
test waters. No biostimulation was observed with
the addition of nitrogen or carbon alone. The test
conditions of the PAAP Bottle Test are designed to
provide adequate interchange of atmospheric CO_2
with the test water; therefore it was not expected
that added carbon should show an effect.

Christie[7] has reported that NTA can act as a
sole source of nitrogen for algal growth when fed
directly or after treatment in an activated sludge
system, albeit not to the same extent as equivalent
quantities of nitrogen fed as nitrate under the
same test conditions. Although most often concern
is expressed over the contribution of detergent
phosphates to the environment, Ryther and Dunstan[12]

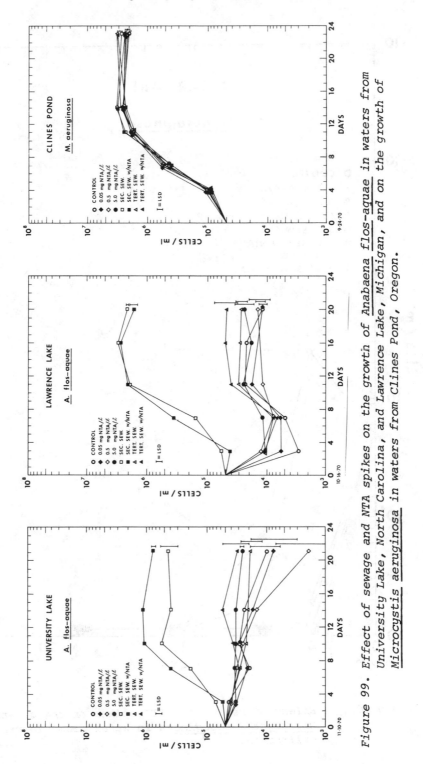

Figure 99. *Effect of sewage and NTA spikes on the growth of Anabaena flos-aquae in waters from University Lake, North Carolina, and Lawrence Lake, Michigan, and on the growth of Microcystis aeruginosa in waters from Clines Pond, Oregon.*

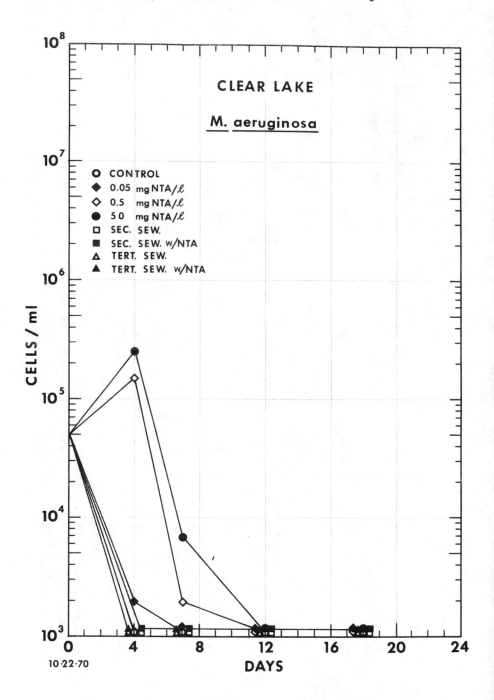

Figure 100. Effect of sewage and NTA spikes on the growth of <u>Microcystis</u> <u>aeruginosa</u> in water from Clear Lake, California.

have questioned the nitrogen contribution of NTA to estuarine waters since they are usually nitrogen-limited. The contribution of NTA to the N and C load in wastewater is negligible. If 10% NTA were to be used in all detergent products, the contribution by NTA to municipal sewage would be 0.27 mg N/l and 1.4 mg C/l. When compared to normal levels of 40 and 200 mg/l, respectively, present in typical raw sewage, NTA input would represent an increase of less than 0.7% N and 0.8% C.

SUMMARY

While data for chronic toxicity studies have not yet been released, information is available from an extensive acute toxicity program which includes several species as well as three degrees of water hardness. The relative non-toxicity of trisodium NTA, even in very soft water where its effects would be the greatest, and the absence of evidence of accumulative toxicity suggest NTA to be compatible with the goal of environmental improvements. While this work is but a small part of the total NTA environmental safety test program, it does produce evidence that 1) NTA exerts no acute toxic effect on the fish or algae species tested at levels well above those anticipated in the environment, 2) Bluegills and fathead minnows exposed to high but sublethal concentrations for a period of 28 days do not exhibit abnormal gill pathology, a further indication of NTA's safety, and 3) NTA stimulatory effects on algae would not be anticipated on the basis of nitrogen contribution and were not found by application of the AAP Bottle Test to a variety of lake waters.

Summary of Algal Assays

Test Water	Trophic* Level	Sampling Date	Inoculum	MAXIMUM STANDING CROP Control 10^6 Cells/ml
A. North Carolina Waters				
Lake James	O-M	8/6/70	Sel.	.005
Lake James		8/26/70	Sel.	.06
			Mic.	.10
			Ana.	.05
Lake James		9/24/70	Sel.	.08
Lake Michie	M	7/28/70	Sel.	.015
Lake Michie		8/19/70	Sel.	.005
			Mic.	.200
University Lake	E	7/16/70	Sel.[a]	20
			Mic.	die-off ---
University Lake		8/14/70	Sel.	.35
University Lake		9/70	Sel.	.005
University Lake		10/70	Sel.	.005
			Mic.	.100
			Ana.	.030
B. Oregon Waters				
Waldo Lake	O	7/23/70	Sel.	.005
Waldo Lake		8/18/70	Sel.	.005
Waldo Lake		9/70	Sel.	.005
Triangle Lake	M	8/5/70	Sel.	.005
Triangle Lake		9/2/70	Sel.	.005
Clines Pond	E	7/14/70	Sel.[a]	18
Clines Pond		8/11/70	Sel.	.100
Clines Pond		9/12/70	Sel.	.700
			Mic.	2.5
			Ana.	.50
C. California Waters				
Clear Lake	E	9/70	Sel.	4.0
			Mic.	die-off ---
			Ana.	.40
Castle Lake	M	9/70	Sel.	.005
			Mic.	die-off ---
			Ana.	.070
D. Michigan Marl Lake Water				
Lawrence Lake	O	10/70	Sel.	.005
			Mic.	.020
			Ana.	.030

*Key: O = Oligotrophic; M = Mesotrophic; E = Eutrophic

MAXIMUM STANDING CROP					
Experimental					
Ratio of Cells, Experimental to Control					
Secondary Sewage		Tertiary Sewage		NTA Alone	P
w/NTA	w/o NTA	w/NTA	w/o NTA		
220X	220X	60X	90X	1X	34X
28X	28X	1.5X	1.5X	1X	7X
6X	6X	0.5X	0.8X	4X	4X
10X	6X	1X	0.4X	1X	1X
21X	21X	1.5X	1.5X	1X	3X
100X	100X	23X	37X	1X	20X
240X	240X	60X	60X	1X	36X
9X	9X	6X	6X	1.5X	1.5X
2.4X	3.2X	1.5X	1.6X	1X	1.5X
6X	6X	2X	2X	1X	3X
300X	300X	1X	1X	1X	180X
400X	500X	2X	2X	1X	180X
9X	9X	3X	1.7X	1X	3X
33X	17X	1X	1X	1X	13X
140X	200X	60X	80X	1X	30X
180X	180X	40X	60X	1X	14X
300X	300X	1X	1X	1X	20X
180X	180X	60X	80X	1X	46X
260X	260X	50X	90X	1X	14X
1.8X	2.6X	1.5X	1.8X	1.2X	1.3X
13X	13X	4X	6X	1X	4X
4X	4X	1X	1X	1X	2.5X
1X	1X	1X	1X	1X	1X
2X	2X	1X	1X	2X	1X
1.1X	1.1X	1.1X	1.1X	1.1X	1X
1X	1X	1X	1X	1X	1X
300X	300X	1X	1X	1X	30X
				4X[b]	die-off
10X	10X	2X	2X	1X	1.4X
700X	700X	1X	1X	1X	180X
200X	200X	1X	2X	2.5X	150X
80X	80X	1.7X	1X	1X	33X

[a]Contamination - Total organic carbon (mg/l) data only.
[b]Increase above initial inoculum concentration - control died-off.

Table 101B

Summary of Algal Assays

Test Water	Trophic* Level	Sampling Date	Inoculum	MAXIMUM STANDING CROP Control 10^6 Cells/ml
E. Ely, Minnesota Area Waters				
Burntside L.	O	2/70	Sel.	.005
			Mic.	.05
			Ana.	.05
Burntside L.		5/70	Sel.	.005
			Mic.	.01
			Ana.	.70
Burntside R.	O	8/69	Sel.	.005
Burntside R.		12/69	Sel.	.005
Burntside R.		2/70	Sel.	.005
			Mic.	.10
			Ana.	.05
Burntside R.		5/70	Sel.	.005
			Mic.	die-off
			Ana.	.07
Shagawa L.	E	12/69	Sel.	.80
Shagawa L.		2/70	Sel.	.90
			Mic.	.70
Shagawa L.		5/70	Sel.	.50
			Mic.	.60
			Ana.	.25

*Key: O = Oligotrophic;   M = Mesotrophic;   E = Eutrophic

Table 101B, Continued

				MAXIMUM STANDING CROP			
				Experimental			
			Ratio of Cells,	Experimental to Control			
Sec. Sew.	NTA Alone	P	N	P+N	Fe + Tr. Nutr.	P & N + NTA	P&N + Fe + Tr. Nutr.
180X	1X	20X	1X	40X	1X	40X	40X
16X	1X	-	-	20X	2X	20X[a]	20X[a]
30X	1X	-	-	10X	-	10X	10X
70X	1X	12X	1X	240X	1X	240X	240X
2X[b]	3X[b]	1X	1X	5X[b]	die-off	35X[b]	2X[bc]
2X	1X	6X	1X	12X	1X	12X[a]	12X
440X	1X	40X	-	-	1X	-	-
-	1X	60X	-	300X	1X	300X	300X
200X	1X	50X	1X	180X	1X	180X	180X
-	1X	-	-	10X	-	10X	10X
30X	1X	-	-	10X	-	10X	10X
70X	1X	20X	1X	260X	1X	260X	260X
die-off	3X[b]	1X[b]	1X[b]	40X[b]	0.5X[b]	40X[b]	40X[b]
2.5X	1X	1X	1X	6X	1X	6X[a]	6X
-	1X	1X	-	2X	1X	2X	2X
-	1X	1X	1.3X	2.5X	1X	-	2.5X
-	-	-	-	2X	-	-	2X
-	1X	1X	1X	7X	-	7X	-
-	1X	1X	1.7X	4X	-	10X	-
-	1X	4X	2X	6X	-	6X	-

[a]Faster growth rate when compared to P & N addition.

[b]Increase above initial inoculum concentration - control died-off.

[c]Without trace nutrients

⁻Not tested.

REFERENCES

1. Pollard, R. R. "NTA: Newest Detergent Builder,"
    Hydrocarbon Process, Petrol. Refiner 45(11):197 (1966).
2. Thompson, J. E. and J. R. Duthie. "The Biodegradability
    and Treatability of NTA," *Journal, Water Pollution
    Control Federation 40*:306 (1968).
3. Swisher, R. D., M. M. Crutchfield, and D. W. Caldwell.
    "Biodegradation of Nitrilotriacetete (NTA) in Activated
    Sludge," *Environmental Science & Technology 1*:820 (1967).
4. Shumate, K. S., J. E. Thompson, J. D. Brookhart, and
    C. R. Dean. "NTA Removal by Activated Sludge--Field
    Study," *Journal, Water Pollution Control Federation 42*:
    631 (1970).
5. Forsberg, C., and G. Lindquist. "On Biological Degrada-
    tion of Nitrilotriacetate (NTA)," *Life Sciences 6*:1961
    (1967).
6. Sprague, J. B. "Promising Anti-Pollutant: Chelating
    Agent NTA Protects Fish From Copper and Zinc," *Nature
    220*:1345 (1968).
7. Christie, A. E. "Trisodium Nitrilotriacetate and
    Algae," *Water and Sewage Works*, 58 (1970).
8. Eyster, H. C. "Response of Aquatic Plants to NTA,"
    Personal Communication (1966).
9. Mount, D. I., and W. A. Brungs. "A Simplified Dosing
    Apparatus for Fish Toxicology Studies," *Water Research
    1*:21 (1967).
10. Algal Assay Procedure: Bottle Test, 1971, Environmental
    Protection Agency, Pacific Northwest Water Laboratory,
    Corvallis, Oregon.
11. Erickson, S. J., T. D. Maloney, and J. H. Gentile.
    "Effect of Nitrilotriacetic Acid on the Growth and
    Metabolism of Estuarine Phytoplankton," *Journal, Water
    Pollution Control Federation 42*:R329 (1970).
12. Ryther, J. H., and W. M. Dunstan. "Nitrogen, Phosphorus
    and Eutrophication in the Coastal Marine Environment,"
    Science 171:1008 (1971).

23. THE MEASUREMENT OF PHOSPHORUS METABOLISM IN
NATURAL POPULATIONS OF MICROORGANISMS

David L. Correll. Radiation Biology Laboratory,
Smithsonian Institution, Rockville, Maryland.

ABSTRACT

Methods and apparatus for making accurate field
measurements of phosphorus metabolism in microbial
populations under relatively undisturbed conditions
are discussed and illustrated. These measurements
should allow a better understanding to be gained of
such phenomena as nutrient cycling and photosynthetic
energy flow. Measurements of the kinetics of incorporation of radiophosphorus into various phosphorus
compounds allow the calculation of turnover numbers.
These compounds are quantitatively isolated after
immersion of the cells in phenol. The product of
the turnover rate and the pool size gives the
metabolic rate. The product of the metabolic rate
and the bond energy involved gives the rate of
energy flow.

Possible applications of microbial phosphorus
metabolism to environmental biology include (a)
measurement of photosynthetic phosphorylation rates
in algae and (b) studies of the mechanisms involved
in phosphate cycling. Studies of algal photosynthetic
phosphorylation can make use of the intracellular
compartmentalization shown schematically in Figure
102. Orthophosphate is taken up from the medium by
active transport systems and arrives in the photosynthetic apparatus without mixing with the large
cytoplasmic pool of orthophosphate. It is then used
in the phosphorylation reactions shown in Figure 103.

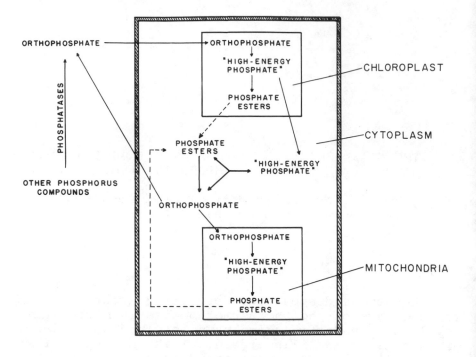

Figure 102. *Some aspects of the intracellular compartmentaliza-tion of algal phosphorus metabolism.*

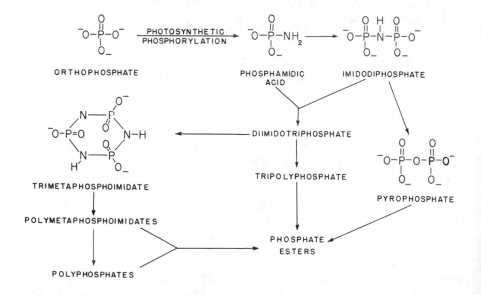

Figure 103. *Algal "high-energy" phosphorus metabolism pathways.*

These reactions and pathways are not meant to represent all reactions or pathways and products. They are the reactions which occur quickly and whose products are compounds with relatively large pool sizes so that they can be readily measured. In order to measure photosynthetic phosphorylation rates both the pool size and the turnover rate of at least one of these "high-energy" compounds or classes of compounds must be measured. The product of the pool size and the turnover rate is the metabolic rate. When this rate is multiplied by the bond energy (ΔF of hydrolysis is about 7.6 Kcal/mole), an estimate of energy flow through the pathway can be determined. Laboratory studies of the chemical pathways and phosphorus compound structures involved in algal photosynthetic phosphorylation have been few.[1-5] Applications to environmental chemistry are rare.[6]

In order to make the necessary measurements of pool sizes and turnover rates, samples of sufficient size must be obtained due to the concentrations of the phosphorus compounds in the microbial cells. A sample of "sufficient" size with present technology is in the range of from 10 to 100 g fresh weight. The cells in the samples must be exposed to labeled phosphate for short, known times of the range of one-half to fifteen minutes. Of course, this must be done under conditions approximating closely as possible the natural growth conditions. The cells of the sample must be separated from the medium and all metabolic activity destroyed at the end of an incorporation experiment without damage to the compounds under study. These requirements can be met in studies of suspended or attached microbial cells in laboratory mass cultures or field populations. In the case of suspended microbial cells the use of a continuous-flow high-speed centrifugal apparatus is recommended (Figure 104). In this system the suspension of cells is pumped through a plankton net in a clear plexiglass housing to remove most zooplankton and large debris. Most phytoplankton, including bacteria, will pass through. Radioisotope is added continuously at a constant ratio for a known time of flow before the liquid is injected into the centrifuge. There the cells are rapidly sedimented into 80% phenol and metabolism is halted. A large enough sample of cells can be obtained in one run for a later laboratory determination and the machinery can be operated from a small boat by a two-man crew.

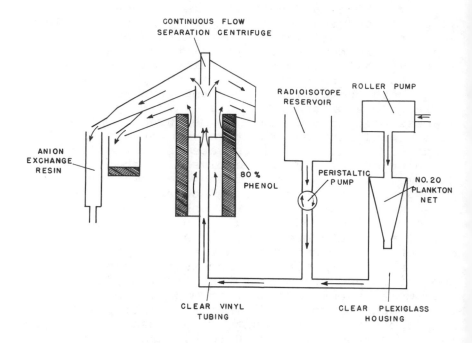

Figure 104. *Schematic diagram of apparatus for continuous-flow*
separation and radioisotopic labeling of
phytoplankton.

In the case of attached cultures of microorganisms, more difficulties are encountered. The
goal here is to study the so-called periphyton
communities of flowing water systems. One approach
is the use of plexiglass artificial substrates.[7]
When these have been colonized, they may be transferred to a nearby apparatus (Figure 105) to carry
out phosphate incorporation experiments. In this
apparatus a suitable medium--for example, filtered
river water with or without the addition of test
chemicals--can be "spiked" with carrier-free P-32
phosphate. The pump can be adjusted to give about
the same flow rate at the location of the artificial
substrate rack as they were exposed to during
colonization. The temperature can be kept the same
and the plexiglass plates can be submerged to the
same depth as they were during colonization, thereby
assuring an exposure to about the same light intensity and quality as they were exposed to in nature.
When the time of incorporation is over, they can be

frozen on dry ice or dropped into 80% phenol for
later laboratory analyses. The criterion most
difficult to satisfy in this method is obtaining a
large enough sample. This can be overcome by using
more area of artificial substrate or perhaps by
using more sensitive assays in the laboratory
analyses.

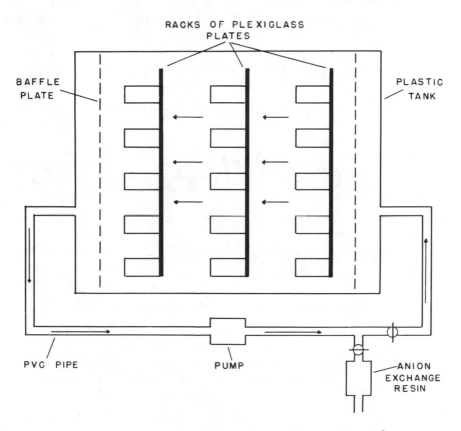

Figure 105. *System for metabolic and growth studies of*
attached microorganisms.

These same two general methods of approach to
the study of metabolism in microbial populations can
be applied to the measurement of such parameters as
net photosynthetic carbon fixation by substitution
of C-14 bicarbonate for P-32 phosphate. They can
be used for the study of mass cultures in the
laboratory or in the field. They avoid any possible
contamination of the environment with radioisotopes,

yet they can provide the researcher with relatively
large samples of cells which have been exposed to
the isotopes under favorable conditions for relatively
short times.

These methods may also be used to study the
mechanisms of phosphate cycling. In Figure 106 some
of the phosphate transformations in a flowing water
system are diagrammed. They could easily be simplified to represent a non-flowing situation. Studies

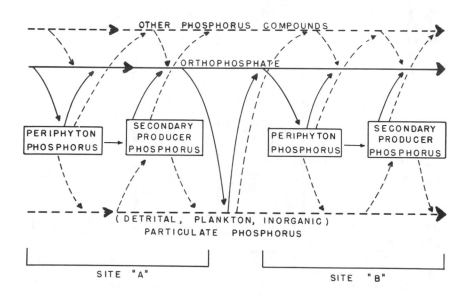

Figure 106. *Schematic diagram of various phosphate cycling
mechanisms by which the downstream movement of
phosphorus is retarded.*

of phosphate cycling in flowing water situations
have been very few,[8] but could be done accurately
and in much more detail with the available methods.
This process of phosphate cycling is important since
it determines the residence time of phosphorus in
the ecological system. Very little is known about
the factors which are important in the control of
the overall rate and what effects man-made changes
in the environment have upon these various steps in
phosphate cycling. Since man will be releasing high
levels of phosphate into our waterways for the
foreseeable future, it is necessary that we develop
methods and apply them to the study of this problem.

REFERENCES

1. Baltscheffsky, H., L. V. Stedingk, H. W. Heldt, and M. Klingenberg. *Science 153:*1120 (1966).
2. Correll, D. L. and N. E. Tolbert. *Plant Physiol. 37:* 627 (1962).
3. Correll, D. L. and N. E. Tolbert. *Plant & Cell Physiol. 5:*171 (1964).
4. Correll, D. L. *Plant & Cell Physiol. 6:*61 (1965).
5. Correll, D. L. *Science 151:*819 (1966).
6. Correll, D. L. *Limn. Ocean. 10:*364 (1965).
7. Grzenda, A. R. and M. L. Brehmer. *Limn. Ocean. 5:*190 (1960).
8. Ball, R. C. *Radioecology, First National Symposium on Radioecology.* (Reinhold, 1961), pp. 217-228.

24. NUTRIENT AVAILABILITY AND ALGAL PRODUCTION

Darrell L. King and Dennis M. Sievers. Department of Civil Engineering, University of Missouri-Columbia, Columbia, Missouri.

ABSTRACT

Using basic chemical equilibrium relationships, a method is derived which permits the calculation of the algal production potential of the available carbon contained within the carbonate alkalinity system. Bioassay data for two algal types are presented which allow comparison of direct measurements of inorganic carbon uptake by the algae with calculated values. The excellent agreement of the two sets of data validates the equations and dissociation constants used in calculating the carbon uptake. Nitrogen and phosphorus availabilities are considered as limits of the amount of available carbon fixed by the algae. Knowledge of the nutrient content of a lake and the use of carbon:nitrogen:phosphorus atomic ratios for algae allows a rough initial prediction of the amount and type of algal production the lake will support. A sample calculation involving carbon and phosphorus is presented illustrating the method.

Although both the natural and cultural, or man-accelerated, aspects of eutrophication have received considerable attention over the past two decades, there has been little effort directed toward development of a model useful in predicting either the rate or extent of algal growth in lakes. In fact, the only criteria generally agreed upon as indicators of a eutrophic lake are mid-summer bluegreen algal blooms and hypolimnetic oxygen

433

deficiencies. Successful retardation of accelerated eutrophication is predicated upon a method for measuring and predicting both the rate and extent of eutrophication in individual lakes. Measurement of the algal growth potential (AGP)[1] is one biological method of arriving at a measure of eutrophication but there is a definite need for a means of predicting algal growth from measurements of physical and chemical properties of lake waters. The following material represents an initial attempt toward that end.

The Carbon System

Although those algae at or near the air-water interface may gain significant amounts of carbon from the atmosphere, the major sources of carbon available to algae are respiratory carbon dioxide resulting from biotic utilization of previously accrued organics and carbon dioxide fixed from the bicarbonate-carbonate alkalinity system of the water. The amount supplied by respiration is a function of the degree of organic enrichment of the water but even in the organic-rich sewage lagoon the bulk of the carbon dioxide used for net algal production is taken from the bicarbonate-carbonate system.[2] This system acting as a battery discharged during the day and recharged at night operates through the first and second dissociations of carbonic acid.[3]

(1) $HOH + CO_2 \rightleftharpoons HCO_3^- + H^+$

(2) $HCO_3^- \rightleftharpoons CO_3^= + H^+$

The amount of carbon present in each of the three carbon-containing species of the alkalinity system can be calculated from the following equations derived from basic equilibrium theory and presented by Harvey[4] and Park.[5]

(3) $[H_2CO_3]$ including free $CO_2 = \dfrac{aH^2}{K_1(H+2K_2)}$

(4) $[HCO_3^-] = \dfrac{aH}{H + 2K_2}$

(5) $[CO_3^=] = \dfrac{aK_2}{H + 2K_2}$

where:
   a = bicarbonate-carbonate alkalinity $[HCO_3^-] + 2[CO_3^=]$
   H = hydrogen ion activity as measured by a pH meter, 10^{-pH}

K_1 = first dissociation constant of carbonic acid

K_2 = second dissociation constant of carbonic acid

Total carbon dioxide within the bicarbonate-carbonate system can be calculated by summing the above three equations to yield the following expression:

$$(6) \quad \Sigma CO_2 = a \left| \frac{\frac{H^2}{K_1} + H + K_2}{H + 2K_2} \right|$$

This equation, a slight rearrangement of Buch's[6] equation, allows estimation of the carbon dioxide potentially available to algae from measurements of bicarbonate-carbonate alkalinity and pH of the water.

Many algae appear to require free carbon dioxide to support photosynthesis[3] and the amount of free carbon dioxide in the water is dependent on the alkalinity and pH of the water as mediated by equilibrium reactions 1 and 2. As the algae fix carbon dioxide the bicarbonate-carbonate system supplies additional carbon dioxide to maintain equilibrium. However, the concentration of free carbon dioxide decreases with the rising pH associated with continued uptake of carbon dioxide by the algae. The amount of free carbon dioxide present under any equilibrium condition can be calculated from Equation 3 and some results of such calculations are presented in Figures 107 and 108. From these figures it can be seen that for a given temperature or alkalinity the amount of free carbon dioxide diminishes rapidly with increasing pH.

When inorganic carbon is taken from the bicarbonate-carbonate system to support algal photosynthesis, there is a concomitant increase in organic carbon associated with increased algal biomass. Measurement of pH and alkalinity at the beginning of an algal growth period, ΔT, and Equation 6 yield an initial value for ΣCO_2. Measurement of the same parameters at the end of the growth period and Equation 6 yield a second value for ΣCO_2 reflecting the equilibrium changes associated with the removal of carbon dioxide by the algae. The difference between these two values, expressed as carbon, is a measure of new algal biomass which can be compared with direct

Figure 107   Figure 108

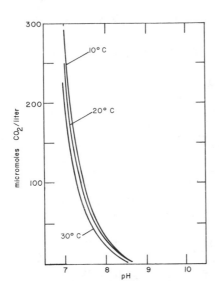

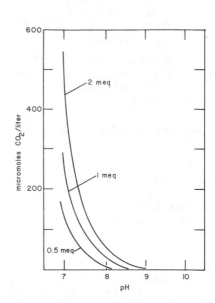

Figure 107.   *Free carbon dioxide concentration for a*
              bicarbonate-carbonate alkalinity of 1 meq under
              various conditions of temperature and pH.

Figure 108.   *Free carbon dioxide concentration at 15°C under*
              various conditions of alkalinity and pH.

measures of increased organic carbon over the time
period ΔT. The validity of such calculations is
dependent directly upon the values used for K_1 and
K_2. The following method is a means of validating
values for these constants such as those presented
by Buch.[6]

Procedures Used to Validate Method

     Pure cultures of *Scenedesmus acutiformis* and
Anacystis nidulans obtained from the R. C. Starr
culture, Indiana University, Bloomington, Indiana,
were grown in batch culture in one liter Erlenmeyer
flasks at 25°C under constant light supplied by
Sylvania Gro-lux bulbs. The growth medium, similar
to that used by Kevern and Ball[7] was inorganic
except for a slight amount of EDTA. Measurements
of organic carbon were obtained with a Beckman
Model IR315 Carbonaceous Analyzer. Measurements

of pH were made with a Fisher Accumet 320 pH meter and all alkalinity determinations were validated by measurement of inorganic carbon with the carbon analyzer.

Validation of the Method

Results of these microcosm studies are presented in Figures 109 and 110. These figures show both the

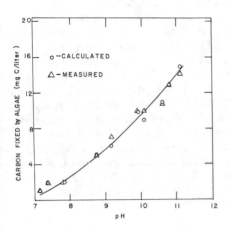

Figure 109. *Calculated and measured carbon fixation by* Anacystis nidulans.

Figure 110. *Calculated and measured carbon fixation by* Scenedesmus acutiformis.

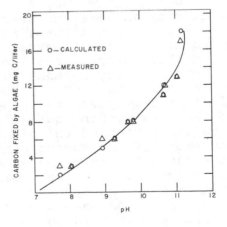

calculated and measured estimates of carbon fixation by the algae. The excellent agreement between the two sets of data validate both the equations and Buch's values for the dissociation constants over a wide pH range for the water used in this study. It appears that measurements of pH and bicarbonate-carbonate alkalinity and Equation 6 could be used as a quantitative assessment of algal growth.

Qualitative Considerations

Using empirical data, King[2,3] proposed that many green algae become carbon limited if the free carbon dioxide concentration is decreased to less than 7.5 μmoles CO_2/liter while the bluegreen algae are not affected seriously until the free carbon dioxide reaches a concentration of about three μmoles CO_2/liter. This suggests that if light and all other required algal nutrients are present in abundance, algal fixation of carbon dioxide from the bicarbonate-carbonate system will eventually reduce the free carbon dioxide concentration to the point where the bluegreen algae are favored. Continued reduction of free carbon dioxide below 7.5 μmoles CO_2/liter increases the probability that the lake will be dominated by bluegreen algae.

Relationship of Carbon and Other Major Nutrients

Although carbon is an extremely important algal nutrient, other required algal nutrients also must be considered since they are more apt to limit algal productivity in less enriched waters. Of these other essential nutrients, nitrogen and phosphorus are required in the greatest amount by algae relative to their general availability in water[8] and algal productivity in many systems is limited by the availability of one or both of these nutrients. Gerloff[9] has proposed that the ratio of nitrogen to phosphorus within aquatic plant biomass reflects the relative need of the plant for these nutrients. Measurement of the carbon content of algal biomass allows the ratio to be expanded to an atomic C:N:P ratio which relates algal demand for carbon to that of nitrogen and phosphorus. Although widely variable C:N:P ratios have been reported for algae, the critical C:N:P ratio for a particular alga sets the base nutrient requirement for that alga in terms of carbon, nitrogen, and phosphorus.

The ability of algae to store excess phosphorus makes it particularly difficult to determine the critical C:P ratio for individual algal species. Results of our laboratory investigations indicate that single species of algae are photosynthetically active at biomass atomic ratios from less than $C_{100}P_1$ to ratios exceeding $C_{1200}P_1$. However, these algae flocculate, settle to the bottom of the flasks and cease to be planktonic at a C:P ratio somewhere around $C_{200}P_1$. In a lake these algae would sink out of the euphotic zone and would be unable to exercise the significant physiological potential and reach $C_{1200}P_1$. Thus the extreme C:P ratios appear to be artifacts of laboratory systems and an atomic ratio somewhere around $C_{200}P_1$ appears to be a reasonable upper estimate for algae in natural systems.

Combination of C:N:P ratios for algal biomass with Equations 3 and 6 offers a means of approximating the quantitative and qualitative algal growth potential of a stratified lake.

If a non-turbid lake with a bicarbonate-carbonate alkalinity of 50 mg $CaCO_3$/liter is at 5°C just after the spring overturn, is in equilibrium with the atmosphere at 16 μmoles free carbon dioxide per liter (pH = 8.31), has an available phosphorus concentration of 4.7 μg P/liter, and warms to 20°C by midsummer, the following values can be calculated to approximate the quantity and quality of algal activity during the spring and summer growth period. Initially the ΣCO_2 of the lake is 12.12 mg carbon/ liter but at a critical atomic ratio for the algae of $C_{200}P_1$ complete utilization of all available phosphorus would allow algal carbon fixation of 0.36 mg C/l leaving ΣCO_2 equal to 11.76 mg carbon/ liter. At 20°C this would occur at pH 8.78 where the free carbon dioxide would be equal to 4 μmoles CO_2/liter, suggesting probable significant bluegreen algal populations at that time.

These calculations assume abundant light and no nutrient limit imposed by any required plant nutrient other than carbon and phosphorus. No consideration is given to nutrient recycle or accrual during the summer months. Nitrogen availability is not considered because it appears that if nitrogen became limiting at any time during the summer growth period, the nitrogen fixing bluegreen algae would begin to dominate the system.

While the proposed method allows a rough prediction of quantity and quality of algal production

within a lake, refinement of these estimates requires
better information on the degree of nutrient recycle
within lakes, more complete knowledge of which forms
of phosphorus are available to algae, better values
of the critical C:N:P ratios for algae, and more
information on the requirements and availability of
trace nutrients, especially as affected by changing
chemical equilibria associated with active algal
photosynthesis.

REFERENCES

1. Oswald, W. J., and C. G. Golueke. *J. Water Pollution Control Federation 38:*964 (1966).

2. King, D. L. Paper presented at the ASLO Symposium--The Limiting Nutrient Controversy, Gull Lake, Mich. (Feb. 1971).

3. King, D. L. *J. Water Pollution Control Federation 42:* 2035 (1970).

4. Harvey, H. W. *The Chemistry and Fertility of Sea Water* (Cambridge, England: Cambridge University Press, 1957).

5. Park, P. K. *Limnol. and Oceanog. 14:*179 (1969).

6. Buck, K. *Havsforskingstitutets Skrift Helginfors No. 151* (1951).

7. Kevern, N. R., and R. C. Ball. *Limnol. and Oceanog. 10:* 74 (1965).

8. Vallentyne, J. In "Special Symposium" Vol. 1, *Nutrients and Eutrophication*, Am. Soc. Limnol. and Oceanog. (1972), p. 107.

9. Gerloff, G. C. In *Eutrophication: Causes, Consequences, Correctives.* (Washington, D.C.: Nat. Acad. Sci., 1969), p. 537.

25

RESPONSE OF PLANKTON AND PERIPHYTON DIATOMS IN
LAKE GEORGE (N.Y.) TO THE INPUT OF NITROGEN
AND PHOSPHORUS

S. L. Williams, E. M. Colon, R. Kohberger, and
N. L. Clesceri.* Rensselaer Fresh Water Institute,
Rensselaer Polytechnic Institute, Troy, N.Y. 12181

Algal assays to determine the effect of present
or potential water chemistries on algal growth in a
lake or other water systems range from relatively
simple batch flask tests performed in the laboratory
to complex *in situ* tests involving large submerged
plastic containers. Complementing these bioassay
techniques in which the lake water is contained or
restricted in some manner, are unrestricted *in situ*
measurements made over relatively long periods of
time, usually one or more years, in which the environmental chemistry and the algal or other biological populations are measured and mathematical
relationships obtained. These mathematical relationships range from simple regression models to
complex phenomenological models relating time,
space and all the interacting variables.

Lake George, a typical soft water oligotrophic
lake in the southeastern Adirondack Region of New
York State (Figure 111) is showing signs of an increasing rate of eutrophication, particularly in
the southern basin of the lake, despite control of
waste discharges to the lake for the past 50 years.
Personnel of the Rensselaer Fresh Water Institute
at Lake George have been conducting a series of
biological, chemical and hydrological measurements

---

*Present Address: Bureau of Water Resources Planning, Area
of Natural Resources, Dept. of Public Works, Box 8218,
San Juan, Puerto Rico, 00910.

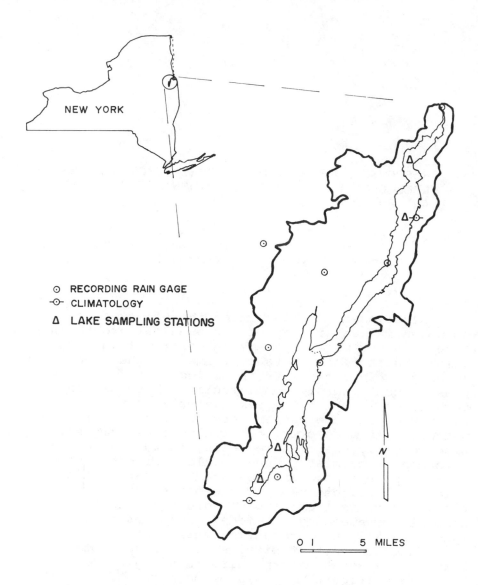

○ RECORDING RAIN GAGE
-○- CLIMATOLOGY
△ LAKE SAMPLING STATIONS

0 1   5 MILES

Figure 111.   Lake George, N.Y.   Location of Sampling Points

over the past few years to determine the relative importance of the various factors which could be contributing to the natural and man-produced aging of this popular recreational lake and the relative value of the corrective actions which could be taken.   Measurements by the FWI personnel and by other workers include flask type algal assays performed in the laboratory, *in situ* plastic bag assay experiments, and long term measurements of algal populations, their environmental chemistry and their zooplankton predators in unrestricted lake water. This report presents some of the results of these latter measurements on unrestricted lake water; specifically the apparent effect through regression analysis of the nitrate and phosphorus present in the lake and entering in the precipitation on the standing crop of plankton and periphyton diatoms. Lake George, which drains to the north, lies in a glacial scoured basin of Precambrian metamorphic plutonic and igneous bedrock. The drainage basin of the lake has an area of 606 Km^2 with 19% of this area covered by the lake. The volume of the lake is approximately 2.1 Km^3. Precipitation is the dominant water input to Lake George with an average annual precipitation of about 93 cm. About 0.1 Km^3 precipitation falls directly on the lake each year and 0.18 Km^3 enters as runoff through many small streams.[1] Based on morphometric and circulation characteristics, the lake has been subdivided into three basins, north, south and a central basin.[2]

EXPERIMENTAL

Lake water samples were generally taken at approximately monthly intervals from four locations—two in the north basin, and two in the south basin. Three depths, 3, 9 and 15 meters, were sampled at each location. Measurements of ammonia- and nitrate-nitrogen, hydrolyzable phosphorus, plankton diatoms, temperature, oxygen, pH, alkalinity, light penetration (photometer) and water clarity (Secchi disc) were made on these samples or at these locations. Aliquots of the water samples were filtered through prewashed 0.45 u membrane filters within 4-8 hours after sampling. The various plankton diatom species were identified and counted on the filters. Ammonia was determined by direct Nesslerization and nitrate by reduction with a cadmium-copper column and EDTA. Soluble acid hydrolyzable phosphorus (filtered sample) and total acid hydrolyzable phosphorus (unfiltered sample) were determined by the phosphomolybdate-stannous chloride method without extraction. Samples of the periphyton diatoms were obtained by exposing glass and plastic substrate slides at the three depths and two locations for the monthly periods and enumerating the various diatom species directly

on the slides. Diatom identification was based on
the description of Hustedt[3] and Patrick and Reimer.[4]
   The precipitation falling on the Lake George
basin was measured at eight locations (Figure 111)
using recording precipitation gauges. Samples of
the precipitation falling at the climatology station
in the northern basin were collected periodically
by event with an automatic precipitation collector
in plastic containers and usually analyzed within
24 hours of collection for ammonia and nitrate-
nitrogen and phosphorus, or frozen and analyzed
subsequently. A few samples of the precipitation
falling at the climatology station at the southern
part of the lake were collected in an open container
and analyzed. Average nutrient loadings and input
corresponding to the monthly sampling intervals
were estimated from the amount of precipitation and
the concentrations of ammonia and nitrate-nitrogen
and phosphorus present in the precipitation samples.
Samples from the principal streams entering Lake
George were also taken and analyzed. The incoming
solar radiation was measured with an integrating
pyranometer and averaged for the same monthly
sampling periods.
   The stepwise multivariate regression analysis
was performed on the data using the BIOMED computer
program BMD 02R.[5] The model utilized in the
analysis was exponential of the form

$$Y = b_j e^{b_i x_i + \ldots b_k x_k}$$, where Y = the

diatom counts and $x_i \ldots x_k$ the following environmental factors: temperature, total incoming solar
radiation and average daily incoming solar radiation for the 28 days preceding the sample, nitrate,
soluble hydrolyzable phosphorus, pH, and alkalinity.
The environmental and diatom data were grouped into
three seasonal periods for the years 1968, 1969 and
1970 as follows: winter/spring--January-May;
summer--June-September and fall--October-December.
Partial F values for the variables in the model
were examined and significant levels at 90% were
established. R^2 values and standard deviations of
the predictive model were also examined for the
overall value of the model. The overriding consideration in the analysis was F-values which test
the significance of an independent variable in
explaining variation in the data.

RESULTS

The seasonal periodicity of the standing crop
of the plankton and periphyton diatoms in the north
basin over a 3-year period is shown in Figure 112.
The pattern in the south basin is generally similar.
The plankton diatoms show a strong spring pulse
which generally starts building up in December and
reaches a peak in April or May of each year. The
periphyton diatoms reach their greatest standing
crop during the summer with no single distinct
peak or maximum occurring. For the most part, the
same diatom species are present in both the plankton
and periphyton samples but there was no relationship by regression analysis between relative or
absolute numbers of the various species found in
the two types of samples.

The concentrations of hydrolyzable phosphorus
and nitrate-nitrogen in the northern and southern
lake basins for the period covered in this report
(September 1969-November 1970) are shown in Figures
113-114. The average concentrations of these two
nutrients in the two basins for the seasonal periods
utilized in the regression analysis are shown in
Table 61. Ammonia nitrogen was occasionally present
but when present it was usually less than 5 μg/L
based on distillation followed by Nesslerization.
Interferences in direct Nesslerization prevented
measurement of low ammonia concentrations when
present in the samples. It can be noted from the
table and graphs that the south basin water on the
average is higher in soluble phosphorus and nitrate,
and lower in pH and alkalinity. An analysis of
variance confirmed that there were significant
seasonal and basin differences in these parameters
as suggested by the averages except that the variation in the monthly nitrate concentrations precluded establishing a significant north-south
difference on a statistical basis even though the
average levels show much higher concentrations in
the south.[6] The dissolved oxygen generally exceeded
7 mg/liter throughout the year except at Station 1
where the dissolved oxygen decreased to 5.6 mg/l
at 15m and 2.0 mg/l at 21m in the fall of 1969
(October 12). Figure 115 shows the total incoming
solar radiation for the monthly periphyton exposure
period and the water temperature and period of ice
cover. The daily total incoming solar radiation
for the monthly period was also included in the

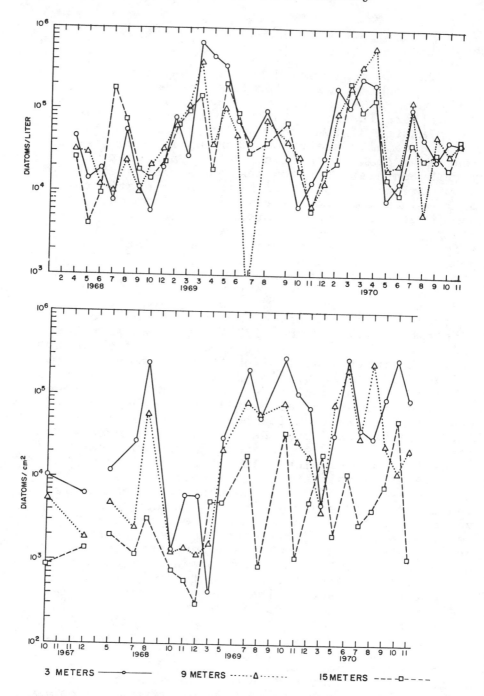

Figure 112. *Standing Crop of Plankton and Periphyton Diatoms in the North Basin, Lake George, N.Y. The Numbers of Periphyton Diatoms Developed Over a 28-Day Exposure Period Prior to the Date Plotted. Plankton Diatoms are Plotted on the Top Graph.*

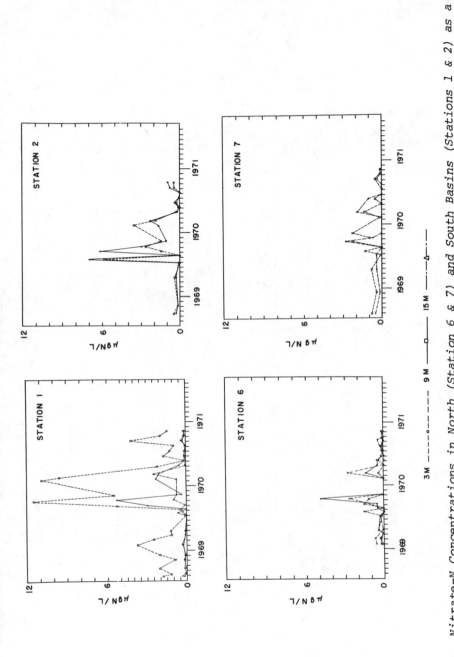

Figure 113. Nitrate-N Concentrations in North (Station 6 & 7) and South Basins (Stations 1 & 2) as a Function of Julian Dates.

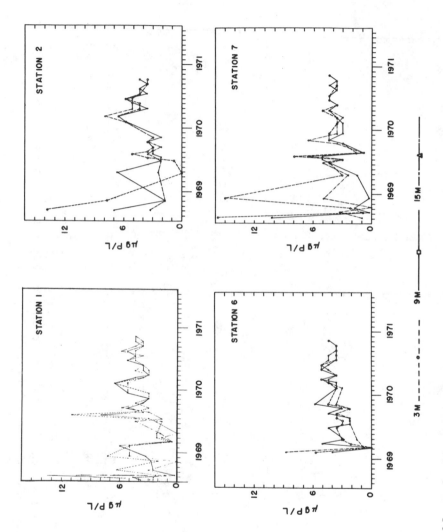

Figure 114. Soluble Hydrolyzable Phosphorus Concentrations in North (Stations 6 & 7) and South (Stations 1 & 2) Basins as a Function of Julian Dates.

Table 61

Average Concentrations of Nutrients in North and South Basins

Seasonal Period	*Soluble Hydrolyzable Phosphorus* µg P/L	*Nitrate-Nitrogen* µg N/L	*Alkalinity* mg CaCO₃/L	*pH*	*Temperature* °C
Winter-Spring South Basin	3.5	10.0	22.6	7.2	2.5
Winter-Spring North Basin	3.4	5.1	21.6	7.2	4.0
Summer South Basin	4.1	11.6	21.2	7.6	11.3
Summer North Basin	3.9	4.0	21.4	7.7	16.7
Fall South Basin	4.3	10.2	23.4	7.5	12.6
Fall North Basin	4.5	7.2	25.3	7.6	15.8

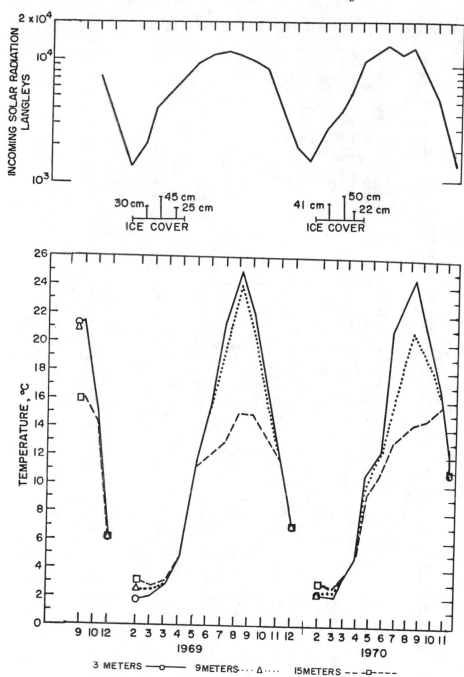

Figure 115.   Water Temperature, Ice Cover and Incoming Solar
Radiation--South Basin (Stations 1 & 2).

analysis and is also shown in Figure 115. The incoming solar radiation was about the same for both
basins of the lake; however, the water clarity and
light penetration, as measured by Secchi disk and
underwater photometer, was significantly greater
in the north basin, as was the temperature.[6]
    Table 62 summarizes the regression analysis
results when the standing crop of plankton and
periphyton diatoms were compared to the independent
variables of nitrate, soluble hydrolyzable phosphorus,
alkalinity, pH, temperature, daily incoming solar
radiation averaged over the 28-day periphyton exposure periods and total incoming solar radiation
for the 28-day periods. In the south basin during
the winter-spring period for 1969 and 1970 the
total plankton diatoms showed a significant relationship with nitrate concentrations. In the summer
period, 1969 and 1970, the total plankton diatoms
in the south showed a correlation with soluble
hydrolyzable phosphorus followed by alkalinity.
In the north basin during the winter-spring period
of 1969 the numbers of plankton diatoms showed a
relationship with the daily average incoming solar
radiation; and, in 1970 there was a relationship
with alkalinity. No correlations were present
during the summer in this basin. In the fall
period in 1969 the plankton diatoms in the north
basin showed a relationship with phosphorus concentrations. The total periphyton diatom standing
crops showed no correlation with nitrate, phosphorus,
or alkalinity during any of the seasonal periods.
    Table 63 lists the species that were the prin-
cipal dominant diatoms (> 50% of the total diatom
population) during the three seasonal periods and
the results of the regression analysis relating
these species to the independent variables.
Synedra tenera which dominated the plankton during
the winter-spring period of 1969 showed a correlation with the phosphorus concentration in the
winter-spring and fall periods. *Stephanodiscus
minutula* which was a subdominant species during
the summer periods also showed a correlation with
phosphorus. The two diatom species which normally
dominated the winter-spring plankton diatom populations, *Asterionella formosa* and *Fragilaria
crotonesis*, both showed a slight relationship to
nitrate but a much greater relationship with lake
temperature. In the periphyton, *Achnanthes
minutissima* which was the dominant diatom most of

Table 62

Results of Regression Analysis Total Diatoms and Environmental Factors

Seasonal Period	Variable	Partial-F	ΔR^2%	F Total	R^2% (total)
PLANKTON					
Winter-Spring (Feb.-May)					
South Basin, 1969	Nitrate	43.3	86.1	43.3	86.1
South Basin, 1970	Nitrate	43.3	75.6	43.3	75.6
North Basin, 1969	IR-AV*	3.5	24.0	3.5	24.0
North Basin, 1970	Alkalinity	5.0	23.7	5.0	23.7
Summer (June-Sept.)					
South Basin, 1969-1970	Phosphorus	8.1	28.8		
	Alkalinity	5.3	26.4	5.6	55.2
North Basin, 1969-1970	**				
Fall (Sept.-Dec.)					
South Basin, 1969-1970	**				
North Basin, 1969	Phosphorus	7.4	55.3	7.4	55.3
1968	**				
1970	**				
PERIPHYTON					
Winter-Spring and Fall					
1968-1970	Temp	29.9	49.3		
North and South Basin	IR-Tot*	8.9	12.5		
	pH	4.3	3.6	25.9	65.4
Summer, 1969-1970					
North and South Basin	**				

*IR-AV - daily incoming solar radiation averaged over 28-day period.
 IR-Tot - total incoming solar radiation over 28-day period.
**No significant correlation with any of the variables.

the year showed a relationship to nitrate in the
winter-spring and fall periods and in the summer a
strong correlation with nitrate, solar radiation
and some variable associated with the different
years not covered in the regression. During the
summer, *Synedra tenera* which was present in periphyton but not dominant showed strong correlations with
both soluble phosphorus and nitrate concentrations
but not at other times of the year.

The precipitation falling on Lake George and
its drainage basin contained much greater concentrations of inorganic nitrogen (ammonia and nitrate)
and slightly greater concentrations of hydrolyzable
phosphorus than found in the lake water. Figure 116
shows the average monthly concentrations of total
hydrolyzable phosphorus, NH_3-N and NO_3-N during
1970. The concentrations were generally independent
of the amount of precipitation collected making it
possible to estimate chemical loadings due to wet
fallout. Concentrations of ammonia and nitrate in
the precipitation appeared to be slightly higher
during the summer period (June, July and August)
while the phosphorus concentration was lowest in
the fall. The amount of precipitation within the
drainage basin was quite variable with the annual
precipitation in the drainage basin tending to decrease to the north.[1] Figure 117 shows the precipitation catch at the rain gauges at the north and
south basin climatology stations. NO_3-N, NH_3-N and
total hydrolyzable phosphorus loadings are also
shown in this figure. Table 64 summarizes the
regression analysis results when the standing crop
of plankton and periphyton diatoms were compared to
the amount of precipitation and loadings of NH_3-N,
NO_3-N and hydrolyzable phosphorus in the precipitation falling directly on the lake. In the case of
the plankton, the amount of precipitation and the
loadings falling on the lake during the 7 days
preceding the plankton sample dates were used in
the regression model with 33% of the precipitation
falling during the ice covered period entering the
lake. The total precipitation and associated
loadings falling on the lake during the 28-day
periphyton exposure periods were used in the regression models involving the periphyton with the
same 33% of the precipitation entering during the
ice covered period.

No significant correlation was found between
the amount of precipitation and diatom counts.
Previous experience[6] has indicated that a logarithmic

Table 63

Relationships of Dominant Diatoms* Species with N & P Concentrations

Dominant Diatoms	Variable	F Partial	$\Delta R^2\%$	F Total	R^2 Total
PLANKTON					
Winter-Spring					
Asterionella formosa	Temp	8.1	51.2		
	pH	9.6	3.6		
	Nitrate	5.6	7.6	22.1	62.4
Synedra tenera	Phosphorus	4.3	18.5	4.3	18.5
Fragilaria crotonensis	Temp	7.3	11.4		
	Nitrate	2.5	5.3	3.9	16.7
Summer					
Cyclotella operculata	Nitrate	4.2	16.0	4.2	16.0
C. glomerata	No correlation (1970)				
Stephanodiscus astraea	Phosphorus	27.7	42.9		
S. minutula	Alkalinity	15.2	37.4	16.3	80.3
Fall					
Tabellaria fenestrata	pH	22.1	25.7	22.1	25.7
A. formosa	IR-Tot	4.7	9.3	4.4	19.7
F. crotonensis	pH	7.1	10.4		
	pH	3.0	7.0	3.0	7.0
S. tenera	Phosphorus	26.2	92.9	26.2	92.9

PERIPHYTON

Winter-Spring-Fall					
Achnanthes minutissima					
(South Basin)	Temp	38.0	52.8		
	IR-Tot	8.5	8.9		
	Nitrate	5.3	5.8	67.5	
(North Basin)	Temp	4.9	67.9	4.9	67.9
S. tenera, 1968	IR-Tot	6.1	19.5		
	Alkalinity	3.4	23.7	3.1	43.2
1969	IR-Tot	9.0	50.4		
	Temp	8.9	21.1	15.1	71.5
1970	Temp	7.7	43.5	7.7	43.5
Summer					
A. minutissima					
(South Basin)	Year	13.4	36.1		
	pH	14.2	52.7	11.9	88.8
(North Basin)	Year	58.6	27.4		
	IR-Tot	48.3	39.6		
	Nitrate	44.7	30.9	41.2	97.9
S. tenera	Phosphate	10.0	38.4		
	Nitrate	9.3	39.3	6.9	69.7

*> 50% of total diatoms in sample.

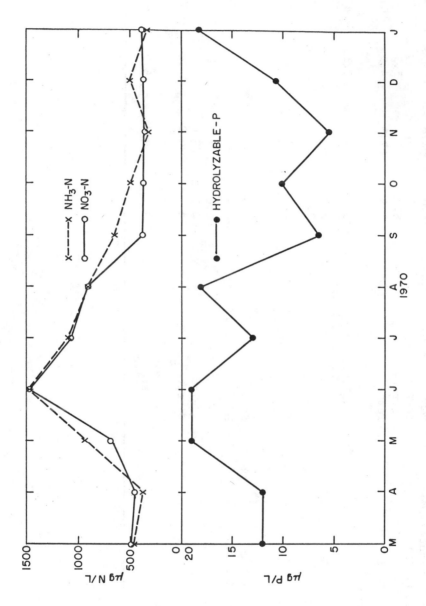

Figure 116. Average Monthly Concentrations of NO_3-N, NH_3-N Total and Hydrolyzable Phosphorus in Precipitation Falling on Lake George.

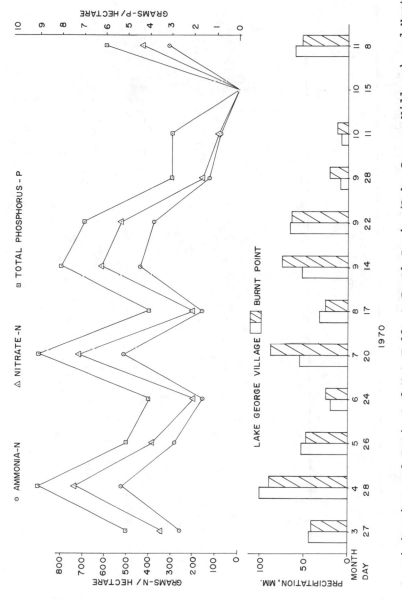

Figure 117. *Precipitation and Estimated Wet Fallout South Basin (Lake George Village) and North Basin (Burnt Point).*

Table 64

Correlation of Dominant Diatom Species and Precipitation Chemistry

	Partial F-Test	Multiple Correlation Coefficient (R)	F-Test (total)	F-test for Significant (95%)
PLANKTON				
Station 1				
Total plankton				
Ammonia	22.28			
Phosphorus	28.02	.5611	14.015	3.15
Station 6				
Asterionella formosa				
Nitrate	11.78			
Phosphorus	8.96	.4608	5.931	3.23
Fragellaria crotonensis				
Nitrate	8.20			
Phosphorus	8.61	.4091	4.423	3.23
Synedra tenera				
Phosphorus	6.589	.3574	6.589	4.08
Total plankton				
Phosphorus	27.75	.6176	27.755	4.08

PERIPHYTON

Station 6

Tabellaria flocculosa
Ammonia   4.02
Phosphorus 7.49
Depth   3.81   .6732   4.15   3.29

Fragellaria crotonensis
Ammonia   40.3
Phosphorus 35.5
Depth   6.10   .867   15.1   3.29

transformation of diatom counts should be used.
The correlations were examined through the use of
a regression model of the form: $y = ae^{bx}$, where
y = diatom counts, x = precipitation (in inches).
The periphyton diatoms were regressed against the
previous 28-day totals of precipitation. The
plankton diatoms were regressed against the previous 28-day totals and 7-day totals of precipitation. Neither period of precipitation showed a
significant correlation with diatom counts.

When the chemical loadings were used in the
regression model instead of precipitation volume,
several significant correlations involving inorganic
nitrogen and phosphorus were shown to be present.
The plankton standing crop in the south basin correlated with the input of ammonia and phosphorus
by precipitation, while the standing crop in the
north basin correlated with the phosphorus only.
The standing crops of the principal dominant diatom
species in the plankton in the north basin showed
a correlation with precipitation nitrate and
phosphorus, or phosphorus in the case of *S. tenera*.
It is interesting to note that this diatom species
also showed a strong correlation with phosphorus
in the lake water (Table 63). The standing crop
of diatoms in the periphyton showed no correlation
with phosphorus or nitrogen in the precipitation
but showed a strong correlation with depth. This
was also true of the most common dominant species
in the periphyton *Achnanthes minutissima* and *S.
tenera*. *F. crotonensis* and *T. flocculosa* which
occasionally were dominant diatoms in the periphyton
showed a correlation with ammonia and phosphorus in
the precipitation as well as with depth in the
north basin but showed no correlation with these
nutrients in the south basin. In all of the regression analyses, including those involving precipitation volume, depth was a significant variable
for all of the periphyton diatoms but was not
significant for the plankton diatoms which move
freely in the water column. Depth did not appear
as a significant variable affecting the periphyton
diatoms in Tables 62 and 63 since the attenuation
of solar radiation and the nutrient concentrations
and temperature at each depth were utilized in the
analyses. A regression analysis was also performed
with precipitation ammonia, nitrate and phosphorus
as independent variables and the concentration of
nitrate and phosphorus in the lake as dependent
variables. As would be expected no correlation

existed since these elements are rapidly lost from
the water column during and after precipitation.

DISCUSSION

The results of the regression analysis shown
in Tables 62 and 63 indicate that the diatom populations are responding at different times to one or
more of several environmental factors. Two of these
are the concentrations of soluble phosphorus and
nitrate in the lake water. These concentrations
are equilibrium or pseudo equilibrium concentrations
related to many factors, one of which is the input
of inorganic N and hydrolyzable P necessary to
balance the losses from biological activity and
lake outflow. A positive correlation between the
standing crop of a particular diatom species or the
total diatom population with nitrogen or phosphorus
can, under some conditions, represent a positive
correlation with the input of these elements, thereby
bearing some relationship to enrichment bioassays in
which phosphorus or inorganic nitrogen are added to
lake water and the response of the standing crop of
algae noted. Stross performed numerous enrichment
experiences on Lake George water during part of the
same time period of our regression analyses.[7] In
June, August and October of 1970 he found an increase in ^{14}C uptake from small phosphorus additions
(half saturation constants K_t of 0.1-0.3 µg P/L)
while in July and December of 1970 and January of
1971 phosphorus additions produced no increase in
carbon uptake. In February of 1971 he reports only
a very slight increase in ^{14}C uptake with phosphorus
additions and in May of 1971 a slightly larger increase with P enrichment but not as large an increase
in the months cited above. In the summer period of
1970, which included June and August when the enrichment experiments showed a good response from P
additions, the regression analysis also showed a
strong correlation between the standing crop of
plankton diatoms and soluble hydrolyzable phosphorus.
During the fall period of 1970 which included
October when the enrichment experiments showed a
response from P additions and December when no
response was noted, the regression analyses showed
no correlation between phosphorus, or any other
environmental variable tested, and the standing
crop of plankton diatoms. A positive correlation
was noted with phosphorus during the fall period
of 1969. No enrichment experiments were reported

461

on Lake George water during the winter of 1970 but, as mentioned above, Stross reported some increase in ^{14}C uptake in February and May in 1971, but none in January.[7] No correlations between P and plankton diatom standing crops were obtained during these months in 1970 in the regression analysis; however, a correlation with nitrate was obtained during this period in 1970 in the south basin but not in the north basin. *Fragilaria crotonensis* showed a weak but significant correlation with nitrate during this period in both basins. Goldman, in enrichment studies on Lake Tahoe water with nitrate-N and ortho-phosphate, found that nitrate additions when considered on a short time basis substantially increased the growth rate of the diatom *F. crotonensis* in February with ortho-phosphate stimulating growth in mid-spring.[8] *F. crotonensis* was the main phytoplankton in the lake water. Schindler *et al.* reported the results of a multiple regression analysis relating phytoplankton production in Clear Lake, eastern Ontario, with nitrate-N, silicate, chlorophyll, solar radiation, total CO_2 (as derived from alkalinity measurements) and water temperature with the first four parameters contributing significantly to the regression.[9] (Clear Lake is very similar in water clarity and nutrient concentrations to Lake George.) These authors also provided a general comparison with four other similar regression analyses in other lakes performed by other workers. This comparison indicated that temperature and light gave significant regressions in four of the five studies and dissolved inorganic nitrogen and carbon in three of the five. Phosphorus was not examined in Schindler's report and was not mentioned in the general comparison. Sakamoto reported a significant increase in carbon uptake from the additions of either phosphorus or nitrate or both to water from lakes 67 and 69 of the western Ontario Experimental Lake Area. Additions of both P and N increased the carbon uptake in Lake 189, while no response occurred from the addition of the elements separately.[10]

The existence of significant correlations between inorganic nitrogen and phosphorus in the precipitation falling directly on the lake and the standing crops of certain plankton and periphyton diatoms suggests that the precipitation is an important source of nutrients and the nutrient limitations in the lake permit utilization of the nitrogen and phosphorus entering in the precipitation.

The regression analyses in Tables 63 and 64 appear internally consistent in that, on occasion, depending on the season, certain of the important diatoms in the plankton and periphyton show correlations with nitrate or phosphorus concentrations in the lake. A closer relationship between Tables 63 and 64 would not be expected since the concentration of nitrate and phosphorus in the lake represents an equilibrium value related to biological activity, lake discharge and nutrient input, while the amount of inorganic nitrogen and phosphorus in the precipitation is true input.

The inorganic nitrogen in the precipitation falling directly on the lake along with the inorganic nitrogen in the dry fallout on the lake is the major source of inorganic nitrogen entering the lake. The amount of inorganic nitrogen in the wet and dry fallout would be expected to be of the same order of magnitude.[11,12] The stream runoff potentially would be expected to be a larger source of inorganic nitrogen than the precipitation falling directly on the lake since precipitation falling on the terrestrial part of the basin and entering the lake is 1.7 times greater than that falling directly on the lake. Chemical analyses performed on the main streams carrying the precipitation runoff to the lake indicate that large quantities of inorganic N and lesser quantities of P are retained by the terrestrial drainage basin and do not enter the lake with the runoff. Several workers have reported similar retentions.[9,13] Our analyses of incoming streams and similar analyses conducted by the New York State Health Department indicate an average concentration of 80 µg/L inorganic nitrogen, as compared to 1328 µg/L in the precipitation. Based on these average concentrations and the amount of runoff precipitation, it is calculated that the inorganic nitrogen introduced by precipitation runoff is only one-tenth of that entering in the precipitation falling on the lake. The amount of inorganic nitrogen removed from the lake by drainage from the lake outlet is estimated to be only 1.6% of the yearly input of inorganic nitrogen from precipitation falling directly on the lake.

The amount of hydrolyzable phosphorus being introduced with the precipitation falling on the lake also represents a significant fraction of the hydrolyzable phosphorus input particularly in the sparsely settled north basin where the major phosphorus inputs are the precipitation and dry

fallout on the lake, and the precipitation runoff.
In the south basin, with its densely populated
shoreline and adjacent lands, a potentially large
input of hydrolyzable phosphorus from septic tanks
and the sewage treatment plant also exists. In the
north basin, the amount of hydrolyzable phosphorus
being introduced with the precipitation falling on
the lake represents about one-third of the input
with the hydrolyzable phosphorus in the dry fallout
on the lake estimated to be another one-third based
on Lake Wingra measurements.[11] The hydrolyzable
phosphorus input from precipitation runoff carried
by streams was calculated to be the same magnitude
as in the wet fallout (or another one-third of the
phosphorus input) based on an average concentration
of 7 µg/L-P in the streams entering the lake. Based
on the average soluble hydrolyzable phosphorus concentration of 4 µg/L-P in the lake, about 22% of
the hydrolyzable phosphorus entering from precipitation, dry fallout and runoff is removed by the
lake outflow.

The correlations between the inorganic nitrogen
(NH_3 or NO_3) and phosphorus in the precipitation
with plankton diatom standing crops appear to be in
agreement with the above observations. In the north
basin where the precipitation on the lake represents
a greater fraction of phosphorus input, a strong
correlation was obtained between total plankton
diatoms and phosphorus alone. In the south basin
both phosphorus and nitrogen (NH_3) showed a correlation with diatom standing crops. The utilization
of precipitation ammonia and nitrate as well as
phosphorus indicated by the regression analyses is
in agreement with the low nitrate concentrations
found in the lake and the low phosphorus half
saturation constant found for Lake George phytoplankton populations. The results also appear to
be in agreement with enrichment tests performed by
Stross on Lake George water. He found a response
from phosphorus and nitrogen together, but not from
nitrogen alone.[7] The plankton standing crops in
the deeper water (9 and 15 m) were equally responding
to the input of nitrogen and/or phosphorus, while
the periphyton diatom standing crops forming on
substrate slides at fixed depths were not. Although
the total periphyton diatom standing crops did not
show a correlation with P or N in the precipitation,
F. crotonensis and *T. flocculosa* show strong correlations. *F. crotonensis* in the plankton also showed
a correlation with precipitation nitrate and

phosphorus. This is the same diatom which Goldman found responded to nitrate and phosphorus additions to Lake Tahoe water.[8]

ACKNOWLEDGMENTS

FWI Report 71-8. This research was supported in part by the Office of Water Resources Research (Contract 14-31-0001-3387) and in part by the Eastern Deciduous Forest Biome, U.S. International Biological Program, funded by the National Science Foundation, under Interagency Agreement AG-199,40-193-69, with the Atomic Energy Commission, Oak Ridge National Laboratory. The research reported represents a portion of a thesis submitted by S. L. Williams in partial fulfillment of the requirements for the Ph.D. degree in Environmental Engineering at Rensselaer Polytechnic Institute. The assistance of Krystyna Mrozinska and Gregory Gustitus in the chemical analyses and of Richard E. Stevens of Ernest F. Fullam, Inc. in the identification of diatoms by scanning electron microscopy is gratefully acknowledged. Dr. Cornelius Weber, National Environmental Research Center, Environmental Protection Agency, Cincinnati, Ohio generously assisted in the identification of the centric diatom species. Dr. Wolfgang Fuhs, Division of Laboratories and Research, New York Dept. of Health, provided data on concentrations of P and N in influent streams.

REFERENCES

1. Colon, E. M. "Hydrological Study of Lake George, New York," D. Eng. Thesis, Rensselaer Polytechnic Institute, Troy, N.Y. (1972).
2. Needham, J. G., C. Juday, E. Moore, C. F. Sibley, and J. W. Titcomb. "A Biological Survey of Lake George, New York," *State of New York Conservation Commission Report*, pp 1-78 (1932).
3. Hustedt, F. "Bacillariophyta (Diatomeae)," In *Die Susswasser Flora Mitteleuropas*, Pascher, A., ed. v. 10 (Gustav Fischer, Jera, 1930).
4. Patrick, R., and C. W. Reimer. *The Diatoms of the United States* (Philadelphia: Acad. Natural Sci., 1966).
5. Dixon, W. J., Ed. *BMD-Bio-Medical Computer Programs* (University of California Press, 1968).
6. Williams, S. L., and R. Kohberger. "Differences in Diatom Populations and Environmental Parameters in the North and South Basins," In *Diatom Populations Changes in Lake George, N.Y.*, Clesceri, N.L., and S. L. Williams, eds. OWRR Contract 14-31-0001-3387 Final Report, 1972. pp 65-85.

7.   Stross, R. G. *Primary Production in Lake George: Its
     Estimation and Regulation*, Eastern Deciduous Forest Biome
     Memo Report No. 71-115. State University of New York at
     Albany, Albany, N.Y. (1971).

8.   Goldman, C. R., and R. Armstrong. "Primary Productivity
     Studies in Lake Tahoe, California," *Verh. Internat.
     Verein. Limnol. Bd. 17*:4 (1969).

9.   Schindler, D. W., and J. E. Nighswander. "Nutrient
     Supply and Primary Production in Clear Lake, Eastern
     Ontario," *J. Fish. Res. Bd. Canada 27*:2009 (1970).

10. Sakamoto, M. "Chemical Factors Involved in the Control
     of Phytoplankton Production in the Experimental Lakes
     Area, Northwestern Ontario," *J. Fish. Res. Bd. Canada
     28*:203 (1971).

11. Kluesener, J. W. "Nutrient Transport and Transformations
     in Lake Wingra, Wisconsin," Unpublished Ph.D. Thesis,
     University of Wisconsin, Madison (1971).

12. Matheson, D. H. "Inorganic Nitrogen in Precipitation and
     Atmospheric Sediments," *Can. J. Technol. 29*:406 (1951).

13. Likens, G. F., F. H. Bormann, and N. M. Johnson.
     "Nitrification Importance to Nutrient Losses from a
     Cutover Forested Ecosystem," *Science 163*:1205 (1969).

26. ZOOPLANKTON REPRODUCTION AND WATER BLOOMS

R. G. *Stross*. Department of Biological Sciences,
State University of New York at Albany, Albany, N.Y.
12222

Food chains are transformed and species replaced
as blooms of bluegreen algae develop in the plankton
of lakes. While the ultimate cause, inadvertent
fertilization[1] coupled with increased nutrient
mobility within the lake[2] is known, the immediate
processes that attend the development of the bloom
are not. The purpose of this paper is to describe
and interpret the disappearance of the grazer
assemblage prior to and during development of the
bloom. In keeping with the constraints of the
symposium, bioassay is involved and knowledge gained
from its application forms an essential precursor to
the conclusions of the paper.

Planktonic populations, like most others, wax
and wane within an annual interval, often with great
seasonal precision. Abrupt numerical changes in
density of a population may or may not be attended
by transformation to a specialized dormant or diapaused stage in the life cycle. In either case
increase and decrease must be the result of changing
rates of birth and death. When an embryonic diapause
is involved, the temporal separation of fecundity
and birth may greatly alter the birth rate while
death rate remains constant. Since diapaused stages
may be involved in changes of abundance, the alternative questions may be: do the seasonal cycles in
abundance result from an environmental forcing of
birth and death rates? Alternatively, are populations
responding to environmental cues which function to
trigger the formation or the germination of resting
stages? Although the resulting effect on rates of
population increase may ultimately be the same, the

alternatives radically affect the explanations for the development of an algal bloom.

Fortuitously, the search for a particular diapause cycle in *Daphnia* resulted in the discovery of an intriguing bluegreen algae-zooplankton grazer relationship in a lake in eastern New York. The grazer assemblage, temporarily at least, goes to near extinction as a dense bloom of bluegreen algae develops. The rapid growth of bluegreen algae in Saratoga Lake, New York may be typical of the way bluegreen blooms develop.[3] In three successive years an explosive increase in density has developed within an interval of one week. Development of the bloom was always associated with and followed the virtual disappearance of the grazing Crustacea.

METHODS

Planktonic organisms were collected in horizontal tows with a Clarke-Bumpus sampler and a #10-mesh net. Duplicate or triplicate tows of 1.0 m^3 or more were made at depth intervals of two meters from a location near the northern end of Saratoga Lake. The sampling program began each year sometime in April or May and continued into late June or early July of 1970 and 1971. The initial experience in 1969 indicated a very rapid change in the plankton at or about June 1. Therefore, the sampling was most intense during late May and early June, declining to a frequency of once per 7 or 10 days thereafter. Prior to preservation in a 5% formalin solution, the Crustacea were narcotized by the addition of CO_2 from a compressed source containing either 50 or 95% carbon dioxide; the treatment, intended to prevent the loss of the brood pouch contents of the Cladocera, was sufficiently effective to determine mean number of embryos per brood.

Numerical densities were determined for all of the more abundant Crustacea in the samples from a minimum of three 1-ml subsamples. Each of the more abundant species was normally subdivided into three or more size or age categories. In the principal species, *Daphnia pulex* Leydig, the distinction was made of embryos destined for immediate development (subitaneous) or for diapause. Development of the blooms of bluegreen algae was indexed by sampling colonies (as in the case of *Microcystis* and *Coelosphaerium*) or of filaments or parts thereof (as in the case of *Aphanizomenon* and *Anabaena*) in the samples of Crustacea. It is to be emphasized that because of the large mesh size, the bluegreen estimates could be indicative only of true changes in numerical density.

Temporal Patterns of Bluegreen and Crustacea Density

The principal grazing population in Saratoga Lake is *Daphnia pulex*, a judgment that invokes numerical density, length of the individuals, and the rule that describes filtration rate as varying with the cube of the body length.[4] The rapid sequence of events immediately before and during the surge of the bluegreen populations is illustrated by the population changes in density and switch in reproduction of *D. pulex*. When the ice melts in April (as in 1971) there are small numbers of adults which begin to release eggs into the brood pouch. Increase is slow in the cold water, and in the second week of May, the mean density for all ages of *Daphnia* is still less than 1.0 per liter in a top 7-meter-deep column of water in Saratoga Lake; the density for all species and ages of Crustacea captured in the net is only 10.0 per liter (Figure 118A). Thereafter increase is rapid. The finite birth rate of *Daphnia* in mid-May is nearly a doubling per day (Figure 118C). Within two more weeks the density of all Crustacea is now at the maximum for the season at 42.0 per liter. The *Daphnia* population increased more slowly and reached a maximum two weeks after species of Copepods had begun to decline or four weeks from the focal point of May 11.

The Copepod populations are first to decline in late May and account for the decline in total Crustacea at that time (Figure 118A). They are followed by the decline and virtual disappearance of *D. pulex* in mid-June 1971. The surviving *Daphnia* is likely to be *D. galeata*, a second species that is rare in mid-June and later. Another Cladoceran in the plankton is a species of *Bosmina* which was rare in mid-June but became increasingly abundant in July (not shown in Figure 118). At the time of its disappearance, the population of *D. pulex* is still giving birth to young, and the population includes both young and juveniles.

Although grazers other than Crustacea may be present, their small size and unimpressive densities indicate that nothing in mid-June substitutes for the loss of the large Crustacea, especially the *Daphnia*.

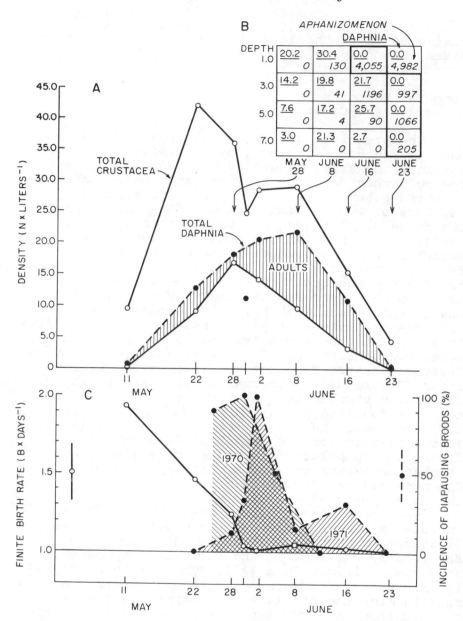

Figure 118. *Temporal sequence preceding and during a bluegreen bloom in Lake Saratoga, N.Y. A. Development and disappearance of Daphnia pulex population relative to that of all Crustacea. B. Spatial-temporal pattern of densities of Daphnia and the bluegreen, Aphanizomenon (individuals or filaments per liter). C. Daily birth rate in 1971 and timing of reproductive switch in the Daphnia pulex population in 1970 and 1971.*

A bluegreen bloom develops within the same
week that the *Daphnia* disappear. The density of
Daphnia at 1.0 meter depth declined from 30.4 on
June 8 to 0.0 per liter on June 16 (Figure 118B).
During the same period the density of the bluegreen
Aphanizomenon flos-aquae increased from 130 to more
than 4,000 filaments per liter. The *Daphnia* remaining at deeper levels were gone by the end of
the following week without substantial change in
density of bluegreens. A similar pattern prevailed
during the preceding years. The transition was
completed some 10 days earlier, however, in 1970.

Aphanizomenon is the primary species of bluegreen in Lake Saratoga during June and early July
although another spore former, *Anabaena*, is present.
Two non-spore forming species, *Coelosphaerium* and
Microcystis, are also present, and their colonies
are visible as the ice melts in April. Density of
the more abundant *Coelosphaerium* does not increase
until the burst of *Aphanizomenon*, increasing from
14.0 to 225 colonies per liter during the June 8-16
interval. An overall increase in algal biomass is
suggested by the threefold increase in total
chlorophyll during that week.

Exploring Relationships Between
Grazers and Bluegreen Algae

Given the information presented above, a ten-
tative conclusion at this stage of the analysis might
be that the bluegreens have forced themselves into
the community from the spore stage as with
Aphanizomenon and from the reproductively inactive
population, as with *Coelosphaerium*. The result
would seem to have been to eliminate the *Daphnia*
population by some unknown mechanism. Evidence
suggests the reverse, however.

The disappearance of the *Daphnia* population
coincident with the burst of bluegreen growth could
suggest a causal relationship. If causal, which
causes which? The evidence, although circumstantial,
would indicate that the bluegreens do not cause the
disappearance of the *Daphnia* population. Rather
the *Daphnia* are apt to have been eliminated by some
predator, such as *Chaoborus*, which is abundant in
the lake and teeming in the surface waters at night,
or by populations of young fishes which are not size
selective. It is to be noted that the Copepods
actually decline two weeks before the *Daphnia* and
therefore well in advance of the bluegreen emergence.

471

The development of the *Bosmina* population, a much
smaller species of Cladocera would be explained by
the predation hypothesis.[4],[5]

A second bit of evidence arguing for a planned
retreat of the *Daphnia* population is its brief and
intensive flight into the production of the diapausing embryos. From previous analysis of diapause
control in *Daphnia pulex*, it seems probable that
Saratoga Lake is inhabited by a distinct cyclic
type, one in which passing from the active to the
diapaused phase of the life cycle is to be expected
in long daylengths. Prior to the decline in June,
1971 (as in 1969 and 1970) reproduction shifted from
the release of young to the release of diapaused
embryos (Figure 118C). The switch was complete on
June 1, but only for the single sampling. During
the week before and after, the reproduction was
mixed. The interval of intense production of
diapausing embryos appeared to be only one week in
1971 and possibly somewhat longer in 1970.

A reproductive switch in *Daphnia* would seem
to be obviously adaptive. Its occurrence in late
spring characterizes it as one of the three cyclic
patterns[6] that may be identified from an analysis
of the environmental stimuli that induce the switch
in reproduction.[7] Cycles are identified by the
timing and number of switches each year. The acycle
remains in a free swimming phase throughout the year,
although it may have a reproductive arrest during
winter (Figure 119). The autumnal monocycle makes
but one switch each year in autumn and spends the
winter in an embryonic diapause. The dicycle is
characterized by two periods of entry into diapause,
one in late spring, a second in autumn. (A fourth,
the spring monocycle, also enters embryonic diapause in the long daylengths of late spring but
separation of it from the dicyclic type on the
basis of diapause initiation is not yet possible).

The validity of each cyclic type and the
adaptive strategy each fulfills is amenable to
analysis. Moreover, their understanding is essential
to one of the conclusions of this paper.

All cyclic types of *D. pulex* which have been
brought into the laboratory and cultured are respon-
sive to photoperiod control[7] as is typical of
arthropods generally.[8] The reproductive switch is
induced at daylengths equivalent to autumn at the
latitude of the source population. Even the dicyclic type may be controlled by photoperiod. Its

ANNUAL CYCLES
DAPHNIA PULEX

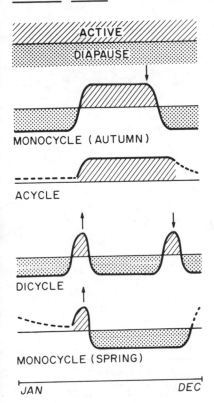

MONOCYCLE (AUTUMN)

ACYCLE

DICYCLE

MONOCYCLE (SPRING)

JAN   DEC

Figure 119. *Annual diapause cycles in Daphnia pulex. Arrows designate the timing of the reproductive switch leading to release of diapausing embryos. Downward pointing arrows show switching in response to inductive photoperiods. Upward pointing arrows show where switching represents an escape from photoperiod control.*

escape can be simulated in the laboratory by manipulating temperatures to conform to that of lakes in late May and June.

Induction of the reproductive shift requires also a second unidentified stimulus which in the bioassays is proportional to the density of the culture.[9,10] The acyclic type is least sensitive (Figure 120). Threshold densities range from 85 to more than 200 individuals per liter. The dicycle is most sensitive and induction in short daylengths is complete at the minimum bioassay density of 50 individuals per liter.

Each cyclic type possesses advantages. The overwintering in a quasi-adult stage, which is characteristic of the acycle, minimizes the time for retooling the reproductive machinery and is

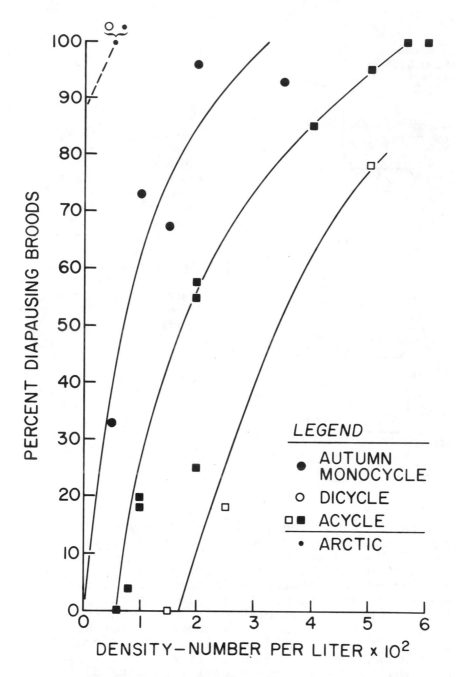

Figure 120. *Reproduccive switch in D. pulex in response to population density. All measurements were done at inductively short photoperiods and at 12°C. An arctic population of D. middendorffiana is included for comparison.*

likely to be the most competitive. Overwintering
in the relative warmth of an ice-blanketed lake
obviously presents few problems provided the reservoir of dissolved O_2 does not become exhausted. The
O_2 is depleted in fertile lakes, however, and a
cycle that produces a diapaused embryo, known to
be resistant to low tensions of O_2, may be the more
successful; there is evidence to indicate reversal
of the above succession of cyclic types, and Frey[11]
suggests that documentary evidence should be available in lake sediments. The adaptive value of a
vernal diapause is well illustrated by the events
in Lake Saratoga.

Bluegreen Blooms and
Food Chains

The rapid growth of bluegreen algae simultaneous
with the loss of a significant component of the
grazing trophic level repeats itself annually in
Saratoga Lake. Temporally the events seem well
ordered, and the presence of spore and diapaused
stage in the life cycles of the principal component
species is by itself strongly suggestive of a well
trained cast moving on and off stage of the
"ecological theater" on cue. Yet it is necessary
to consider alternatives to a non-interactive kind
of species succession. Retreading the initial
question, does the bluegreen population merely
appear after the retreat of the grazing component
of the food chain or does it force itself into
prominence?
    A series of alternative "models" are presented
in Figure 121. They may be classed as causal or
selective. Bluegreen blooms may be "caused" by the
loss of grazers as the simple result of nothing to
eat them. Alternatively, the relationship may be
indirect with the loss of grazers resulting in reduced rates of nutrient recycling, which one supposes
is a necessary condition to maintain an excluding
effect by other algae on the bluegreens. The larger
standing crop of algae is consistent with the
shortening of a food chain.[12] The indirect
hypothesis is supported by a frequently held con-
tention that bluegreen blooms develop in organically
rich but temporarily impoverished inorganic environments.[13] A third alternative is the selective
process mentioned above in which grazers disappear
and bluegreens appear as a result of completely
independent causes.

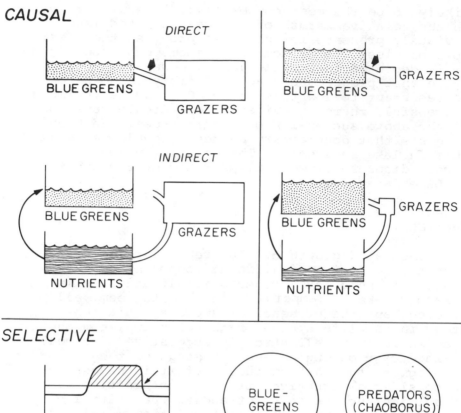

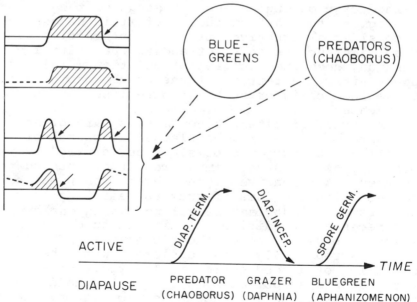

Figure 121. Alternative models describing the relationship between disappearance of the grazing species from the plankton and the simultaneous "blooming" of bluegreen algae.

Conclusions are possible when field observations are combined with the life cycle bioassay for *Daphnia*. The oscillation in density of the *Daphnia* is probably forced and the predatory destruction of the population is too rapid for the birth rate to readjust to greater food availability. Destruction of the population seems to have "selected" for a cyclic type of *Daphnia pulex* in which reinoculation can be achieved through the hatching of the diapaused embryos, although there is no indication that the population does indeed depend upon that source of supply.

The evidence is insufficient to attribute a cause and effect reaction between the disappearance of the grazer and the rapid growth of the bluegreen algae. It is clear that the two events are closely correlated. Since this paper was presented, the 1972 data have been collected and partially analyzed. As expected, the *Daphnia* disappeared and *Aphanizomenon* appeared at the expected times. Development of the algal bloom beyond the initial density was delayed, however, and for approximately one month. A sufficient explanation for the bloom is provided with the so-called indirectly causal hypothesis. It is believed that greatly increased rainfalls and cloudy skies of June, 1972, increased nutrient input from outside the lake while decreasing demand, thus substituting for nutrients normally excreted by the grazing population. In effect, the 1972 data argue for both forced and phased oscillations of population densities to explain blooms of bluegreen algae.

Acknowledgments

Generous assistance was given by Penny Chisholm, J. M. O'Connor, J. W. Sutherland, Timothy A. Downing, and Peter Starkweather.

REFERENCES

1. Hasler, A. D. "Eutrophication of Lakes by Domestic Drainage," *Ecology, 28*:383 (1947).
2. Mortimer, C. H. "Physical Factors with Bearing on Eutrophication in Lakes in General and in Large Lakes in Particular," In *Eutrophication*, NAS Symp. Proc., Rohlich, G., ed. (1969), pp. 340-368.
3. Hammer, U. T. "The Succession of "Bloom" Species of Bluegreen Algae and Some Causal Factors," *Verh. Int. Ver. Limnol. 15*:829 (1964).
4. Brooks, J. L. and S. I. Dodson. "Predation, Body Size, and Composition of Plankton," *Science, 150*:28 (1965).

5. Brooks, J. L. "Eutrophication and Changes in the Com-
    position of the Zooplankton," In *Eutrophication,* NAS
    Symp. Proc., Rohlich, G., ed. (1969), pp. 236-255.
6. Berg, K. *Studies on the Genus Daphnia. O.F. Muller:
    with Especial Reference to the Mode of Reproduction.*
    (Cophenhagen: Bianco Luno A/S, 1931), 222 pp.
7. Stross, R. G. "Photoperiod Control of Diapause in
    Daphnia III. Two Stimulus Control of Long-day, Short-
    day Induction," *Biol. Bull. 137*:359 (1969).
8. Danilevskii, A. S. *Photoperiodism and Seasonal Develop-
    ment of Insects.* (Edinburgh: Oliver and Boyd, 1965),
    283 pp.
9. Stross, R. G. and J. C. Hill. "Diapause Induction in
    Daphnia Requires Two stimuli," *Science, 150*:1462 (1965).
10. Stross, R. G. and J. C. Hill. "Photoperiod Control of
    Winter Diapause in the Fresh-water Crustacean, *Daphnia,*"
    Biol. Bull. 134:176 (1968).
11. Frey, D. G. "Evidence for Eutrophication from Remains
    of Organisms in Sediments," In *Eutrophication,* NAS Symp.
    Proc., Rohlich, G., ed. (1969), pp. 594-612.
12. Smith, F. E. "Effects of Enrichment in Mathematical
    Models," In *Eutrophication,* NAS Symp. Proc., Rohlich, G.,
    ed. (1969), pp. 631-645.
13. Lange, W. "Cyanophyte-Bacteria Systems: Effects of
    Added Carbon Compounds and Phosphate on Algal Growth at
    Low Nutrient Concentrations," *J. Phycol. 6*:230 (1970).

EFFECT OF WASTEWATER ORGANIC FRACTIONS ON THE GROWTH OF SELECTED ALGAE

Gerald C. McDonald, Executive Director, Albany County Sewer District, Albany, New York; *Nicholas L. Clesceri,* Director, Rensselaer Fresh Water Institute at Lake George, Associate Professor, Bio-Environmental Engineering Division, Rensselaer Polytechnic Institute, Troy, New York

With the rise of the Industrial Revolution the accelerated aging of the natural water resource has become increasingly evident. This phenomenon has, in part, been attributed to the unchecked discharge of man-made wastes.[1-5] The agents within such wastes responsible for the increased algal productivity, indicative of such phenomena, have been investigated extensively. Nitrogen and phosphorus have received the main attention, but the results elicited to date have not fully explained the occurrences of increased productivity solely in light of these entities.[6-11] The investigations have produced grounds for speculation and hypotheses requiring further serious study. To date, a specific aspect of wastewaters, namely, the possible stimulatory effects produced by organic components, has received little attention. The work outlined herein describes an attempt to determine the existence and extent of algal growth enhancement brought about by the addition of wastewater organic fractions to representative algal cultures.

MATERIALS AND METHODS

The objective of this investigation was to ascertain the effects of organic wastewater fractions on the growth rate of selected algae. To this end a sample of effluent from a conventional

479

activated sludge facility located at Batavia, New York was subjected to fractionation using gel chromatographic techniques. Prior to such fractionation the sample was membrane filtered (0.45 μ) and concentrated by freeze drying to insure its usefulness for further experimentation. The latter technique was carried out using a 10 liter Virtis large port freeze dryer specially equipped with sixteen 3/4" ports to allow for an increased rate of sublimation (Virtis Co., Gardiner, New York). Preliminary studies with locally available wastewater clearly demonstrated that this concentration technique did not cause a change in the basic character of the wastewater, *i.e.*, did not result in selective organic carbon losses. Figure 122 displays chromatographic data relative to one such study.

Subsequent to filtration and freeze drying, the concentrated wastewater was separated into organic fractions through the use of gel chromatography. The chromatographic apparatus used was utilized in a 4° C cold room. The gel columns, 2.5 by 100 cm (K25/100 Pharmacia Fine Chemicals, Inc., Piscataway, New Jersey), were fitted with up-flow adapters to optimize the resolution of the separation procedure. Five ml eluent fractions were collected and analyzed for total organic carbon (TOC) content with a Beckman infra-red carbonaceous analyzer, model 315.

The gels used for separation of the organic fractions were Sephadex G-10, G-25 and G-50 (Pharmacia Fine Chemicals, Inc.). Separation ranges for such gels, in terms of molecular weight are 0-700 (G-10), 1000-5000 (G-25) and 1000-30,000 (G-50). Before introducing the wastewater concentrate the columns were standardized with compounds of known molecular weight. Based on these standardizations apparent molecular weights (AMW) were assigned to the fractions from the wastewater concentrate.[12-14]

The chromatographic procedure involved the application of the wastewater concentrate to the G-10 column in 10 ml passes and the subsequent compositing of similar fractions produced in each of the runs. The composite G-10 frontal peak (AMW > 700) was then reconcentrated using the freeze drying technique and applied to the G-25 column. The fractions resulting from this operation were composited in the same manner as the G-10 fractions.

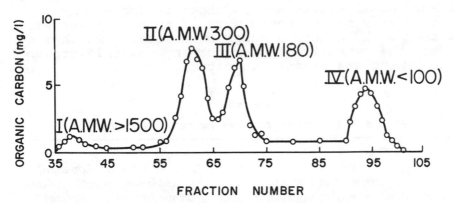

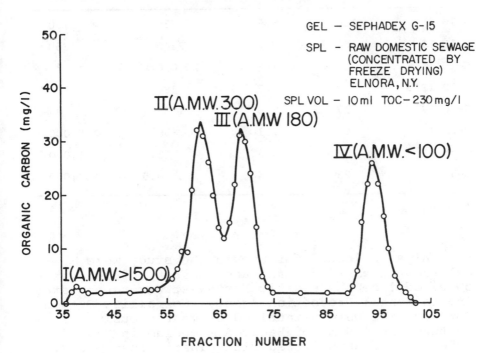

Figure 122. *Elution diagrams of the fractionation of*
unconcentrated and concentrated samples of
raw domestic sewage from Elnora, New York on
Sephadex G-15.

The G-25 composite frontal peak was in turn concentrated and used as the G-50 column sample. Displayed in Figures 123, 124, 125 are typical chromatograms produced during these chromatographic procedures. The composite fractions were examined for selected physical and chemical properties. The data relative to such examinations are as shown in Table 65. Conductivity measurements were of special interest as they would afford a means of assessing whether or not the causative factor for any possible growth enhancement was organic in nature.

Table 65

Selected Parameters for Wastewater Fractions

Fraction	Total Organic Carbon Concentration mg/l	pH	Conductivity μmhos/cm	Apparent Molecular Weight
G-10 I	120	5.6	850	> 700
G-10 II	100	8.2	2400	250
G-10 IIIa	60	7.1	20,000	*
G-10 IIIb	30	7.7	380	*
G-10 IV	17	6.3	34	*
G-25 I	64	7.1	56	> 5,000
G-25 II	39	6.4	5.7	1,000
G-50 I	23	6.4	8.5	>30,000

* indefinable

Algae selected for use in the study were *Selenastrum capricornutum* and *Anabaena flos-aquae* obtained from the Pacific Northwest Water Laboratory, Environmental Protection Agency, Corvallis, Oregon. Both species were maintained as continuously stirred cultures in 100 ml of Basic ASM medium in 250 ml Erlenmeyer flasks fitted with plastic foam plugs. This medium was the modification of the ASM of McLachlan and Gorham[16] originally proposed for use in the Provisional Algal Assay Procedure. The medium was modified[15] through a reduction in the concentration of K_2HPO_4 from 17.4 mg/l to 3.48 mg/l,

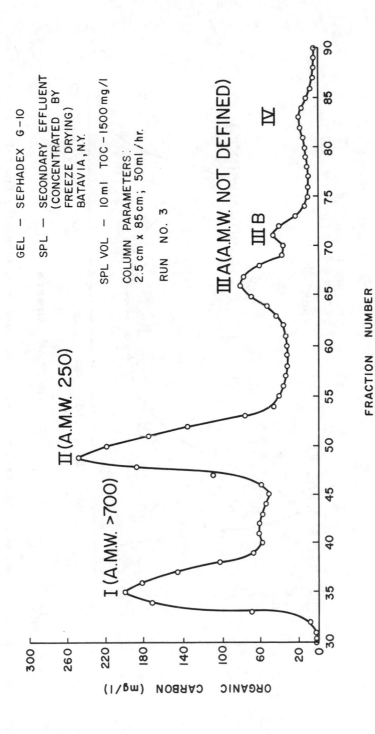

Figure 123. Elution diagram of the fractionation of concentrated Batavia effluent on Sephadex G-10.

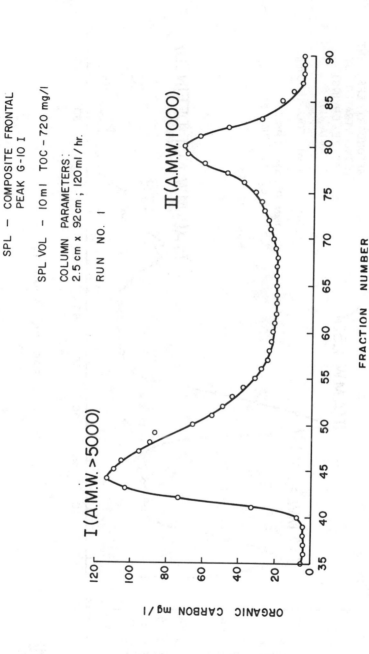

GEL — SEPHADEX G-25

SPL — COMPOSITE FRONTAL
PEAK G-10 I

SPL VOL — 10 ml TOC — 720 mg/l

COLUMN PARAMETERS:
2.5 cm x 92 cm ; 120 ml / hr.

RUN NO. 1

I (A.M.W. >5000)

II (A.M.W. 1000)

Figure 124.   Elution diagram of the fractionation of concentrated composite frontal peak G-10 I on Sephadex G-25.

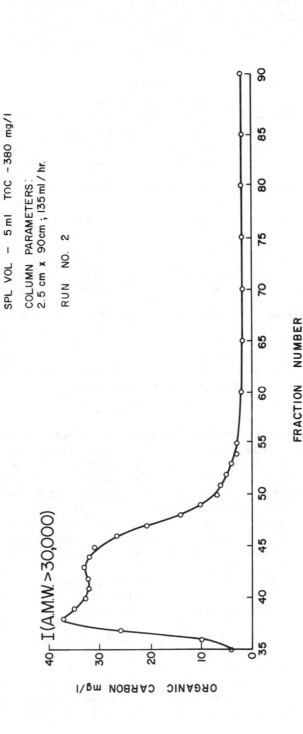

GEL – SEPHADEX G-50

SPL – COMPOSITE FRONTAL
PEAK G-25 I

SPL VOL – 5 ml   TOC –380 mg/l

COLUMN PARAMETERS:
2.5 cm x 90cm ; 135ml / hr.

RUN NO. 2

I (A.M.W. >30,000)

Figure 125.   *Elution diagram of the fractionation of concentrated composite frontal peak G-25 I on Sephadex G-50.*

of Na_2EDTA from 7.4 mg/l to 1.0 mg/l, and the addition of 50 mg/l Na_2CO_3 to insure adequate buffering capacity. Trace metals were not added to the medium. The constituents are shown in Table 66.

Table 66

Basic ASM Medium

Compound	Stock Solution (g/l)	For 1 Liter of Solution Use (ml)	Final Concentration in Medium (mg/l)
$NaNO_3$	8.50	10	85.0
K_2HPO_4	0.348	10	3.48
$CaCl_2 \cdot 2H_2O$	1.47	10	14.7
Na_2CO_3	5.00	10	50.0
$MgSO_4 \cdot 7H_2O$	4.90	10	49.0
$MgCl_2$	1.90	10	19.0
$FeCl_3$	0.032	10	0.32
$Na_2 \cdot EDTA$	0.100	10	1.0

Glass Distilled Water - Dilute to 1 Liter

The culture room temperature was kept at 23 ± 1° C and illumination levels provided were 550 ft-candles[17,18] for *Selenastrum capricornutum* and 150 ft-candles for *Anabaena flos-aquae*. Prior to inoculation, the Erlenmeyer flasks and medium were sterilized by autoclaving. Inocula for both species were transferred from stock cultures to the control and test flasks at set periods. The stock cultures were maintained under the same conditions outlined for the test cultures.

Five-day-old cultures were used as inocula for *Selenastrum capricornutum* and seven-day-old cultures as inocula for *Anabaena flos-aquae*. In the case of *Selenastrum capricornutum* this transfer schedule produced an initial cell concentration of 35,000 cells/ml in the test cultures.

Growth of both species was determined by absorbance measurements at 750 nm with a Beckman DU-2 spectrophotometer.[19] The procedure had previously been correlated with cell count for

Selenastrum capricornutum and dry weight determinations for *Anabaena flos-aquae*. For each culture those points on the growth curve delineating the log phase of growth were subjected to a regression analysis from which the growth rate constant (K_{10}) was calculated. Growth rate constants from similar cultures were then averaged and a mean growth rate constant obtained for a specific set of cultures.

RESULTS

Significant and marked effects on the growth rate of the algal forms were noted upon addition of several of the organic wastewater fractions to the respective algal cultures.

Data relative to the effects on *Selenastrum capricornutum* are shown in Table 67. Analysis of these data indicates that a marked increased growth rate response was noted on addition of fraction G-50 I (AMW > 30,000) and the concentrated effluent. Fractions G-25 II (AMW = 1000) and G-10 IIIb produced somewhat lesser responses. The concentrated effluent and G-50 I produced growth rate constants of 0.96 and 0.72, respectively, as opposed

Table 67

Effect of Organic Fractions on the Growth Rate of Selenastrum capricornutum

Fraction	Concentration of Fraction mg/l Organic Carbon	Mean Growth Rate K_{10} (day^{-1})	95% Confidence Interval
Control*	---	0.43	0.42 - 0.44
G-10 II	2.0	0.43	0.41 - 0.45
G-10 IIIa	0.6	0.42	0.40 - 0.44
G-10 IIIb	0.3	0.49	0.48 - 0.50
G-10 IV	0.2	0.43	0.39 - 0.47
G-25 II	0.4	0.50	0.49 - 0.51
G-50 I	1.3	0.72	0.70 - 0.74
Concentrated Effluent	7.0	0.96	0.89 - 1.03

*Basic ASM medium, without fraction addition

to 0.43 for the control. The data also indicate that the responses noted were selective, in that no correlation between increased growth rate and the organic carbon concentration in the culture vessel is evident. The organic carbon concentrations of the fractions noted in Table 67 are those, calculated on the basis of chromatographic data, that would be contained in the original wastewater effluent.

Reduced concentrations of the active fractions were also investigated. The data developed in these experiments are shown in Table 68. It is evident that, in order to eliminate growth rate enhancement, the strongly active G-50 I fraction and the concentrated effluent must be reduced to a point where their respective concentrations are only 10% of those existing in the original effluent; whereas a 50% reduction in the concentrations of fractions G-25 II and G-10 IIIb eliminates all effects.

Table 68

Effect of Organic Fractions in Reduced Concentrations
on the Growth Rate of Selenastrum capricornutum

Fraction	Concentration of Fraction mg/l Organic Carbon	Mean Growth Rate K_{10} (day^{-1})	95% Confidence Interval
Control	---	0.43	0.42 - 0.44
G-10 IIIb	0.2	0.43	0.39 - 0.47
G-25 II	0.2	0.45	0.40 - 0.50
G-50 I	1.3	0.72	0.70 - 0.74
	0.7	0.63	0.60 - 0.66
	0.1	0.43	0.40 - 0.46
Concentrate	3.5	0.86	0.79 - 0.93
	0.7	0.43	0.40 - 0.46

Data were also developed regarding the possible interactive effects of the fractions. The addition, in concert, of the three active fractions, G-10 IIIb, G-25 II and G-50 I, to a *Selenastrum* culture allowed for the development of a growth rate constant of 0.85. Comparison of this constant with that found using only the concentrated effluent ($K_{10} = 0.96$) suggests that the greater portion of the concentrated effluent response is due principally to these fractions.

Anabaena flos-aquae was not similarly affected by the wastewater fractions. As shown in Table 69, only fraction G-50 I and the concentrated effluent enhanced the growth rate of this alga. This growth rate response, measured as a percent increase, was much less than that exhibited on addition of these same fractions to the *Selenastrum* cultures. Data

Table 69

Effect of Organic Fractions on the Growth Rate
of <u>Anabaena flos-aquae</u>

Fraction	Concentration of Fraction mg/l Organic Carbon	Mean Growth Rate K_{10} (day^{-1})	95% Confidence Interval
Control	---	0.34	0.32 - 0.36
G-10 II	2.0	0.34	0.30 - 0.38
G-10 IIIa	0.6	0.38	0.34 - 0.42
G-10 IIIb	0.3	0.33	0.29 - 0.37
G-10 IV	0.2	0.34	0.29 - 0.39
G-25 II	0.4	0.37	0.33 - 0.41
G-50 I	1.3	0.41	0.38 - 0.44
Concentrated Effluent	7.0	0.54	0.50 - 0.58

collected during investigations at reduced concentrations, Table 70, show that a 50% reduction in the concentration of G-50 I did not allow for any enhanced growth while a 90% reduction in the concentration of the whole effluent was required to achieve this same end.

Investigations with phosphorus at the level found in the concentrated effluent (1.5 mg/l) indicated growth rate enhancement for *Selenastrum capricornutum* alone, but the growth rate constant achieved ($K_{10} = 0.50$) was significantly less than

Table 70

Effect of Organic Fractions in Reduced Concentrations
on the Growth Rate of <u>Anabaena flos-aquae</u>

Fraction	Concentration of Fraction mg/l Organic Carbon	Mean Growth Rate K_{10} (day^{-1})	95% Confidence Interval
Control	---	0.34	0.32 - 0.36
G-50 I	1.3	0.41	0.38 - 0.44
	0.7	0.37	0.33 - 0.41
	0.1	0.33	0.30 - 0.36
Concentrated Effluent	3.5	0.41	0.39 - 0.43
	0.7	0.33	0.31 - 0.35

that produced by the concentrated effluent or fraction G-50 I, those organic entities exhibiting the greatest effect on the algal species.

Limited information relative to the effect on final cell yield was also developed. Figures 126 and 127 depict typical *Selenastrum* and *Anabaena* growth curves developed from control and concentrated effluent data. Final yield (200 hours) was consistently higher in those cultures spiked with full concentrated effluent than in the control cultures.

DISCUSSION

Reviewing the data in Table 65 relative to the conductivity of each of the organic fractions allows for a preliminary assessment of the cause of the demonstrated effects. In view of this information, it may be assumed that any algal growth rate enhancement produced by fractions G-25 II and G-50 I, with conductivities of 5.7 and 8.5 μmhos/cm respectively, was due either to indirect or direct action of an organic agent. In the case of G-10 IIIb (380 μmhos/cm) such a clear determination is not possible. The high conductivity of this fraction may well indicate that the growth enhancement found resulted from inorganic factors or possibly organic ions within the fraction.

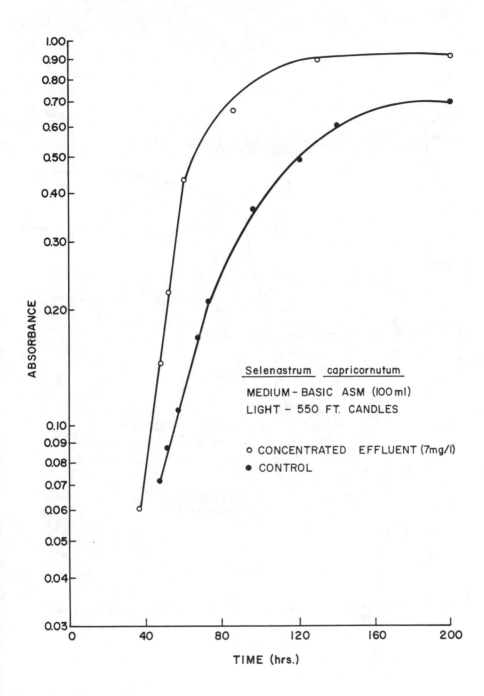

Figure 126. <u>*Selenastrum capricornutum*</u> *growth curves.*

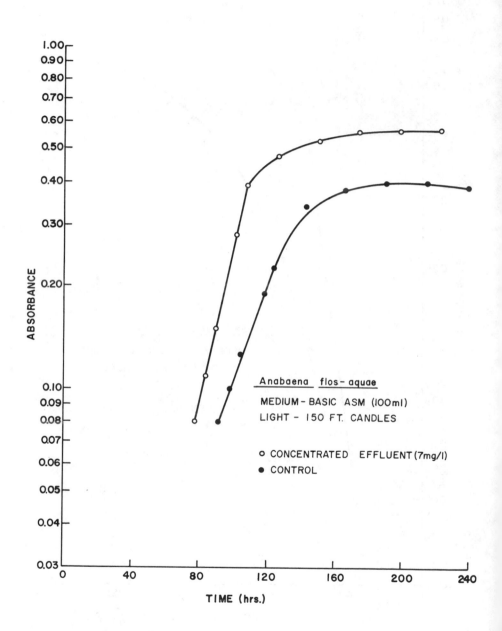

Figure 127.   *Anabaena flos-aquae* growth curves.

Though no direct experimental evidence was produced relative to the question of carbon limitation, the data developed indicates that removal of carbon limitation was not a factor. The specificity of the effects coupled with significant increases in growth rate achieved by the addition of smaller quantities of carbon, *e.g.*, comparing fraction G-10 II to fraction G-50 I, would not be indicative of this explanation. If carbon alone was the operative factor, then the fraction which adds the greatest amount would be expected to produce the largest increase in growth rate.

As noted previously, experiments with phosphorus at the level found in the concentrated effluent (1.5 mg/l) elicited a growth rate increase with *Selenastrum capricornutum* but none with *Anabaena flos-aquae*. However, the growth rate achieved ($K_{10} = 0.50$) was significantly lower than that produced by the concentrated effluent ($K_{10} = 0.96$). These data in combination with the absence of a measureable phosphorus concentration in all fractions, except the concentrated effluent, show that the effects produced upon addition of the fractions cannot be attributed to the action of phosphorus.

Several possible factors may be responsible for the effects exhibited. First among these would be the provision of a required trace element not available in the Basic ASM medium. The conductivity for G-10 IIIb would allow that such trace elements could be provided by this fraction directly. In the case of fractions, G-25 II and G-50 I, based on conductivity measurements, an indirect effect may be suggested. This indirect effect would involve previous chelation by a compound or compounds within such fractions of required trace elements and their availability or release in the culture vessel as a result of bio-degradation.

Another possible explanation geared to the assumption of an organic causative agent would be the provision of an accessory growth factor or substance in the fractions. Examples of each would be the vitamin B_{12} as the factor and 3-indoleacetic acid as the substance. Both entities have been reported as being present in secondary effluents, though only vitamin B_{12} would have a molecular weight to match that of a particular active fraction, namely G-25 II. But it is noted that a vitamin requirement has yet to be established for *Selenastrum capricornutum* and that *Anabaena flos-aquae* was unaffected by fraction G-25 II. Thus, the relationship between such growth factors and substances and the effects found are tenuous.

An explanation in line with that proposed by Prakash and Rashid[20] for the effect of humic substances on marine algal growth may also be offered. Such explanation supposes that a factor or factors within the active fractions act as sensitizing agents for cellular transport. It is believed that such agent or agents allow for increased rate of cellular transport resulting in enhancement of nutrient uptake from the basal medium and thus leading to increased growth.

Final consideration must be given to the provision of a proteinaceous material resulting from bacterial breakdown of the active fractions. Since the cultures were not maintained in the axenic state, it may be that biological breakdown of organic compounds within the active fractions provides protein degradation products such as amino acids and peptides. Subsequent algal uptake of such materials may allow the organism to bypass normal synthetic processes, resulting in increased growth.

Although the dilution used to obtain the above data would not normally be expected to occur in the natural environment, the effects noted are still important. The Batavia Activated Sludge Plant effluent is not unique. Effluents from activated sludge systems serving similar areas would generally be the same. Thus, if the conventional system were built on a water course not allowing for maximum dilution, problems resulting from algal growth enhancement would be expected. Examples of such receiving waters are lakes and tidal rivers, both of which, under certain circumstances, could allow for a concentration buildup of the organic fractions responsible for accelerated growth rate. Future investigations with other effluents will be undertaken to assess the generality of the results produced in this study.

CONCLUSION

Organic compounds contained in wastewater fractions have been found to exert a growth-enhancing effect on two algae species. The exact nature of the causative factors and the pathway by which they achieve their effects are as yet unknown, and further investigative efforts are being conducted to assess the generality of these results. This undertaking is important as increased algal productivity may yet continue to occur if such factors are general constituents of wastewater effluents, notwithstanding the removal of nitrogen and phosphorus from such effluents.

ACKNOWLEDGMENT

Financial support for this work was in part provided by United States Public Health Service Training Grants 2T1RH4-06 and 3T1RH4-0651 (67) and the New York State Department of Health. Major support was forthcoming under United States Environmental Protection Agency Grant No. 16010DHN.

REFERENCES

1. David, C. C. "An Approach to Some Problems of Secondary Production in Western Lake Erie," *Limnology and Oceanography 3*:15 (1958).
2. Duffer, W. R., and T. C. Dorris. "Primary Productivity in a Southern Great Plains Stream," *Limnology and Oceanography 11*:143 (1966).
3. Sylvester, R. O., and G. C. Anderson. "A Lake's Response to Its Environment," *Journal of Sanitary Engineering Division, ASCE 90*, SA1 Proc. Paper 3786, 1 (1964).
4. Oswald, W. J., and C. G. Golueke. "Eutrophication Trends in the United States--A Problem," *Journal Water Pollution Control Federation 38*:964 (1966).
5. Hasler, A. D. "Eutrophication of Lakes by Domestic Drainage," *Ecology 28*:383 (1947).
6. Dugdale, R. C., and J. J. Goering. "Uptake of New and Regenerated Forms of Nitrogen in Primary Productivity," *Limnology and Oceanography 12*:196 (1967).
7. Gerloff, G. C., and F. Skoog. "Nitrogen as a Limiting Factor for the Growth of *Microcystis aeruginosa* in Southern Wisconsin Lakes," *Ecology 38*:556 (1967).
8. Ketchum, B. H. "Mineral Nutrition of Phytoplankton," *Annual Review of Plant Physiology 5*:55 (1954).
9. Kratz, W. A., and J. Myers. "Nutrition and Growth of Several Blue-Green Algae," *American Journal of Botany 42*:282 (1955).
10. Krauss, R. W. "Physiology of the Fresh-Water Algae," *American Journal of Botany 9*:207 (1958).
11. Oglesby, R. T., and W. T. Edmondson. "Control of Eutrophication," *Journal Water Pollution Control Federation 38*:1452 (1966).
12. Andrews, P. "Estimation of the Molecular Weights of Proteins, by Sephadex Gel Filtration," *Biochem. Journal 91*:222 (1964).

13. Andrews, P.   "The Gel-Filtration Behaviour of Proteins Related to Their Molecular Weights Over a Wide Range," *Biochem. Journal 96:*595 (1965).
14. Determann, H.   *Gel Chromatography* (New York: Springer Verlag, 1968).
15. Maloney, T. E.   Personal Communication, Provisional Algal Assay Procedure Medium, 1968.
16. McLachlan, J., and P. R. Gorham.   "Growth of *Microcystis aeruginosa* Kutz in a Precipitate-Free Medium Buffered with Tris," *Canadian Journal of Microbiology 7:*869 (1961).
17. Skulberg, O. M.   "Algae Problems Related to the Eutrophication of European Water Supplies and a Bio-Assay Method to Assess Fertilizing Influence of Pollution on Inland Waters," In *Algae and Man*, Jackson, D. F., ed. (New York: Plenum Press, 1964) p 262.
18. Skulberg, O. M.   "Algal Cultures as a Means to Assess the Fertilizing Influence of Pollution," *Journal Water Pollution Control Federation 38:*319 (1966).
19. Yentsch, C. S.   "A Non-Extractive Method for the Quantitative Estimation of Chlorophyll in Algal Cultures," *Nature 179:*1302 (1957).
20. Prakash, A., and M. A. Rashid.   "The Influence of Humic Substances on the Growth of Marine Phytoplankton Dinoflagellates," *Limnology and Oceanography 13:*598 (1968).

INDEX